143
Advances in Biochemical Engineering/Biotechnology

For further volumes:
http://www.springer.com/series/10

Aims and Scope

This book series reviews current trends in modern biotechnology and biochemical engineering. Its aim is to cover all aspects of these interdisciplinary disciplines, where knowledge, methods and expertise are required from chemistry, biochemistry, microbiology, molecular biology, chemical engineering and computer science.

Volumes are organized topically and provide a comprehensive discussion of developments in the field over the past 3–5 years. The series also discusses new discoveries and applications. Special volumes are dedicated to selected topics which focus on new biotechnological products and new processes for their synthesis and purification.

In general, volumes are edited by well-known guest editors. The series editor and publisher will, however, always be pleased to receive suggestions and supplementary information. Manuscripts are accepted in English.

In references, Advances in Biochemical Engineering/Biotechnology is abbreviated as *Adv. Biochem. Engin./Biotechnol.* and cited as a journal.

Holger Zorn · Peter Czermak
Editors

Biotechnology of Food and Feed Additives

With contributions by

Gert-Wolfhard von Rymon Lipinski · Dieter Elsser-Gravesen
Anne Elsser-Gravesen · Marco Alexander Fraatz · Martin Rühl
Holger Zorn · Zoltán Kovács · Eric Benjamins · Konrad Grau
Amad Ur Rehman · Mehrdad Ebrahimi · Peter Czermak
Lex de Boer · Hans-Peter Hohmann · Hendrich Quitmann
Rong Fan · Peter Czermak · Andreas Karau · Ian Grayson

Springer

Editors
Holger Zorn
Institute of Food Chemistry
and Food Biotechnology
Justus Liebig University Giessen
Giessen
Germany

Peter Czermak
Institute of Bioprocess Engineering and
Pharmaceutical Technology
University of Applied Sciences Mittelhessen
Giessen
Germany

ISSN 0724-6145 ISSN 1616-8542 (electronic)
ISBN 978-3-662-43760-5 ISBN 978-3-662-43761-2 (eBook)
DOI 10.1007/978-3-662-43761-2
Springer Heidelberg New York Dordrecht London

Library of Congress Control Number: 2014941091

Printed on acid-free paper

Springer is part of Springer Science+Business Media (www.springer.com)

Preface

Already millenniums before the chemical industry invented "white biotechnology", food has been produced in biotechnological ways. Wine, beer, soy sauce, tempeh, sauerkraut, and many more traditional foods impressively show that biotechnological processes today are securely controlled and operated on a large scale. This knowledge, which has already been achieved by executing biotechnological processes, provides an optimal basis for us to overcome the big challenges involved in supplying the steadily increasing world population with high-quality food in the future. These challenges focus on four main aspects.

- Of central importance is to supply people globally with enough nutrients. In particular, the provision of proteins of high biological value is limiting. Here new concepts, e.g., approaches based on insects or mycoproteins, are currently discussed worldwide.
- Even if in the developed states, sufficient amounts of food is available, the avoidance of loss, e.g., due to spoilage or over-storage, is a central social task. The "biopreservation" of food can help us use the available food resources in a more sustainable way.
- The third trend is the enrichment of food with functional ingredients which improve, e.g., the tolerability or can support digestion. Examples are, among others, galacto- and fructo-oligosaccharides which can be produced by enzymatic synthesis. The tolerability of food can also be improved by degradation of the proteins which elicit allergies for certain target groups significantly.
- The fourth main focus of research in Food Biotechnology concentrates on replacing existing chemical processes with more ecologically friendly biotechnological processes. In comprehensive ecological efficiency analyses, new processes must definitely show their benefit in comparison to old chemical processes.

This volume focuses on the biotechnology of food and feed additives to enhance the production of food and feed while ensuring the quality of ingredients. Another aim is to improve the properties of food e.g., for a balanced diet, for natural based preservation, for stable colors and alternative sweeteners.

Avoidance of Food Loss

According to a recent study of the "Food and Agriculture" organization (FAO) of the United Nations, only about two thirds of the food produced worldwide is currently consumed. One third, yearly about 1.3 billion tons, is disposed of by the consumer directly or is lost either during the agricultural process or on the way from the producer to the consumer. In the long term, this can lead to a shortage of food in poorer countries [1]. Modern processes of "biopreservation" offer fascinating possibilities to protect food against spoilage and minimize losses . The spectrum of possibilities includes the production of bacteriocins by starter cultures and protective cultures and the addition of so-called "fermentates". This method involves employing bacterial diversity and functionality in biotechnological food processes using specific metabolic qualities of the starter cultures and protective cultures, e.g., from lactic acid bacteria. This approach supports the discovery of new molecules which not only suppress undesirable micro-organisms, but also show functional qualities and contribute to the flavor profile and texture attributes of the food [2]. The application of bacteriophages, in particular, is efficient and specific [3]. In the USA, the use of bacteriophages to control e.g., *Listeria monocytogenes*, *E. coli*, *Xanthomonas campestris*, *Pseudomonas syringae* and *Salmonellae* is already permitted. Chapter 2 of this volume discusses the production and the possibilities of "Biopreservatives" and gives definitions and applications. Furthermore, Chap. 4 "Acidic Organic Compounds in Beverage, Food, and Feed Production" also deals with this topic.

Food with Functional Ingredients

Prebiotica, which are indigestible food components for humans, have a positive influence on the balance in the intestine by stimulating growth and the activity of the bacterial flora. This is due to their role as a substrate for the metabolism of the so-called "positive" intestinal bacteria. Currently, there are only two substance groups that fulfill all criteria for prebiotica: (i) fructans (fructo-oligosaccharides, FOS) including lactulose and the fructo-polysaccharides inulin and (ii) galacto-oligosaccharides (GOS) [4, 5]. The prebiotica FOS, GOS, inulin, and lactulose are accredited in Europe as food ingredients and are classified as safe (GRAS—generally recognized ace safe). Other oligosaccharides will most certainly follow, as for example xylo-oligosaccharides (XOS), gluco-oligosaccharides (glucoOs), and isomalto-oligosaccharides (IMO). These substances are also of interest for fat-reduced and dietary products for the improvement of food texture. Sugar, as an example, can be substituted by FOS and in combination with e.g., Aspartam or Acesulfam K, additional synergistic effects can be reached. The bioprocess technologies on the enzymatic synthesis and recovery of FOS and GOS show considerable similarities. Besides a higher yield of OS and continuous processes, research also focusses on the purity of the OS fractions. Today, up to 45 % of GOS and FOS, depending on the total content of sugar, can be reached with easy enzymatic systems. This gives high yields regarding time-and-reaction volume in continuous Enzyme-Membrane-(Bio) reactor systems (EMR). In future, concepts with mixed enzyme systems and selective fermentations will serve to remove by-products, which inhibit the reaction, as well as mono and disaccharide from the OS. However, efficient and well-matched enzyme systems and microorganisms still have to be found and bioprocesses have to be optimized, especially focusing on lifetime/standing time of biocatalyzed reactions. Chapter 8 of the book gives an overview on "Recent Developments in Manufacturing Oligosaccharides with Prebiotic Functions"

Numerous interesting options for the production of food and feed ingredients arise by the cultivation of photoautotrophic algae. Algae of the type Chlorella are valued for their content of proteins and unsaturated fatty acids. In addition, algae contain a high portion of vitamins of the B group, and various carotenes and xanthophylls. Prominent examples will be discussed in Chap. 3 "Biotechnological Production of Colorants". Food or food ingredients can be generated for special dietary purposes by precise and very specific decomposition of the proteins which elicit food allergies or intolerances (as for example coeliac disease). Therefore, however, suitable peptidases with high substrate specificity are required. Promising sources for such enzymes are, for example, eatable mushrooms from the phylum Basidiomycota or insects that, as grain or stock pests, have specialized in the degradation of herbal storage proteins. In Chap. 7 "Food and Feed Enzymes" of the present book the degradation of proteins is discussed besides other enzyme applications for the improvement of resource efficiency, for the biopreservation of food, and for the treatment of food intolerances.

Substitution of Chemical by Biotechnological Processes

Successful examples of the integration of environmentally friendly and sustainable biotechnological steps in the synthesis of e.g., sweeteners (Isomalt, Aspartam, Xylit, Erythrit etc.), amino acids, or vitamins (among others ascorbic acid and rioboflavin) are manifold. In Chap. 1 "Sweeteners" of the book the biotechnological production of e.g., polyols, isomalt or intensive sweeteners like Aspartame as a non-cariogenic alternative to sucrose is discussed for the application in beverages, sugar-free sweets and confections for dietetic nutrition. Chapter 5 focuses on the bioprocesses for the "Industrial Production of L-Ascorbic Acid (Vitamin C) and D-Isoascorbic Acid", and Chap. 6 is dedicated to the industrial production of amino acids.

Though the biotechnological production of food and feed ingredients may not be discussed exhaustively, this volume provides numerous interesting insights into current industrial processes and impressively illustrates the huge potential for future markets. New options still arise from the discovery of new enzymes and the clarification of whole metabolic pathways for the optimization of existing processes or for the development of alternative processes.

Giessen, August 2013

References

1. Gustavsson J et al (2011) Global food losses and food waste. FAO. http://ucce.ucdavis.edu/files/datastore/234-1961.pdf
2. Ravyts F et al (2012) Bacterial diversity and functionalities in food fermentations. Eng Life Sci 12:356–367
3. Garcia P et al (2010) Food biopreservation: promising strategies using bacteriocins, bacteriophages and endolysins. Trends Food Sci Technol 21:373–382
4. Torres DPM et al (2010) Galacto-oligosaccharides: production, properties, applications, and significance as prebiotics. Compr Rev Food Sci Food Saf 9:438–454
5. Patel S et al (2011) Functional oligosaccharides: production, properties and applications. World J Microbiol Biotechnol 27:1119–1128

Contents

Adv Biochem Eng Biotechnol (2014) 143: 1–28
DOI: 10.1007/10_2013_222

Published Online: 27 July 2013

Sweeteners

Gert-Wolfhard von Rymon Lipinski

Abstract Polyols as sugar substitutes, intense sweeteners and some new carbohydrates are increasingly used in foods and beverages. Some sweeteners are produced by fermentation or using enzymatic conversion. Many studies for others have been published. This chapter reviews the most important sweeteners.

Keywords Aspartame • Erythritol • Fermentation • Isomalt • Maltitol • Mannitol • Production • Sorbitol • Steviol glycosides • Tagatose • Thaumatin

Contents

G.-W. von Rymon Lipinski (✉)
MK Food Management Consulting GmbH, 61118 Bad Vilbel, Germany
e-mail: gwvrl@t-online.de

1 Summary

Sweeteners, sweet substances other than sugar and related carbohydrates, are polyols or intense sweeteners. Most of these substances are produced by chemical synthesis. Among the group of polyols, erythritol and part of mannitol are produced by fermentation. Immobilized cells or enzymes are used in the production of isomalt and maltose, an intermediate for maltitol. Many papers on the production of sorbitol and xylitol by fermentation are available. Among the intense sweeteners, the building blocks of aspartame, aspartic acid and phenylalanine, are produced by fermentation, and enzymatic coupling was used in practice by one producer. Stevioside and glycyrrhizin can be modified enzymatically, and possibilities to express the genes for thaumatin were reported in several papers. Tagatose, a reduced-calorie carbohydrate, can be produced by enzymatic conversion of galactose. Important papers describing organisms, enzymes, and fermentation conditions used in practice and in studies are reviewed in this chapter.

2 Introduction

Sweet-tasting substances other than sugar have become increasingly important in food production in the course of the last decades. In certain areas such as soft drinks, the quantity of products sweetened with these substances has almost equalled the conventional, sugar-sweetened products in some countries including the United States. In others, such as in some European countries, the percentage of these beverages has increased steadily after a harmonized approval for all Member States of the European Community in 1995. In other fields of application such as sugar-free sweets and confections, polyols have been established as a noncariogenic alternative to sucrose.

Many sweet-tasting substances are known. This chapter focuses on products used in foods and beverages. Several others can be produced by fermentation, but are of no practical importance.

3 Definitions and General Aspects

The general field of sweet-tasting substances can be divided in two main sectors. One comprises sugar (sucrose) and other nutritive carbohydrates including glucose, fructose, and products obtained from hydrolyzed starch such as high-fructose corn syrup. The other sector covers products generally called sweeteners. They are noncarbohydrate alternatives such as polyols and intense sweeteners. A third group of still rather limited commercial importance comprises sweet carbohydrates of physiological characteristics different from the standard carbohydrates normally used in food production.

3.1 Sweetness

All substances covered in this chapter are sweet. They are, however different in their sweetness intensity and characteristics of their sweetness.

Several substances show sweetness intensity in the same range as the sweetness of sucrose. These are generally polyols and also the carbohydrates described here. Others are distinguished by much a higher sweetness intensity and therefore are normally called intense or high-intensity sweeteners.

In addition to the sweetness intensity, other characteristics are important for the assessment of sweeteners, such as side-tastes, for example, bitter or licorice-like aftertastes and delayed or lingering sweetness or cooling effects. Although polyols normally have a more or less clean sweetness, most of them have a cooling effect when ingested as the dry substance. Intense sweeteners may have aftertastes, a bitter aftertaste like saccharin, a licorice-like taste like steviol glycosides, a delayed sweetness onset like thaumatin or a lasting sweetness like aspartame and sucralose. They are therefore often used in combinations balancing their taste properties.

3.2 Physiology

Most polyols are metabolized, but absorbed only slowly. Partial absorption and fermentation in the intestine result in some contribution to the calorie content of foods. The European Union uses 2.4 kcal/g or 10 kJ/g for all polyols except for erythritol which is noncaloric [10]. Other countries use other, mostly similar, but not always the same, values for polyols. Osmotic effects and microbial metabolization of polyols in the intestine can result in laxative effects causing intestinal discomfort after ingestion of larger amounts.

Most intense sweeteners are not metabolized in the human body and are therefore calorie-free. Others such as aspartame are fully metabolized but, owing to their intense sweetness, are only used in minute quantities that do not make any significant contribution to the caloric content of foods or beverages.

The caloric values of the carbohydrates covered here vary from zero calories for tagatose to the full energy value for, as an example, isomaltulose.

Polyols and intense sweeteners are suitable for diabetics within a suitable diet, whereas for the fully metabolized carbohydrates the rules for the diet should apply, although they may not be absorbed as quickly as sucrose or glucose and therefore trigger a lower blood glucose level than sucrose.

As intense sweeteners and polyols are either not or only very slowly metabolized by the bacteria of the oral cavity to acids, they are generally considered noncariogenic [89].

3.3 Applications

Polyols have a similar sweetness level to that of sugar and are therefore used in similar quantities. Important applications are sweets and confections, chewing gum, tablets, or carriers for sugar-free powders. Owing to the rather low sweetness of some polyols, they are often combined with intense sweeteners to adjust the sweetness to the customary sucrose level.

Intense sweeteners are used in too small a quantity to have any of the technological functions sugar has in many foods. Therefore their main fields of application are beverages, table-top sweeteners and dairy products, but also combinations with some polyols, for example, in confectionery products.

3.4 Regulatory Aspects

Several polyols and intense sweeteners are approved as food additives in the European Union [11]. Change of their manufacturing processes (e.g., replacement of synthetic production by fermentation) requires an additional approval [9]. The reduced-calorie and other carbohydrates are normally not food additives in the EU regulatory framework. New substances would require approval as novel food; approved substances produced by a new fermentation process would also require this approval, but could be notified as substantially equivalent to existing substances if no significant deviation from the existing product could be demonstrated [4].

In the United States, intense sweeteners with the exception of steviol glycosides are regulated as food additives; polyols are either Generally Recognized As Safe (GRAS) or approved as food additives (Anonymous). Substances occurring in nature are GRAS eligible. For these substances, submission of a GRAS notice to the US Food and Drug Administration (FDA) is possible. They are considered acceptable unless the FDA objects or asks questions within 90 days after submission [5].

Generally, a high purity is required for food uses. The specifications laid down in legislation, are, however, slightly different among the EU, USA, and international proposals.

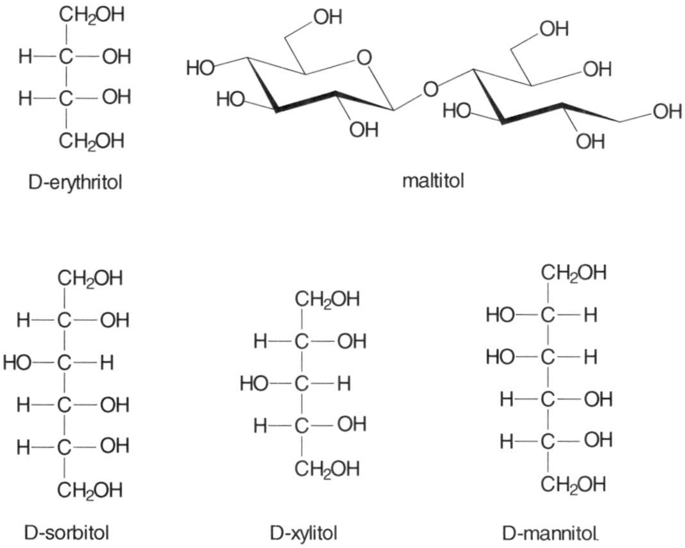

Fig. 1 Structures of commercially produced polyols

4 Polyols

4.1 Erythritol

4.1.1 General Aspects and Properties

Erythritol (*meso*-erythritol, *meso*-1,2,3,4-Tetrahydroxybutan; Fig. 1) has been known for a long time. Its potential use as a bulk sweetener was, however, recognized rather late.

Erythritol is a natural constituent of several foods and beverages in levels sometimes exceeding 1 g/kg. Its solubility in water is approximately 370 g/L at room temperature and increases with increasing temperature. Erythritol melts at 121 C and is stable up to more than 160 C and in a pH range from 2 to 10.

Depending on the concentration used, erythritol is approximately 60 % as sweet as sucrose. It is noncariogenic and not metabolized in the human body which means that it is more or less calorie-free [26].

In the European Union, erythritol is approved as E 968 for a large number of food applications [11]. It is GRAS in the United States [6, 8, 12] and also approved in many other countries.

4.1.2 Microorganisms Producing Erythritol

Microorganisms producing erythritol have been known for many years [140]. Papers describing microorganisms producing yields of 35–40 % of the sugar used in the medium were published as early as 1960 and 1964, and the need carefully to control nitrogen and phosphorus levels in the medium were also highlighted [39, 139]. Further research resulted in the discovery of a variety of organisms. Among these are *Aspergillus niger* [102], *Aurobasidium sp.* [49], *Beauveria bassiana* [145], *Candida magnoliae* [158], *Moniliella sp.* [87], especially *Moniliella pollinis* [29], *Penicillium sp.* [80], *Pseudozyma tsukubaensis* [55], *Torula corallina* [77], *Trigonopsis variabilis* [65], *Trichosporonoides sp.* [90], and especially *Trichosporonoides megachiliensis* [131], *Ustilagomycetes sp.* [44], and *Yarrowia lipolytica* [122]. Patent applications specify a number of different species.

4.1.3 Biochemical Pathways

Different types of microorganisms use different pathways for the biosynthesis of erythritol.

For *C. magnoliae*, transaldolases and transketolases are involved [139]. For mutant strains of *C. magnoliae*, up-regulated enzymes of the citric acid cycle with resulting higher NADH and ATP formation, down-regulated enolase, and up-regulated fumarase with improved conversion of erythritol-4-phosphate to erythritol were held responsible for the higher yields of erythritol [73]. The enolase, erythrose reductase, is an NAD(P)H-dependent homodimeric aldose reductase [78, 79]. Reduction of fumarate production resulted in higher yields of erythritol inasmuch as fumarate is a strong inhibitor of erythrose reductase, the enzyme converting this substance to erythritol [77].

Trichosporonoides megachiliensis mainly uses the pentose phosphate way for the production of erythritol. Transketolase activity was correlated with erythritol yields under various production conditions. It is therefore concluded that transketolase appears to be a key enzyme for formation of erythritol in this organism [131].

In *Y. lipolytica*, glucose is supposed to be converted to erythrose-4-phosphate via the pentose phosphate pathway and reduced by erythrose reductase to erythritol-4-phosphate with subsequent hydrolysis of the ester bond [121].

4.1.4 Production

The synthesis of erythritol is rather difficult. One of the possibilities is the catalytic reduction of tartaric acid with Raney nickel, which does, however, also produce threitol, a diastereomere of erythritol that requires separation of both. Threitol may be isomerized which increases the yields of erythritol. Another chemical synthesis starts from butane-2-diol-1.4 which is reacted with chlorine in aqueous alkali to yield erythritol-2-chlorohydrin and can be hydrolyzed with sodium carbonate solution. Synthesis from dialdehyde starch in the presence of a nickel catalyst at high temperatures is also possible [16].

Owing to the special physiological properties of erythritol, commercial interest increased with the discovery of an increasing number of microorganisms able to produce this substance. Today, the commercial production of erythritol is apparently only based on fermentation.

Erythrytitol fermentations mostly use osmophilic yeasts. Based on regulatory submissions for commercial production, *T. megachiliensis*, *M. pollinis* [7], and *Y. lipolytica* [12] are used. It is also claimed that *P. tsukubaensis* and *Aureobasidium sp.* are used for commercial production [95].

Erythritol-producing microorganisms often produce other polyols such as ribitol. Nevertheless, some strains had a rather high yield of erythritol. A two-step fermentation of *C. magnoliae* on 400 g/L glucose resulted in a 41 % conversion rate and a productivity of 2.8 g/Lh [124]. *M. pollinis* cultivated on glucose and several nitrogen sources yielded erythritol concentrations up to 175 g/L with a conversion rate of 43 %. Oxygen limitation resulted in ethanol formation, and nitrogen limitation in strong foaming. A mutant gave even better yields [17].

Aerobically on glucose cultured *P. tsukubaensis* KN 75 produced 245 g/L of erythritol with an especially high yield of 61 %. The productivity was 2.86 g/Lh. Scale-up from 7-L laboratory fermenter to 50,000-L industrial scale resulted in productivities similar to the laboratory value [55].

Several factors influence productivity and conversion rates. Investigated were, among others, supplementation of the medium with Mn^{2+} and Cu^{2+} for *Torula sp.* Supplementation with Mn^{2+} resulted in lower intracellular concentrations of erythritol, whereas Cu^{2+} increased the activity of erythrose reductase [75]. Phytic acid, inositol, and phosphate also had a positive effect on the yields in *Torula sp.* by increasing the cell growth and increasing the activity of erythrose reductase [76].

A further increase in productivity was obtained by using mutant strains. Examples are an osmophilic mutant strain of *C. magnoliae* with a yield of 200 g/L, a conversion rate of glucose of 43 %, and a productivity of 1.2 g/Lh [70]. Among several mutants of *Moniliella sp.* 440 fermented in 40 % glucose and 1 % yeast extract, the highest yields were 237.8 g/L [88].

Many aspects of fermentation of an osmophilic fungus are described in a thesis by [16]. A survey covers the most important aspects of fermentation [58].

Owing to the commercial importance of erythritol, much information on production conditions is laid down in patent applications. They describe new strains or species producing erythritol and new mutants that have no commercial importance or none as yet. Also specific compositions of the media, methods to reduce viscosity of the media and specific processing, purification, and crystallization conditions are claimed.

Strains not producing polysaccharides eliminate problems caused by increasing viscosity of the medium such as reduced oxygen transfer rates with increasing formation of ethanol and difficulties in filtration during processing of the medium [147].

The use of inorganic nitrogen sources, especially nitrates, as the main nitrogen source for fermentation of *M. pollinis* was claimed to facilitate the adjustment of the pH, the purification, and also to increase the erythritol yields [30].

Common isolation and purification steps are filtration or centrifugation to remove the microorganisms, demineralization with anion exchangers, other types of chromatographic separation, decolorization with activated carbon, and crystallization and recrystallization [125].

4.2 Isomalt

4.2.1 General Aspects and Properties

Isomalt is a more or less equimolar mixture of 1-*O*-α-D-glucopyranosy-D-mannitol-dihydrate and 6-*O*-α-D-glucopyranosyl-D-sorbitol. Different production conditions, however, allow variations in the ratio of the two products. The solubility in water is about 24.5 % (w/w) at room temperature, but varies with the composition and increases with increasing temperature. In addition to the dry isomalt, a syrup is available.

Isomalt is, depending on the concentration, approximately 45–60 % as sweet as sucrose, stable under normal processing conditions of foods, and noncariogenic [132].

In the European Union, isomalt is approved as E 953 for a large number of food applications [11]. It is GRAS in the United States and also approved in many other countries.

Owing to its low glycemic index, isomaltulose, an intermediate of the production, has found increasing interest as a food ingredient in recent years.

4.2.2 Microorganisms Transforming Sucrose into Isomaltulose

For commercial production of isomalt, the sucrose starting material has to be transformed into isomaltulose. The enzyme for this transformation is a glycosyltransferase (sucrosemutase). An organism producing this enzyme suitable for commercial use is commonly named *Protaminobacter rubrum*. It is, however, claimed that it should be *Serratia plymuthica* [36]. Several other organisms have a similar enzymatic activity. Among these are *Erwinia sp* D 12 [59], *E. rhapontici* [155], and *Klebsiella terrigena* JCM 1687 [143].

A variety of enzymes from other sources and cloning into other organisms has been described in the literature. However, they seem to have no commercial importance or none as yet.

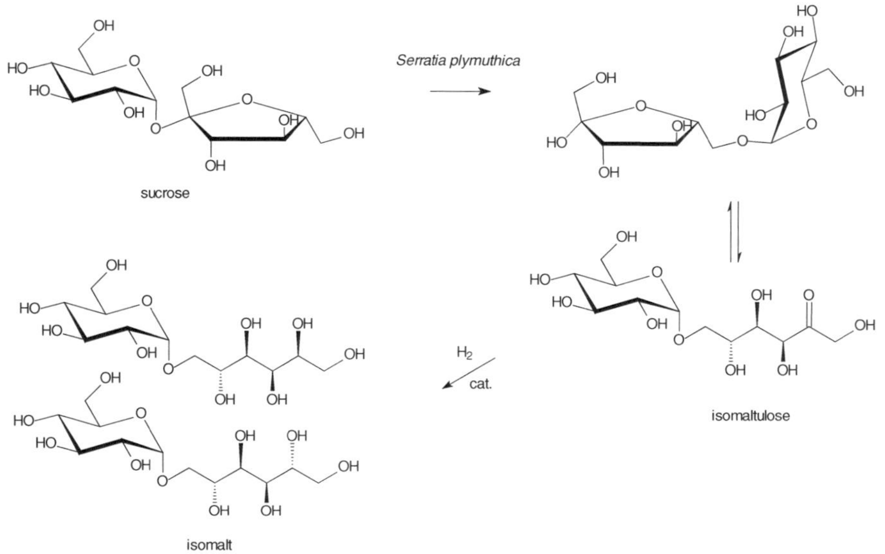

Fig. 2 Production of isomalt from sucrose

4.2.3 Production

For the production of isomalt sucrose is converted to isomaltulose which is then hydrogenated to yield a mixture of the two components of isomalt (Fig. 2). Although the production of isomalt itself from isomaltulose is a chemical hydrogenation, transformation of sucrose into isomaltulose requires enzymatic transformation.

The enzyme sucrosemutase is sensitive to glutaraldehyde, therefore cross-linking is not possible. For industrial use it is, however, not necessary to isolate the enzyme, as immobilized cells of the organism can be used. Addition of sodium alginate to the cultivated cells and subsequent addition of calcium acetate immobilizes the cells. This allows for the use of the cells in a bed reactor, and also facilitates the separation of the product from the reaction mixture.

The long-term stability of the immobilized organism is high and can exceed 5,000 h, even if high sucrose concentrations of 550 g/L are used. The yields are about 80–85 % with 9–11 % of trehalulose and small quantities of other saccharides as by-products.

Prior to hydrogenation, free sucrose has to be removed. This is carried out by nonviable cells of *Saccharomyces cerevisiae*. Remaining by-products of the reaction are converted to the respective sugar alcohols.

Although the hydrogenation of isomaltulose theoretically should yield an equimolar mixture of the two constituents of isomalt, the share of each component may vary between 43–57 % depending on the conditions of hydrogenation [120].

An alternative possibility is the direct cultivation of suitable microorganisms such as *P. rubrum* on sucrose-containing juices obtained during the production of beet and cane sugar. It is claimed that glucose and fructose produced during the transformation are consumed by the microorganisms which results in lower amounts of by-products [24].

4.3 Maltitol

4.3.1 General Aspects and Properties

Maltitol is α-D-glucopyranosyl-1.4-glucitol. The solubility in water is approximately 1,750 g/L at room temperature. Maltitol is stable under the common processing conditions of foods. In addition to dry maltitol several types of syrups are available.

Maltitol is, depending on the concentration, approximately 90 % as sweet as sucrose and noncariogenic [60].

In the European Union, maltitol is approved as E 965 for a large number of food applications. It is GRAS in the United States and also approved in many other countries.

4.3.2 Production

Maltitol is produced by chemical hydrogenation of maltose, which can be obtained by enzymatic degradation of starch under conditions similar to those used for other starch hydrolysates such as glucose. The Starting material can be the different commercially available starches including corn, potato, and others. A partially degraded starch, which can be obtained by treatment with diluted hydrochloric or sulphuric acid and subsequent neutralization or with heat-stable α-amylase, is then subjected to enzyme treatment for further degradation to maltose-rich products.

Enzymes used for maltose production are β-amylases, fungal α-amylases, α-1.6-glucosidases, maltogenic amylases, and debranching enzymes, preferably with high temperature optimum.

Examples can be found in patent applications for processes for production of maltose and maltitol [33, 34, 41, 97, 109, 141].

4.4 Mannitol

4.4.1 General Aspects and Properties

D-mannitol (D-mannohexan-1.2.3.4.5.6-hexaol) is a constituent of several plants including the Manna ash, several edible plants, and seaweed. Parts of the latter contain up to 10 % mannitol by weight. The solubility in water is approximately

230 g/L at room temperature and it increases with increasing temperature. Mannitol is stable under the common processing conditions of foods.

Mannitol is approximately 50 % as sweet as sucrose and non-cariogenic [52].

In the European Union, maltitol is approved as E 421 for a large number of food applications. In the United States, mannitol produced by hydrogenation of glucose or fructose solutions or by fermentation by *Zygosaccharomyces rouxii* or *Lactobacillus intermedius* is approved for several food applications. It is also approved in many other countries.

4.4.2 Microorganisms Producing Mannitol

Several microorganisms are able to produce mannitol, some of which have been known for a long time [105]. Among these are several species of *Aspergillus* [135], *C. magnoliae* [137], several species of *Lactobacillus* [153], especially *L. intermedius*, [128], *Leuconostoc* [20], *Penicillium* [148], or *Torulopsis* [104] and *Z. rouxii* [101].

4.4.3 Biochemical Pathways

Several heterofermentative lactic acid bacteria produce mannitol in large amounts, using fructose as an electron acceptor. Under anaerobic conditions, acetylphosphate produced in the metabolization of glucose would normally be converted to ethanol. In the presence of fructose it is used as an electron acceptor and converted to mannitol by mannitol dehydrogenase. The enzyme requires $NADH_2$ or $NADPH_2$, which is regenerated during hydrogenation of fructose. The now possible conversion of acetylphosphate to acetic acid is energetically advantageous for the organism [136]. *C. magnoliae* also uses mannitol dehydrogenase [13]. *Aspergillus sp.* uses glucose as the starting material and reduces to fructose-6-phosphate instead of fructose [81].

4.4.4 Production

The by far largest quantity of mannitol is produced by chemical hydrogenation of fructose which yields a mixture of mannitol and sorbitol. The mixture is subjected to fractionated crystallization. As direct sorbitol production is less costly, the processing costs have mostly to be borne by mannitol which makes it more expensive than sorbitol. Production from seaweed seems to be of limited importance.

Possibilities to produce mannitol by fermentation were studied using several organisms. They mostly use fructose as an acceptor for hydrogen and glucose as a source of carbon. In a fed-batch culture of *C. magnoliae* with 50 g/L of glucose as the initial carbon source and increasing levels of fructose up to 300 g/L in 120 h,

248 g/L of mannitol were obtained from 300 g/L of fructose equivalent to a conversion rate of 83 % and a productivity of 2.07 g/Lh [138].

High yields were obtained from *Lactobacillus fermentum* grown in a batch reactor. The conversion rates increased from 25 to 35 C to 93.6 % with average and high productivities of 7.6 and 16.0 g/Lh [153]. A fast mannitol production of 104 g/L within 16 h was obtained from *L. intermedius* on molasses and fructose syrups in a concentration of 150 g/L with a fructose-to-glucose rate of 4:1 [126]. High productivity (26.2 g/Lh) and conversion rates (97 mol%) were obtained in a high cell density membrane cell recycle bioreactor. Increase of the fructose concentration above 100 g/L reduced the productivity [154]. A fed-batch process with *L. intermedius* yielded 176 g/L of mannitol from 184 g/L fructose and 94 g/L glucose within 30 h. The productivity of 5.6 g/Lh could be increased to more than 40 g/Lh at the expense of reduced mannitol yield and increased residual substrate concentrations [112].

As mannitol is more expensive than sorbitol, production by fermentation may become an alternative to hydrogenation of fructose.

4.5 Sorbitol

4.5.1 General Aspects and Properties

The solubility of D-sorbitol (D-glucitol, is D-glucohexan-1.2.3.4.5.6-hexaol) in water is up to approximately 2,350 g/L at room temperature. Sorbitol is stable under the common processing conditions of foods. In addition to the dry sorbitol, syrups are available.

Sorbitol is, depending on the concentration, approximately 60 % as sweet as sucrose and noncariogenic [52].

In the European Union, sorbitol is approved as E 420 for a large number of food applications, in the United States as GRAS, and is also approved in many other countries.

Sorbitol is generally produced by chemical hydrogenation of glucose or, together with mannitol, by chemical hydrogenation of fructose.

4.5.2 Fermentation

Several microorganisms are known to produce significant amounts of sorbitol, especially after genetic engineering.

Zymomonas mobilis grown on glucose, fructose, or sucrose produced sorbitol in addition to the main product, ethanol. Strain ZM31 gave the highest concentrations of 43 g/L when grown on 250 g/L of sucrose. As the mechanism, inhibition of fructokinase by free glucose and reduction of fructose by a dehydrogenase is assumed [14]. In a hollow fiber membrane reactor, a productivity of 10–20 g/Lh was found for *Z. mobilis* on 100 g/L each of glucose and fructose. Gluconic acid was produced simultaneously with similar productivities [107]. Immobilized cells of *Z. mobilis* in combination with immobilized invertase produced sorbitol with a productivity of 5.11 g/Lh and gluconic acid with a productivity of 5.1 g/Lh on 20 % sucrose in a recycle packed-bed reactor [117]. Immobilized and permeabilized cells of *Z. mobilis* reached more than 98 % conversion of equimolar concentrations of glucose and fructose to sorbitol and gluconic acid and maximum concentrations of 295 g/L each [115].

A high conversion rate of 61–65 % was found in a *Lactobacillus plantarum* strain with a high expression of two sorbitol-6-phosphate dehydrogenase genes grown on glucose. Small amounts of mannitol were also detected [72].

A high conversion of fructose with 19.1 g/L of sorbitol from 20 g/L of fructose with methanol as the energy source was reported for small-scale fermentation of *Candida boidinii* No. 2201 [144].

Inasmuch as glucose as the starting material and hydrogenation leads to a low-cost production process it seems unlikely that production of sorbitol by fermentation will play a significant role, at least in the near future.

4.6 Xylitol

4.6.1 General Aspects and Properties

The solubility of D-xylitol (D-xylopentan-1.2.3.4.5-pentaol) in water is approximately 1,690 g/L at room temperature. Xylitol is stable under the common processing conditions of foods.

Xylitol is, depending on the concentration, similarly or slightly sweeter than sucrose and noncariogenic [159].

In the European Union, xylitol is approved as E 967 for a large number of food applications. In the United States, it is approved for use in foods following Good Manufacturing Practice and it is also approved in many other countries.

4.6.2 Microorganisms Producing Xylitol

Xylitol can be formed through reduction of xylose by a xylose reductase, in many organisms a NADPH-dependent enzyme [2].

Microorganisms producing xylitol have been studied extensively. Many organisms are able to produce xylitol. Among these are *C. boidinii* [150], *Candida guilliermondii* [103], *C. magnoliae* [69], *Candida maltosa* [37], *Candida mogii* [146], *Candida parapsilosis* [99], *Candida peltata* [127] *Candida tropicalis* [133], *Corynebacterium sp.* [113], especially *Corynebacterium glutamicum* [130], *Debaryomyces hansenii* [106], *Hansenula polymorpha* [129], *Mycobacterium smegmatis* [50], *Pichia sp.*, especially *Pichia caribbica*, *Issatchenkia sp.*, and *Clavispora sp.* [142].

Not only mutants of *C. tropicalis* [35, 54, 56, 114] and *C. magnoliae* [69], but also genetic engineering was used in several organisms to improve xylitol production. Genetic engineering was used to replace the xylose reductase in some organisms in which this enzyme is significantly repressed in the presence of glucose [53].

Strains of *C. tropicalis* with a disrupted gene for xylitol dehydrogenase which catalyzes the oxidation of xylitol to xylose were studied [68]. In one strain, genes were co-expressed that respectively encode glucose-6-phosphate dehydrogenase and 6-phosphogluconate dehydrogenase, under the control of glyceraldehyde-3-phosphate dehydrogenase promoter [2]. In another strain, a highly efficient xylose reductase from *Neurospora crassa*, which is not expressed as such in *C. tropicalis*, was modified and placed in a strain under control of a constitutive glyceraldehyde-3-phosphate dehydrogenase of *C. tropicalis*. This allowed for the use of glucose as a co-substrate with xylose [53]. A gene for an NADH-dependent xylose reductase from *C. parapsilosis* was transferred to *C. tropicalis* which resulted in dual co-enzyme specificity [79].

Higher productivities in *C. glutamicum* were especially achieved when the possible formation of toxic intracellular xylitol phosphate was avoided by elimination of genes encoding xylulokinase (XylB) and phosphoenolpyruvate-dependent fructose phosphotransferase (PTSfru) to yield the strain CtXR7 [130].

Further examples comprise the modification of *Escherichia coli* W3110 to produce xylitol from a mixture of glucose and xylose [61] and *E. coli* containing xylose reductase genes from several sources [23]. Xylitol-phosphate dehydrogenase genes were isolated from *Lactobacillus rhamnosus* and *Clostridium difficile* and expressed in *Bacillus subtilis* [108]. D-xylose reductase from *Pichia stipitis* CBS 5773 and the xylose transporter from *Lactobacillus brevis* ATCC 8287 were expressed in active form in *Lactococcus lactis* NZ9800 [98], and *S. cerevisiae* was supplemented with a xylose reductase gene from *P. stipitis* [82].

4.6.3 Production

Xylitol is mostly produced by chemical hydrogenation of xylose which is obtained by hydrolysis of xylans of plants such as birch and beech trees, corn cobs, bagasse, or straw, but also by fermentation of xylose, for example, using *Candida* species.

Xylose, especially for hydrogenation, requires a high purity. It may be obtained from wood extracts or pulp sulfite liquor, a waste product of cellulose production, by fermentation with a yeast that does not metabolize pentoses. Some strains of *S. cerevisiae*, *Saccharomyces fragilis*, *Saccharomyces carlsbergensis*, *Saccharomyces pastoanus*, and *Saccharomyces marxianus* are suitable for this purpose [51]. Hydrolysates of xylan-rich material are often treated with charcoal and ion-exchangers to remove by-products causing problems in hydrogenation or fermentation.

Many studies of xylitol production by fermentation have been published. Different organisms, substrates, and conditions were investigated. As the starting material, xylose or xylose in combination with glucose was used. Fermentation was carried out in batch reactors as well as continuously [134].

Among the variations studied was cell recycling in a submerged membrane bioreactor for *C. tropicalis* with a high productivity of 12 g/Lh, a conversion rate of 85 % and a concentration of 180 g/L [71]. Many studies addressed the immobilization of cells such as *S. cerevisiae* [119], *C. guilliermondii* [19], or *D. hansenii* [28], especially with calcium alginate.

In some studies, high xylitol concentrations, conversion rates and productivities were achieved. For *C. tropicalis,* concentrations of 290 g/L, a conversion rate of 97 %, and a productivity of more than 6 g/Lh [66], and 180 g/L, 85 % conversion, and 12 g/Lh were reported [71]. For *C. guilliermondii,* a concentration of 221 g/L (conversion rate of 82.6 %; [92]), for *C. glutamicum,* a concentration of 166 g/L at 7.9 g/Lh [130], and for *D. hansenii*, a concentration of 221 g/L and a conversion rate of 79 % [27] were reported. With *S. cerevisiae*, productivities of up to 5.8 g/Lh were observed [119].

4.7 Others

Polyols can generally be produced by hydrogenation of sugars and some also by fermentation. Most of the other polyols are, however, of no commercial interest for the food industry. The only other polyol of some importance is lactitol (E 966), produced by chemical hydrogenation of lactose, a constituent of milk. It seems that no possibilities for production of lactitol by fermentation have been investigated.

5 Intense Sweeteners

5.1 Aspartame

5.1.1 General Aspects and Properties

Aspartame (*N*-L-aspartyl-L-phenylalanine-1-methyl ester, 3-amino-*N*-(α-carbomethoxy-phenethyl)-succinamic acid-*N*-methyl ester) is an intense sweetener widely used in foods and beverages. Its solubility in water is approximately 10 g/L at room temperature. Aspartame is not fully stable under common processing and storage conditions of foods and beverages with the highest stability around pH 4.3 [1].

Aspartame is about 200 times sweeter than sucrose with a clean, but slightly lingering sweetness. It is used as the single sweetener, but often also in blends with

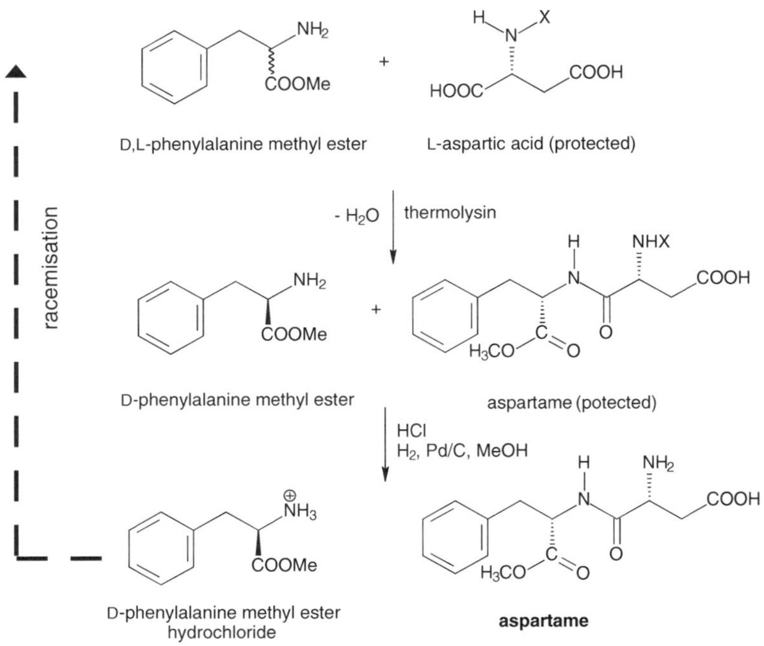

Fig. 3 Production scheme of aspartame

other intense sweeteners owing to synergistic taste enhancement and taste quality improvement often seen in such blends.

In the European Union, aspartame is approved as E 951 for a large number of food applications. In the United States, it is approved as a multipurpose sweetener for food and beverage uses and it is also approved in many other countries.

5.1.2 Production

Aspartame is produced from L-aspartic acid and L-phenylalanine and methanol or alternatively L-phenylalanine methyl ester. The standard process uses common chemical methods of peptide synthesis. Enzymatic coupling of the two amino acids is also possible. *N*-formyl-L-aspartic acid and L- or D.L-phenylalanine methyl ester can be condensed to aspartame by thermolysin-like proteases [43] The formylated aspartame can be deformylated chemically or with a formylmethionyl peptide deformylase to yield the sweetener [111].The enzymatic coupling does not require L-phenylalanine but can start from the racemic product obtained in chemical synthesis, and the remaining D-phenylalanine can be racemized again [151] (Fig. 3).

Production processes based on fermentation are available for the two main components, aspartic acid and phenylalanine [40, 83]

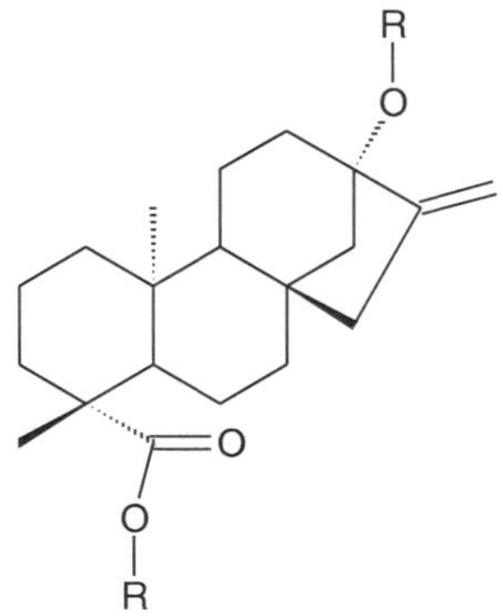

Fig. 4 Structure of steviol glycosides; R = mono- or disaccharide residues

5.2 *Steviol Glycosides*

5.2.1 General Aspects and Properties

Steviol glycosides (Fig. 4) are a family of related substances occurring in *Stevia rebaudiana*, a plant originating in South America, but now also cultivated in Asian countries especially. Main components are typically stevioside and rebaudioside A. The ratio of the different components varies, depending on the product. It may be changed by breeding, which aimed especially at an increase in Rebaudioside A, the product with the best sensory properties. Depending on the composition, steviol glycosides are 200–300 times as sweet as sugar but leave a more or less pronounced bitter and licorice aftertaste. They are stable under normal processing conditions of foods and beverages, but only poorly soluble in water [18].

In the European Union, steviol glycosides are approved as E 960 for a large number of food applications. In the United States, several preparations are GRAS. Steviol glycosides are also approved in many other countries, especially in Asia and South America.

Steviol glycosides are extracted from the leaves of the Stevia plant. The extracts are purified further by flocculation and treatment with ion exchangers before crystallization of the steviol glycosides.

5.2.2 Enzymatic Modifications

To overcome the taste disadvantages of steviol glycosides and their limited solubility, enzymatic modifications were studied. Transglycosylations were used to improve taste quality and solubility. Among the different products obtained, α-glucosyl stevioside seems to be the most interesting. Glucosylated steviosides can be obtained from stevioside and α-glucosyl oligosaccharides including maltose, maltooligosaccharides, or sucrose in the presence of glucosyltransferases [93]. Effective transglycosylation was also achieved with dextrin dextranase of *Acetobacter capsulatus* in a mixture of stevioside and a starch hydrolysate with isoamylase [157]. Glucosyl stevioside has a less pronounced aftertaste than stevioside, is better soluble, of similar sweetness as stevioside, and approved in Japan but neither in Europe nor in the United States.

Transglycosylations of the other steviol glycosides are also possible but apparently of lower, if any, practical importance.

5.3 Thaumatin

5.3.1 General Aspects and Properties

Thaumatin is a mixture of sweet proteins occurring in the arils of the fruits of the African plant *Thaumatococcus daniellii*. Thaumatins I and II are the main components, but four more thaumatin molecules are known [67]. The proteins may be extracted with water. Thaumatin is about 2,000–2,500 times sweeter than sucrose but has a lingering sweetness. In addition to its sweet taste, it has flavor-enhancing properties. It is freely soluble in water and of fairly good stability [42].

In Europe, thaumatin is approved as E 957 for use as a sweetener. It is also approved in a variety of other countries, but in the United States, GRAS as a flavor enhancer only.

5.3.2 Fermentation

Genes encoding thaumatin, mostly thaumatin II, were expressed in several organisms. Among the organisms heterologously producing thaumatin are *Aspergillus awamori* [32, 96], *A. oryzae* [38], *E. coli* [25], *Penicillium roqueforti* [31], *Pichia pastoris* [91], and *Streptomyces lividans* [48]. Thaumatin I was produced in *P. pastoris*, too [47]. The thaumatins A and B, but not Thaumatin I were secreted by engineered *S. cerevisiae* [74]. Proteolytic activities of the production organism may impair the yields, as steps to eliminate this activity significantly improved the yields [96]. When specifically investigated, the secreted products were sweet [25, 32, 47, 91].

The recombinant expression in plant cells was studied, too. The secretion of small levels of thaumatin by recombinant hairy root cells of tobacco could be achieved. However, the yields decreased with increasing amounts of proteases in the medium [110].

5.4 Others

Most intense sweeteners are synthetic products [151]. Approved for food use are acesulfame K (E 950), cyclamate (E 952), neohesperidin dihydrochalcone (E 959), saccharin (E 954), sucralose (E955), and neotame (E 961). Aspartame–acesulfame salt (E 962) is produced by a reaction of acesulfame acid with aspartame.

Neotame (E 961), (*N*-(*N*-(3,3-dimethylbutyl)-L-α-aspartyl-L-phenylalanine 1-methyl ester, is obtained by reacting aspartame with 3,3-dimethylbutyraldehyde.

Advantame, *N*-[*N*-[3-(3-hydroxy-4-methoxyphenyl) propyl-α-aspartyl]-L-phenylalanine 1-methyl ester, monohydrate, is synthesized from aspartame and 3-(3-hydroxy-4-methoxyphenyl)-propionaldehyde. It is not yet approved in the European Union and the United States.

A variety of other sweet-tasting compounds was discovered in plants [67]. Most of these have no commercial importance. Dried aqueous extracts of *Siraitia grosvenori* (formerly *Momordica grosvenori*) are called Luo Han Guo and are regarded GRAS in the United States, but not approved in Europe.

Glycrrhizin, triterpene glycoside salts occurring in the roots of licorice, is not approved as a sweetener, but as a flavoring. Extracts from the roots are up to 100 times sweeter than sucrose. The *3-O-β*-D-monoglucuronide can be prepared using an enzyme from *Cryptococcus magnus*. It is more than 900 times sweeter than sucrose [94], but not approved in Europe and the United States.

6 Carbohydrates

6.1 Isomaltulose

Isomaltulose, 6-*O*-α-D-Glucopyranosyl-D-fructofuranose, is a carbohydrate that has found interest owing to its low glycemic index and noncariogenicity [152]. It is approved as a novel food in the European Union and GRAS in the United States. Production details are given above under isomalt.

6.2 Tagatose

6.2.1 General Aspects and Properties

D-tagatose is a carbohydrate occurring in small amounts in several foods. The solubility in water is approximately 580 g/L at room temperature. As a ketohexose, tagatose reacts in foods in browning reactions like other ketohexoses, for example, fructose (Fig. 5).

Tagatose is, depending on the concentration, approximately 92 % as sweet as sucrose and noncariogenic. The caloric value of tagatose is generally set to 1.5 kcal/g [149].

In the European Union, tagatose is approved as a novel food. In the United States, tagatose has GRAS status and it is also approved in many other countries.

Fig. 5 Fischer projection of the keto hexose tagatose

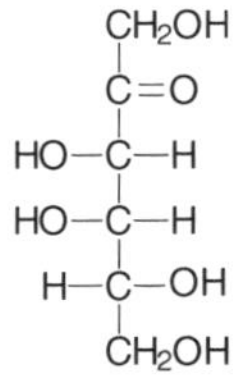

D-tagatose

6.2.2 Microorganisms Producing Tagatose

Enzymatic transformation of galactose into tagatose is possible with L-arabinose isomerase which is found in many microorganisms. Enzymes stable at high temperatures were found in *Acidothermus cellulolytics* [21], *Anoxybacillus flavithermus* [84], *Geobacillus thermodenitrificans* [100], *Thermoanaerobacter mathranii* [85], *Thermotoga maritime*, *Geobacillus stearothermophilus* [46], *Thermotoga neapolitana* [86], and *Thermus sp.* [63]. A thermostable galactose isomerase was isolated from bacteria [64].

Mutations were induced to increase the production rates of tagatose, for example, in *G. thermodenitrificans* [100] or *G. stearothermophilus* [62].

Genetic engineering to improve the performance of fermentation and to use common organisms was reported in several studies. The overexpression of genes of *T. mathranii* [57], *Bacillus stearothermophilus* [21], *T. neapolitana* [45], or *A. cellulolytics* [22] in *E. coli* was described.

6.2.3 Production

Tagatose is produced from galactose, which can be obtained by enzymatic hydrolysis of lactose, the main carbohydrate of milk. Galactose is separated from glucose by chromatography and either isomerized by treatment with calcium hydroxide, subsequent precipitation of calcium carbonate with carbon dioxide, filtration, demineralization with ion exchangers and crystallization [15], or converted enzymatically.

Especially high conversion rates of 96.4 % were obtained with an enzyme extract of an engineered *E. coli* [118], and of 60 % at 95 C for *A. flavithermus* in the presence of borate [84]. Conversion rates of 58 % were reported for an enzyme obtained from a mutant of *G. thermodenitrificans* [100], of 54 % at 60 C for a recombinant enzyme of *Thermus sp.* expressed in *E. coli* [63], and of more than 50 % at 75 C for *E. coli* containing an enzyme of *A. cellulolytics* [21, 22].

Immobilized enzymes or whole cells were used for practical applications. In some studies, high yields and productivities were achieved.

Immobilized L-arabinose isomerase in calcium alginate produced 145 g/L of tagatose with 48 % conversion of galactose and a productivity of 54 g/Lh in a packed-bed reactor [123]. An enzyme of *T. mathranii* immobilized in calcium alginate had its optimum at 75 C with a conversion rate of 43.9 % and a productivity up to 10 g/Lh with, however, lower conversion. After incubation of the resulting syrup with *S. cerevisiae,* purities above 95 % were achieved [85]. The enzyme of *T. neapolitana* immobilized on chitopearl beds gave a tagatose concentration of 138 g/L at 70 C [86].

Lactobacillus fermentum immobilized in calcium alginate had a temperature optimum of 65 C. A conversion rate of 60 % and a productivity of 11.1 g/Lh were obtained in a packed-bed reactor after addition of borate [156].

Direct production of tagatose in yogurt was possible by expressing the enzyme of *B. stearothermophilus* in *Lactobacillus bulgaricus* and *Streptococcus thermophilus* [116].

6.3 Others

A variety of other reduced-calorie or caloric sweeteners was studied in the course of the last years. Properties, production cost, or lack of advantages over established sweet-tasting carbohydrates resulted in no market success [152].

References

1. Abegaz EG, Mayhew DA, Butchko HH et al (2012) Aspartame. In: ÓBrien Nabors L (ed) Alternative sweeteners, 4th edn. CRC, Boca Raton
2. Ahmad I, Shim WY, Jeon WY et al (2012) Enhancement of xylitol production in *Candida tropicalis* by co-expression of two genes involved in pentose phosphate pathway. Bioprocess Biosyst Eng 35(1–2):199–204
3. Anonymous. US Code of Federal Regulations 21§172, 180 and 184 http://www.gpo.gov/fdsys/pkg/CFR-2012-title21-vol3/pdf/CFR-2012-title21-vol3-chapI-subchapB.pdf
4. Anonymous (1997a) Regulation (EC) 258/97 concerning novel foods and novel food ingredients. Off J L 43:1–6
5. Anonymous (1997b) Substances generally recognized as safe; proposed rule. Fed Regist 62:18938–18964
6. Anonymous (2001) GRAS Notice 76 http://www.accessdata.fda.gov/scripts/fcn/fcnDetailNavigation.cfm?rpt=grasListing&id=76
7. Anonymous (2003) European Commission. Scientific Committee on Food (2003) Opinion of the Scientific Committee on Food on Erythritol. SCF/CS/ADD/EDUL/215 Final. http://ec.europa.eu/food/fs/sc/scf/out175_en.pdf
8. Anonymous (2007) GRAS Notice 208 http://www.accessdata.fda.gov/scripts/fcn/fcnDetailNavigation.cfm?rpt=grasListing&id=208
9. Anonymous (2008) Regulation (EC) 1331/2008 establishing a common authorisation procedure for food additives, food enzymes and food flavourings. Off J L 354:1–6
10. Anonymous (2011a) Regulation (EU) 1169/2011 on the provision of food information to consumers. Off J L 304:18–63
11. Anonymous (2011b) Regulation (EU) 1129/2011 amending Annex II to Regulation (EC) No 1333/2008. Off J L 295:1–177

12. Anonymous (2011c) GRAS Notice387 http://www.accessdata.fda.gov/scripts/fcn/gras_notices/GRN000382.pdf
13. Baek H, Song KH, Park SM et al (2003) Role of glucose in the bioconversion of fructose into mannitol by *Candida magnoliae*. Biotechnol Lett 25(10):761–765
14. Barrow KD, Collins JG, Leight DA et al (1984) Sorbitol Production by *Zymomonas mobilis*. Appl Microbiol Biotechnol 20(4):225–232
15. Beadle JR, Saunders JP (1991) Process for manufacturing tagatose. US 5002612 (Biospherics)
16. Bippes MT (2007) Verfahrensentwicklung zur mikrobiellen Herstellung von Erythrit. Thesis Technische Unversität Braunschweig
17. Burschäpers J, Schustolla D, Schügerl K et al (2002) Engineering aspects of the production of sugar alcohols with the osmophilic yeast *Moniliella tomentosa var pollinis*: Part 2. Batch and fed-batch operation in bubble column and airlift tower loop if reactors. Process Biochem 38(4):559–570
18. Carakostas M, Prakash I, Kinghorn AD et al (2012) Steviol glycosides. In: ÓBrien Nabors L (ed) Alternative sweeteners, 4th edn. CRC, Boca Raton.
19. Carvalho W, Canilha L, Silva SS (2008) Semi-continuous xylose-to-xylitol bioconversion by Ca-alginate entrapped yeast cells in a stirred tank reactor. Bioprocess Biosyst Eng 5:493–498
20. Carvalheiro F, Moniz P, Duarte LC et al (2011) Mannitol production by lactic acid bacteria grown in supplemented carob syrup. J Ind Microbiol Biotechnol 38(1):221–227
21. Cheng L, Mu W, Jiang B (2010a) *Bacillus stearothermophilus* IAM 11001 for *D*-tagatose production: gene cloning, purification and characterisation. J Sci Food Agric 90(8): 1327–1333
22. Cheng L, Mu W, Zhang T et al (2010b) An L-arabinose isomerase from *Acidothermus cellulolytics* ATCC 43068: cloning, expression, purification, and characterization. Appl Microbiol Biotechnol 86(4):1089–1097
23. Cirino PC, Chin JW, Ingram LO (2006) Engineering *Escherichia coli* for xylitol production from glucose-xylose mixtures. Biotechnol Bioeng 95(6):1167–1176
24. Crueger W, Drath L, Munir D (1979) Process for the continuous isomerisation if saccharose to isomaltulose by means of microorganisms. EP0001099 (Bayer)
25. Daniell S, Mellits KH, Faus I et al (2000) Refolding the sweet-tasting protein thaumatin II from insoluble inclusion bodies synthesised in *Escherichia coli*. Food Chem 71(1):105–110
26. De Cock P (2012) Erythritol. In:ÓBrien Nabors L (ed) Alternative sweeteners, 4th edn. CRC, Boca Raton
27. Dominguez JM, Gong CS, Tsao GT (1997) Production of xylitol from D-xylose by *Debaryomyces hansenii*. Appl Biochem Biotechnol 63–65:117–127
28. Dominguez JM, Cruz JM, Roca E et al (1999) Xylitol production from wood hydrolyzates by entrapped *Debaryomyces hansenii* and *Candida guilliermondii* cells. Appl Biochem Biotechnol 81(2):119–130
29. Dooms GL, Hennebert GL, Verachtert H (1971) Polyol synthesis and taxonomic characters in the genus *Monilia*. Antonie van Leeuwenhoek 37(1):107–118
30. Edlauer R, Trimmel S (2008) Process for the producing erythritol using *Moniliella tomentosa* strains in the presence of inorganic neutral nitrates, such as potassium nitrate, ammonium nitrate or sodium nitrate, as nitrogen source. AT504230 (Jungbunzlauer Austria)
31. Faus I, Patio C, del Rio JL (1997) Expression of a synthetic gene encoding the sweet-tasting protein thaumatin in the filamentous fungus *Penicillium roquefortii*. Biotechnol Lett 19(12):1185–1191
32. Faus I, del Moral C, Androer N et al (1998) Secretion of the sweet-tasting protein thaumatin by recombinant strains of *Aspergillus niger var. awamori*. Appl Microbiol Biotechnol 49(4):393–398
33. Fouache C, Delobeau D, Quenon B (1999) Production of maltitol-rich syrup. EP0905256 (Roquette)
34. Gibs G (1984) Maltose and maltitol syrups and their preparation. EP0171964 (Lonza)

35. Gong CS, Chen LF, Tsao GT (1981) Quantitative production of xylitol from D-xylose by a high-xylitol producing yeast mutant *Candida tropicalis* HXP2. Biotechnol Lett 3(3):125–130
36. Goulter HC, Hashimi SM, Birch RG (2012) Microbial source isomerases. Enzyme Microb Technol 50(1):57–64
37. Guo C, Zhao C, He P et al (2007) Screening and characterization of yeasts for xylitol production. J Appl Microbiol 101(5):1096–1104
38. Hahm YT, Batt CA (1990) Expression and secretion of thaumatin from *Aspergillus oryzae*. Agric Biological Chem 54(10):2513–2520
39. Hainy GJ, Smith JH, Garver JC (1964) Erythritol production by a yeastlike fungus. Appl Environ Microbiol 12(3):240–246
40. Herrmann T (2003) Industrial production of amino acids by coryneform bacteria. J Biotechnol 104(1–3):155–172
41. Hidaka H, Kohno T, Eida T (1981) Method of producing high purity maltose. US4294623 (Meiji Seika)
42. Higginbotham JD (1986) Talin protein (thaumatin). In: Gelardi RC, ÓBrien Nabors L (eds) Alternative sweeteners. Marcel Dekker, New York
43. Hirata A (1999) Enzymatic method for producing *N*-formyl-alpha-*l*-aspartyl-L-phenylalanine methyl ester. WO 9816546 (Holland Sweetener)
44. Hirata Y, Igarashi K, Ezaki S et al (1999) High-level production of erythritol by strain 618A–01 isolated from pollen. J Biosci Bioeng 87(5):630–635
45. Hong YH, Lee DW, Lee SJ et al (2007) Production of D-tagatose at high temperatures using immobilized *Escherichia coli* cells expressing *l*-arabinose isomerase from *Thermotoga neapolitana*. Biotechnol Lett 29(4):569–574
46. Hong YH, Lee DW, Pyun YR et al (2011) Creation of metal-independent hyperthermophilic L-arabinose isomerase by homologous recombination. J Agric Food Chem 59(24): 12939–12947
47. Ide N, Kaneko R, Wada R et al (2007) Cloning of the thaumatin I cDNA and characterization of recombinant thaumatin I secreted by *Pichia pastoris*. Biotechnol Prog 23(5):1023–1030
48. Illingworth C, Larson G, Hellekant G (1989) Secretion of the sweet-tasting plant protein thaumatin by Streptomyces lividans. J Ind Microbiol Biotechnol 4(1):37–42
49. Ishizuka H, Wako K, Kasumi T et al (1989) Breeding of a mutant of *Aureobasidium sp.* with high erythritol production. J Ferment Bioeng 68(5):310–314
50. Izumori K, Tuzaki K (1988) Production of xylitol from D-xylulose by *Mycobacterium smegmatis*. J Ferment Technol 66(1):33–36
51. Jaffe GM, Szkrybalo W, Weinert PH (1971) Manufacture of xylitol. US 3627636 (Hoffmann La Roche)
52. Jamieson PR (2012) Sorbitol and mannitol. In: ÓBrien Nabors L (ed) Alternative sweeteners, 4th edn. CRC, Boca Raton
53. Jeon WY, Yoon BH (2012) Xylitol production is increased by expression of codon-optimized *Neurospora crassa* xylose reductase gene in *Candida tropicalis*. Bioprocess Biosyst Eng 35(1–2):191–198
54. Jeon YJ, Shin HS, Rogers PL (2011) Xylitol production from a mutant strain of *Candida tropicalis*. Lett Appl Microbiol 53(1):106–113
55. Jeya M, Lee KM, Tiwari MK et al (2009) Isolation of a novel high erythritol-producing *Pseudozyma tsukubaensis* and scale-up of erythritol fermentation to industrial level. Appl Microbiol Biotechnol 83(2):225–231
56. Jin CS, Cruz J, Jeffries TW (2005) Xylitol production by a *Pichia stipitis* D-xylulokinase mutant. Appl Microbiol Biotechnol 68(1):42–45
57. Jœrgensen F, Hansen OC, Stougaard P (2004) Enzymatic conversion of d-galactose to d-tagatose: heterologous expression and characterisation of a thermostable L-arabinose isomerase from *Thermoanaerobacter mathranii*. Appl Microbiol Biotechnol 64(6):816–822

58. Kasumi T (1995) Fermentative production of polyols and utilization for food and other products in Japan. Jpn Agric Res 29:49–55
59. Kawaguti HY, Manrich E, Sato HH (2006) Production of isomaltulose using *Erwinia sp.* D12 cells: culture medium optimization and cell immobilization in alginate. Biochem Eng J 29(3):270–277
60. Kearsley MW, Boghani N (2012) Maltitol. In: ÓBrien Nabors L (ed) Alternative sweeteners, 4th edn. CRC, Boca Raton
61. Khankal R, Luziatelli F, Chin JW et al (2008) Comparison between *Escherichia coli* K-12 strains W3110 and MG1655 and wild-type *E. coli* B as platforms for xylitol production. Biotechnol Lett 30(9):1645–1653
62. Kim HJ, Kim JH, Oh HJ et al (2006) Characterization of a mutated *Geobacillus stearothermophilus* L-arabinose isomerase that increases the production rate of D-tagatose. J Appl Microbiol 101(1):213–221
63. Kim JW, Kim YW, Roh HJ et al (2003) Production of tagatose by a recombinant thermostable L-arabinose isomerase from *Thermus sp.* IM6501. Biotechnol Lett 25(12):963–967
64. Kim P, Yoon SH, Seo MJ et al (2001) Improvement of tagatose conversion rate by genetic evolution of thermostable galactose isomerase. Biotechnol Appl Biochem 34(2):99–102
65. Kim SY, Lee K-H, Kim JH et al (1997) Erythritol production by controlling osmotic pressure in *Trigonopsis variabilis*. Biotechnol Lett 19(8):727–729
66. Kim SY, Oh DK, Jung SR (1999) Fermentation process for preparing xylitol using *Candida tropicalis*. US5998181 (Bolak)
67. Kinghorn AD, Compadre CM (2012) Less common high-potency sweeteners. In: ÓBrien Nabors L (ed) Alternative sweeteners, 4th edn. CRC, Boca Raton
68. Ko BS, Kim J, Kim JH (2006) Production of xylitol from d-xylose by a xylitol dehydrogenase gene-disrupted mutant of *Candida tropicalis*. Appl Environ Microbiol 72(6):4207–4213
69. Ko BS, Kim DM (2011) Enhancement of xylitol production by attenuation of intracellular xylitol dehydrogenase activity in *Candida tropicalis*. Biotechnol Lett 33(6):1209–1213
70. Koh ES, Lee TH, Lee DY et al (2003) Scale-up of erythritol production by an osmophilic mutant of *Candida magnoliae*. Biotechnol Lett 25(24):2103–2105
71. Kwon SG, Park SW, Oh DK (2006) Increase of xylitol productivity by cell-recycle fermentation of *Candida tropicalis* using submerged membrane bioreactor. J Biosci Bioeng 101(1):13–18
72. Ladero V, Ramos A, Wiersma A et al (2007) High-level production of the low-calorie sugar sorbitol by *Lactobacillus plantarum* through metabolic engineering. Appl Environ Microbiol 73(6):1864–1872
73. Lee DY, Park YC, Kim HJ et al (2003) Proteomic analysis of *Candida magnoliae* strains by two-dimensional gel electrophoresis and mass spectrometry. Proteomics 3(12):2330–2338
74. Lee JH, Weickmann JL, Koduri RK et al (1988) Expression of synthetic thaumatin genes in yeast. Biochem 27(14):5101–5107
75. Lee JK, Ha SK, Kim SY et al (2000) Increased erythritol production in *Torula sp.* by Mn^{2+} and Cu^{2+}. Biotechnol Lett 22(12):983–986
76. Lee JK, Ha SK, Kim SY et al (2001) Increased erythritol production in *Torula sp.* with inositol and phytic acid. Biotechnol Lett 23(7):497–500
77. Lee JK, Koo BS, Kim SY (2002) Fumarate-mediated inhibition of erythrose reductase, a key enzyme for erythritol production by *Torula corallina*. Appl Environ Microbiol 68(9):4534–4539
78. Lee JK, Kim SY, Ryu YW et al (2003a) Purification and characterization of a novel erythrose reductase from *Candida magnoliae*. Appl Environ Microbiol 69(7):3710–3718
79. Lee JK, Koo BS, Kim SY et al (2003b) Cloning and characterization of the xyl1 gene, encoding an NADH-preferring xylose reductase from *Candida parapsilosis* and its functional expression in *Candida tropicalis*. Appl Environ Microbiol 69(10):6179–6188

80. Lee KJ, Lim JY (2003) Optimized conditions for high erythritol production by *Penicillium sp.* KJ-UV29, mutant of *Penicillium sp.* KJ81. Biotechnol. Bioprocess Eng 8(3):173–178
81. Lee WH (1967) Carbon balance of a mannitol fermentation and the biosynthetic pathway. Appl Environ Microbiol 15(5):1206–1210
82. Lee WJ, Ryu Seo JH (2000) Characterization of two-substrate fermentation processes for xylitol production using recombinant *Saccharomyces cerevisiae* containing xylose reductase gene. Process Biochem 35(10):1199–1203
83. Leuchtenberger W, Huthmacher K, Drauz K (2005) Biotechnological production of amino acids and derivatives: current status and prospects. Appl Microbiol Biotechnol 69(1):1–8
84. Li Y, Zhu Y, Liu A et al (2011) Identification and characterization of a novel L-arabinose isomerase from *Anoxybacillus flavithermus* useful in D-tagatose production. Extremophiles 15(3):441–450
85. Liang M, Chen M, Liu X (2012) Bioconversion of D-galactose to D-tagatose: continuous packed bed reaction with an immobilized thermostable L-arabinose isomerase and efficient purification by selective microbial degradation. Appl Microbiol Biotechnol 93(4): 1469–1474
86. Lim BC, Kim HJ, Oh DK (2008) Tagatose production with pH control in a stirred tank reactor containing immobilized L-arabinose rom *Thermotoga neapolitana*. Appl Biochem Biotechnol 149(3):245–253
87. Lin SJ, Wen CY, Liau JC et al (2001) Screening and production of erythritol by newly isolated osmophilic yeast-like fungi. Process Biochem 36(12):1248–1259
88. Lin SJ, Wen CY, Wang PM et al (2010) High-level production of erythritol by mutants of osmophilic *Moniliella sp.* Process Biochem 45(6):973–979
89. Maguire A (2006) Dental health. In: Mitchell H (ed) Sweeteners and sugar alternatives in food technology. Blackwell, Oxford
90. Marina Ayoki MAY, Pastore GM, Park YK (1993) Microbial transformation of sucrose and glucose to erythritol. Biotechnol Lett 15(4):383–388
91. Masuda T, Tamaki S, Ryosuke K et al (2004) Cloning, expression and characterization of recombinant sweet-protein thaumatin II using the methylotrophic yeast *Pichia pastoris*. Biotechnol Bioeng 85(7):761–769
92. Meyrial V, Delgenes JP, Moletta R et al (1991) Xylitol production from D-xylose by *Candida guillermondii*: fermentation behaviour. Biotechnol Lett 13(4):281–286
93. Miyake T (1980) Process for producing a sweetener. US4219517 (Hayashibara)
94. Mizutani K, Kuramoto T, Tamura Y et al (1994) Sweetness of glycyrrhetic acid 3-*O*-beta-*D*-monoglucuronide and the related glycosides. Biosci Biotechnol Biochem 58(3):554–555
95. Moon HJ, Jeya M, Kim IW et al (2010) Biotechnological production of erythritol and its applications. Appl Microbiol Biotechnol 86(4):1017–1025
96. Moralejo FJ, Cardosa EJ, Gutierrez S et al (2002) Silencing of the aspergillopepsin B (pepB) gene of *Aspergillus awamori* by antisense RNA expression or protease removal by gene disruption results in a large increase in thaumatin production. Appl Environ Microbiol 68(7):3550–3559
97. Norman BE, Vang Hendriksen H (2002) Method for producing maltose syrup by using a hexosyltransferase.WO2002010427 (Novozymes)
98. Nyyssölä A, Pihlajaniemi A, Palva A et al (2005) Production of xylitol from d-xylose by recombinant *Lactococcus lactis*. J Biotechnol 118(1):55–66
99. Oh DK, Kim SY, Kim JH (1998) Increase of xylitol production rate by controlling redox potential in *Candida parapsilosis*. Biotechnol Bioeng 58(4):440–444
100. Oh HJ, Kim HJ, Oh DK (2006) Increase in D-tagatose production rate by site-directed mutagenesis of *l*-arabinose isomerase from *Geobacillus thermodenitrificans*. Biotechnol Lett 28(3):145–149
101. Ojamo H, Koivikko H, Heikkil H (2000) Process for the production of mannitol by immobilized micro-organisms.US 6602691 (Xyrofin)

102. Omar SH, Honecker S, Rehm HJ (1992) A comparative study on the formation of citric acid and polyols and on morphological changes of three strains of free and immobilized *Aspergillus niger*. Appl Microbiol Biotechnol 36(4):518–524
103. Onishi H, Suzuki T (1969) Microbial production of xylitol from glucose. Appl Environ Microbiol 18(6):1031–1035
104. Onishi H, Suzuki T (1968) Production of D-mannitol and glycerol by yeasts. Appl Environ Microbiol 16(12):1847–1852
105. Onishi H, Suzuki T (1971) Production of mannitol and glycerol by yeasts. Appl Environ Microbiol 16(12):1847–1852
106. Parajó JC, Dominguez H, Dominguez JM (1997) Improved xylitol production with *Debaryomyces hansenii* Y-7426 from raw or detoxified wood hydrolysates. Enzyme Microb Technol 21(1):18–24
107. Paterson SL, Fane AG, Fell CJD et al (1988) Sorbitol and gluconate production in a hollow fibre membrane reactor by immobilised *Zymomonas mobilis*. Biocatal Biotransformation 1(3):217–229
108. Povelainen M, Miasnikov AN (2007) Production of xylitol by metabolically engineered strains of *Bacillus subtilis*. J Biotechnol 128(1):24–31
109. Pedersen S, Vang Hendriksen H (2002) Method for production of maltose and/or enzymatically modified starch. EP1214442 (Novozymes)
110. Pham NB, Schäfer S, Wink M (2012) Production and secretion of recombinant thaumatin in tobacco hairy root cultures. Biotechnol J 7(4):537–545
111. Quaedflieg PJLM, Sonke T, Wagner AFV (2000) Synthesis and recovery of aspartame involving enzymatic deformylation step. EP 1013663 (Holland Sweetener)
112. Racine FM, Saha BC (2007) Production of mannitol by *Lactobacillus intermedius* NRRL B-3693 in fed-batch and continuous cell-recycle fermentations. Process Biochem 42(12):1609–1613
113. Rangaswamy S, Agblevor F (2002) Screening of facultative anaerobic bacteria utilizing D-xylose for xylitol production. Appl Microbiol Biotechnol 60(1–2):88–93
114. Rao RS, Jyothi CP (2006) Strain improvement of *Candida tropicalis* for the production of xylitol: biochemical and physiological characterization of wild-type and mutant strain CT-OMV5. J Microbiol 44(1):113–120
115. Rehr B, Wilhelm C, Sahm H (1991) Production of sorbitol and gluconic acid by permeabilized cells of *Zymomonas mobilis*. Appl Microbiol Biotechnol 35(2):144–148
116. Rhimi M, Chouayekh H, Gouillouard I et al (2011) Production of D-tagatose, a low caloric sweetener during milk fermentation using L-arabinose isomerase. Bioresour Technol 102(3):3309–3315
117. Ro HS, Kim HS (1991) Continuous production of gluconic acid and sorbitol from sucrose using invertase and an oxidoreductase of *Zymomonas mobilis*. Enzyme Microbial Technol 13(11):920–924
118. Roh HJ, Kim P, Park YC et al (2000) Bioconversion of *D*-galactose into *D*-tagatose by expression of L-arabinose isomerase. Biotechnol Appl Biochem 31(1):1–4
119. Roca E, Meinander N, Hanh Hägerdal B (1996) Xylitol production by immobilized recombinant *Saccharomyces cerevisiae* in a continuous packed-bed bioreactor. Biotechnol Bioeng 51(3):317–326
120. Rose T, Kunz M (2002) Production of Isomalt. Landbauforsch Völkenrode (2002) SH 241:75–80
121. Rymowicz W, Rywinska A, Gladkowski W (2008) Simultaneous production of citric acid and erythritol from crude glycerol by *Yarrowia lipolytica Wratislavia* K1. Chem Pap 68(3):239–246
122. Rymowicz W, Rywinska A, Marcinkiewicz M (2009) High-yield production of erythritol from raw glycerol in fed-batch cultures of *Yarrowia lipolytica*. Biotechnol Lett 31(3): 377–380

123. Ryu SA, Kim CS, Kim HJ et al (2003) Continuous d-tagatose production by immobilized thermostable *l*-arabinose isomerase in a packed-bed bioreactor. Biotechnol Prog 19(6):1643–1647
124. Ryu YW, Park CY, Park JB et al (2000) Optimization of erythritol production by *Candida magnoliae* in fed-batch culture. J Ind MicrobiolBiotechnol 25(2):100–103
125. Saburo C, Toshiro O (1999) Process for producing erythritol by using microorganism. EP 935002 (Nikken Chemicals)
126. Saha BC (2006) A low-cost medium for mannitol production by Lactobacillus intermedius NRRL B-3693. Appl Microbiol Biotechnol 72(4):676–680
127. Saha BC, Bothast RC (1999) Production of xylitol by *Candida peltata*. J Ind Microbiol Biotechnol 22(6):633–636
128. Saha BC, Nakamura LK (2003) Production of mannitol and lactic acid by fermentation with *Lactobacillus intermedius* NRRL B-3693. Biotechnol Bioeng 82(7):864–871
129. Sánchez S, Bravo V, Castro E et al (1998) The production of xylitol from D-xylose by fermentation with *Hansenula polymorpha*. Appl Biotechnol Microbiol 50(5):508–611
130. Sasaki M, Jojima T, Inui M et al (2010) Xylitol production by recombinant *Corynebacterium glutamicum* under oxygen deprivation. Appl Microbiol Biotechnol 86(4):1057–1066
131. Sawada K, Taki A, Yamakawa T et al (2009) Key role for transketolase activity in erythritol production by *Trichosporonoides megachiliensis* SN-G42. J Biosci Bioeng 108(5):385–390
132. Sentko A, Bernard J (2012) Isomalt. In: ÓBrien Nabors L (ed) Alternative sweeteners, 4th edn. CRC, Boca Raton
133. Silva SS, Afschar AS (1994) Production of xylitol from D-xylose using *Candida tropicalis*. Bioprocess Biosyst Eng 11(4):129–134
134. Silva SS, Felipe MG et al (1998) Factors that affect the biosynthesis of xylitol by xylose-fermenting yeasts: a review. Appl Biochem Biotechnol 70–72:331–339
135. Smiley KL, Cadmus MC, Liepins P (1967) Biosynthesis of D-mannitol from D-glucose by *Aspergillus candidus*. Biotechnol Bioeng 9(3):365–374
136. Soetart W, Vanhooren PT, Vandamme EJ (1999) The production of mannitol by fermentation. In: Bucke C (ed) Carbohydrate biotechnology protocols, methods in biotechnology, vol 10. Springer, Heidelberg
137. Song KH, Lee JK, Song JY et al (2002) Production of mannitol by a novel strain of *Candida magnoliae*. Biotechnol Lett 24(1):9–12
138. Song KH, Baek H, Park SM et al (2003) *Candida magnoliae* producing mannitol and fermentation method for producing mannitol. US20010978559 (Biongene and Bolak)
139. Spencer JFT, Gorin PAJ (1960) The biosynthesis of erythritol and glycerol by *Torulopsis magnoliae*. Studies with C14-labelled glucose. Can J Biochem Physiol 38(2):157–164
140. Spencer JFT, Sallans HR (1956) Production of polyhydric alcohols by osmophilic yeasts. Can J Microbiol 2(2):72–79
141. Spinner C, Wagner A (2008) Method for production of a syrup with a high maltitol content. WO 2008029033 (Syral)
142. Sreenivas Rao R, Bhadra B, Shivaji S (2007) Isolation and characterization of xylitol-producing yeasts from the gut of *Colleopteran* insects. Curr Microbiol 55(5):441–446
143. Susumu H, Kuriiwa NMSK, Iwasaki S (1990) Process for producing isomaltulose EP0392556 (Meito Sangyo)
144. Tani V, Vongsuvanlert V (1987) Sorbitol production by a methanol yeast, *Candida boidinii* (Kloeckera sp.) No. 2201. J Ferment Technol 65(4):405–411
145. Tarocco F, Lecuona RE, Cuoto AS et al (2005) Optimization of erythritol and glycerol accumulation in conidia of *Beauveria bassiana* by solid-state fermentation, using response surface methodology. Appl Microbiol Biotechnol 68(4):481–488
146. Tochampa W, Sirisansaneeyakul S, Vachnisriratana W et al (2005) A model of xylitol production by the yeast *Candida mogii*. Bioprocess Biosyst Eng 28(3):175–183
147. Troostembergh JCM, Debonne IA, Obyn WR (2001) Process of producing erythritol from culture medium containing the same. EP1106698 (Cerestar Holding)

148. Vang Hendriksen H, Mathiasen TE, Adler-Nissen J et al (1988) Production of mannitol by *Penicillium* strains. J Chem Technol Biotechnol 43(3):223–228
149. Vastenavond CM, Berstelsen H, Juhl Hansen S et al (2012) Tagatose (d-tagatose). In: ÓBrien Nabors L (ed) Alternative sweeteners, 4th edn. CRC, Boca Raton
150. Vongsuvanlert V, Tani V (1989) Xylitol production by a methanol yeast, *Candida boidinii* (Kloeckera sp.) No. 2201. J Ferment Bioeng 67(1):35–39
151. von Rymon Lipinski GW (1995) Sweeteners. In: Ullmanńs encyclopedia of industrial chemistry, 5th edn. VCH, Weinheim
152. von Rymon Lipinski GW (2006) Reduced-calorie sweeteners and caloric alternatives. In: Spillane WJ (ed) Optimising sweet taste in foods Woodhead. Cambridge and CRC, Boca Raton
153. von Weymarn N, Hujanen M, Leisola M (2002a) Production of D-mannitol by heterofermentative lactic acid bacteria. Process Biochem 37(11):1207–1213
154. von Weymarn N, Kiviharju K, Leisola M (2002b) High-level production of D -mannitol with membrane cell-recycle bioreactor. J Ind Microbiol Biotechnol 29(1):44–49
155. Wu L, Birch RG (2004) Characterization of *Pantoea dispersa* UQ68 J: producer of a highly efficient sucrose isomerase for isomaltulose biosynthesis. J Appl Microbiol 97(1):93–103
156. Xu Z, Ki S, Fu F et al (2012) Production of D-tagatose, a functional sweetener, utilizing alginate immobilized *Lactobacillus fermentum* CGMCC2921 cells. Appl Biochem Biotechnol 166(4):961–973
157. Yamamoto K, Yoshikawa K, Okada S (1994) Effective production of glycosyl-steviosides by alpha-1,6 transglucosylation of dextrin dextranase. Biosci Biotechnol Biochem 58(9):1657–1661
158. Yang SW, Park JB et al (1999) Production of erythritol from glucose by an osmophilic mutant of *Candida magnoliae*. Biotechnol Lett 21(10):887–890
159. Zacharis C, Stowell J (2012) Xylitol. In: ÓBrien Nabors L (ed) Alternative sweeteners, 4th edn. CRC, Boca Raton

Adv Biochem Eng Biotechnol (2014) 143: 29–49
DOI: 10.1007/10_2013_234

Published Online: 2 November 2013

Biopreservatives

Dieter Elsser-Gravesen and Anne Elsser-Gravesen

Abstract Food producers of today are met with inherently contradictory demands as seen from a microbiological point of view: producing foods that are less stable (due to nutritional and taste requirements) by processes that confer less control of the detrimental microflora (due to trends of convenience, minimal processing, and reducing or removing additives including preservatives). How should food producers manage to develop such products with a sufficiently long shelf-life and at a competitive price? Some of the most promising tools to this end are the so-called biopreservatives, which are various types of products derived from lactic acid bacteria and other suitable microorganisms, namely bacteriocins and other antimicrobials, fermentates, bioprotective cultures, and bacteriophages. This chapter provides an overview of the scientific background and functionality, as well as food applications and further commercial aspects of each of these categories of biopreservatives.

Contents

D. Elsser-Gravesen (✉) · A. Elsser-Gravesen
ISI Food Protection ApS, Aarhus, Denmark
e-mail: deg@isifoodprotection.com

1 Introduction

The exploitation of biopreservation is by no means a new concept. Biotechnological processes for preserving food have already been used for thousands of years, even though the underlying mechanisms were not understood. Today, biopreservation of foods is as relevant as ever before because it is one of the few possible answers to what at first glance appears to be totally contradictory trends and demands:

- Health trends: The levels of salt, sugar and fat in foods are under pressure to be reduced. These changes are beneficial for human health, but they also all confer an increase in water activity, which provides a friendlier environment for microorganisms.
- Taste preferences: In many products, trends are towards a milder (i.e. less acidic) taste, which results in a higher pH that again is less adverse for microorganisms.
- Perception of "natural": This results in milder or minimal processing, which results in a fresher appearance of the food but also less inactivation of unwanted microorganisms. Furthermore, it increases the demand for "preservative-free" products.
- Convenience trends ("practically homemade"): There are two main risks associated with this trend—namely, more extensive processing, which results in more steps in which contamination with detrimental microorganisms can occur, and the need for proper handling by the consumer (e.g. sufficient heating), which may be neglected.
- Durability and open shelf-life: Market access and economically viable logistics require a long shelf-life. Furthermore, a sufficient open shelf-life is required to ensure customer loyalty.
- Ethical issues: Concerns such as corporate social responsibility, carbon dioxide (CO_2) footprint, and fair-trade and organic products put restrictions on which solutions a food producer can employ.

All in all, these trends lead to food formulations that provide better growth conditions for microorganisms, milder processing that results in less initial reduction, more processing steps that increase the risk of contamination, a need for longer shelf-life, and pressure to reduce food waste. In addition, many of the conventional preservatives are deemed to be unacceptable by trendsetters and consumers. Everyone wants preservative-free food, but most will agree that we cannot maintain our present society and standard of living—and certainly cannot reduce the global food waste problems—with food that is not preserved.

There is thus a strong market need for natural food protection solutions that can ensure both food safety (i.e. reduce the number and/or outgrowth of pathogenic microorganisms) and food shelf-life (i.e. delayed development of the spoilage microflora). One of the few possible solutions is biopreservation based on the concept of using food-grade microorganisms as so-called cell factories (Fig. 1).

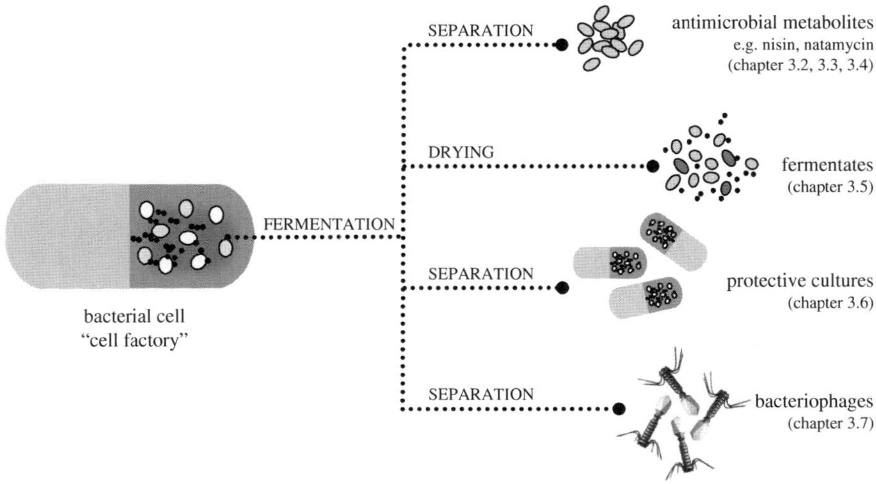

Fig. 1 Overview of the main categories of biopreservatives that can be produced by using lactic acid bacteria and other suitable microorganisms as "cell factories"

Food-grade microorganisms can form a multitude of different substances that are inhibitory to other microorganisms. These mechanisms are part of the natural balance in complex microbial ecosystems. By exploiting the fittest of the naturally occurring microorganisms in organoleptically appealing food products, it is conceivable to design preservation systems that ensure an adequate safety and shelf-life while maintaining the desired quality of the food product.

The biopreservation principles from food-grade microorganisms can be categorized according to the antimicrobial compound (e.g. bacteriocin, other metabolites, bacteriophages, enzymes) as well as product format (purified antimicrobial, fermentate, protective culture), as illustrated in Fig. 1. Not all microbial inhibitory mechanisms are fully understood, and not all antimicrobial metabolites from food-grade microorganisms have yet been discovered. It is highly likely that the near future will bring new understanding and discoveries, which will further expand the options for natural food biopreservation systems.

When investigating biopreservation systems, one should not search for "the silver bullet". As described by Roller [76]: "Antimicrobial compounds in nature rarely function in isolation; combination systems such as those found in the hen's egg are far more common." This principle is central for developing sound biopreservation solutions. Targeted intelligent strategies based on multifactorial systems are the most likely to succeed for protecting food against detrimental microorganisms.

This chapter gives an overview of the current knowledge on biopreservative compounds and concepts, covering solutions that are actually being used industrially today, and points out perceivable directions for future solution development.

2 Nisin

Nisin is a cationic, amphiphilic peptide produced by various strains of *Lactococcus lactis*, which has a relatively broad target spectrum that inhibits a wide range of gram-positive bacteria. The antimicrobial property of nisin was first observed in 1928, when it was reported that inhibition of a dairy starter culture was caused not by phages but by a strain of *L. lactis* (formerly called lactic streptococci and group N streptococci) [75]. The inhibitory compound was further studied the following years and given the name nisin, alluding to its origin as a "group **N** streptococci **I**nhibitory **S**ubstance" [65]. The application of nisin for preservation of dairy products was suggested already in 1951 for inhibiting blowing of Swiss-type cheese [49]. Soon after, the first commercial preparation was made by Aplin and Barret in 1953. The use of nisin as a food preservative was approved by the Food and Agriculture Organization of the United Nations and World Health Organization (WHO) in 1969, by the European Union (EU) in 1983 (E 234), and granted Generally Recognized As Safe (GRAS) status by the U.S. Food and Drug Administration (FDA) in 1988.

Thus, nisin has a long history of safe use in food. It is the only purified bacteriocin that is widely approved as food additive—a fact which presumably also reflects its early discovery. Through the years, a substantial number of scientific papers have described the biosynthesis, chemical and physical properties, mode of action, and practical applications of nisin, and furthermore the accrued knowledge has been extensively reviewed. (For comprehensive reviews, see [13, 25, 87, 89]. A short summary is given below, with focus on aspects that affect industrial applications.

Nisin belongs to the lanthionine-containing bacteriocins, which are designated as class I bacteriocins (Table 1). Production of bacteriocins containing the unusual lanthionine residues, which are formed by posttranslational modifications, is not uncommon amongst lactic acid bacteria; linear, globular, and two-peptide variants have been characterized. Many of these peptides are effective at low concentrations against a wide range of gram-positive bacteria, which has been attributed to a common mode of action: nisin and other lantibiotics bind with high affinity to a docking molecule in the cell envelope of target bacteria, lipid II, an intermediary molecule for building bacterial cell walls [52]. Nisin is a linear lantibiotic that exerts its antibacterial action through inhibiting cell wall formation as well as forming membrane pores; it is furthermore active against spores. Several variants of nisin occur naturally; the two that are currently available as commercial products, nisin A and nisin Z, differ in one amino acid, which confers a difference in charge and solubility.

Lipid II is an essential and highly conserved molecule, providing the broad target spectrum of lantibiotics against gram-positive bacteria. However, in gram-negative bacteria, lipid II is protected under the outer membrane. These organisms are therefore only sensitive to lantibiotics in cases where their outer membrane has been disrupted. The producer organisms, being gram-positive bacteria themselves,

Table 1 Classification of bacteriocins produced by lactic acid bacteria as suggested by Cotter et al. [13]

Classification	Remarks/suggestions	Examples
Class I		
Lanthionine-containing bacteriocins/lantibiotics	Includes both single- and two-peptide lantibiotics; up to 11 subclasses have been proposed	Single-peptide: nisin, mersacidin, lacticin 481 Two-peptide: lacticin 3147, cytolysin
Class II		
Non-lanthionine-containing bacteriocins	Heterogeneous class of small peptides; includes pediocin-like (subclass a bacteriocins), two-peptide (subclass b bacteriocins), cyclic (subclass c; formerly class V), non-pediocin single linear peptides (subclass d)	Class IIa: pediocin PA1, leucocin A Class IIb: lactacin F Class IIc: enterocin AS48, reuterin 6 Class IId: lactococcin A, divergicin A
Bacteriolysins (Suggested that these are no longer considered bacteriocins)		
Non-bacteriocin lytic proteins	Large, heat-labile proteins, often murein hydrolases	Lysostaphin, enterolysin A

are protected by a dedicated immunity system that is encoded in conjunction with the biosynthesis genes [2].

A general concern with the prolonged use of an antimicrobial is the potential development of resistance in the target microorganisms. A moderate reduction in nisin sensitivity due to various modifications in the cell envelope has been described for laboratory-acquired mutants [55]; a prevalent mechanism involved enhanced expression of a penicillin-binding protein [41], which could reduce the accessibility or the affinity of nisin to lipid II. However, high level or complete resistance to nisin has not been observed, presumably due to the essential nature of the docking molecule. Transfer of the immunity system from producer to target organisms has similarly not been reported.

The tested and actual applications of nisin are numerous. Initially, nisin was used in conjunction with heat treatment to prevent spoilage of processed cheese by heat-resistant spores. Since then, effective use of nisin has been demonstrated both for shelf-life and safety purposes in various types of food, including dairy products and processed meats and vegetables [34, 81]. Nisin is particularly effective in heat-treated low pH products.

Technical limitations to be aware of relate to the characteristics of the nisin molecule as well as the mode of action. Nisin is sensitive to degradation by peptidolytic enzymes (e.g. in raw meat) and can be sequestered in food matrices (e.g. in the fat phase). In addition, nisin is relatively heat-stable at low pH but not at neutral or higher pH. Furthermore, if used in fermented products, nisin will inhibit gram-positive starter cultures.

The efficacy and application range of nisin, like any other antimicrobial, can be expanded by use in a multifactorial system. Nisin can be protected from peptidolytic enzymes or sequestering by incorporation in liposomes [63] or incorporation in edible coatings or films [11]. Efficacy and target range can be enhanced by combination with plant extracts or essential oils or with physical treatments, such as high hydrostatic pressure [48, 91].

As mentioned previously, nisin was first approved for food applications in 1969. The initial approvals were based on toxicity testing results from 1962 [33, 46]. Recently, two independent studies have shown that even a very high daily intake was not toxic [44, 45]. In the EU, nisin is currently approved as additive in ripened and processed cheese, clotted cream, puddings such as semolina or tapioca, mascarpone, and pasteurized liquid egg. In the United States and Australia/New Zealand, further approvals have been granted, such as for use in sauces, soups, salads, dressings, and ready-to-eat and processed meat products.

Nisin-containing products on the market are manufactured by a batch fermentation process followed by concentration, drying, milling, and standardization [21]. For many years, Nisaplin, which contains 2.5 % nisin A and is standardized with salt, was the main product on the market (developed by Aplin and Barrett, now DuPont). In recent years, other producers have emerged, and both nisin A and nisin Z are now commercially available. However, toxicology studies were performed with nisin A.

In addition to the products consisting of standardized nisin A or Z, some combination products are also available, such as those combined with rosemary extract. Furthermore, nisin-producing cultures are available (see Sect. 5).

3 Natamycin

Natamycin was discovered in the 1950s. As described by Struyk et al. [84], "A new crystalline antibiotic, pimaricin, has been isolated from fermentation broth of a culture of a *Streptomyces* species, isolated from a soil sample obtained near Pietermaritzburg, State of Natal, Union of South Africa. This organism has been named *Strepyomyces natalensis*". The original name "pimaracin" can be found in earlier publications but it is no longer accepted by the WHO [24]. Natamycin is classified as a macrolide polyene antifungal and is characterized by a macrocyclic lactone-ring with a number of conjugated carbon–carbon double bonds (Fig. 2). The full chemical name is 22-(3-amino-3,6-dideoxy-β-D-manno pyranosol) oxy-1,3,26 trihydroxy-12-methyl-10-oxo-6,11,28-trioxiatri [22.3.1.05.7] o catosa-8,14,16,18,20-pentanene-25-carboxylic acid.

Natamycin has a low solubility in water (approximately 40 ppm), but the activity of neutral aqueous suspensions is very stable. Natamycin is stable to heat and it is reported that heating processes for several hours at 100 °C lead to only slight activity losses. Natamycin is active against almost all foodborne yeasts and molds but has no effect on bacteria or viruses. The sensitivity to natamycin in vitro (minimal inhibitory concentration) is in most cases below 20 ppm (see Table 2).

Natamycin acts by binding irreversibly with ergosterol and other sterols, which are present in the cell membranes of yeasts and vegetative mycelium of molds. It

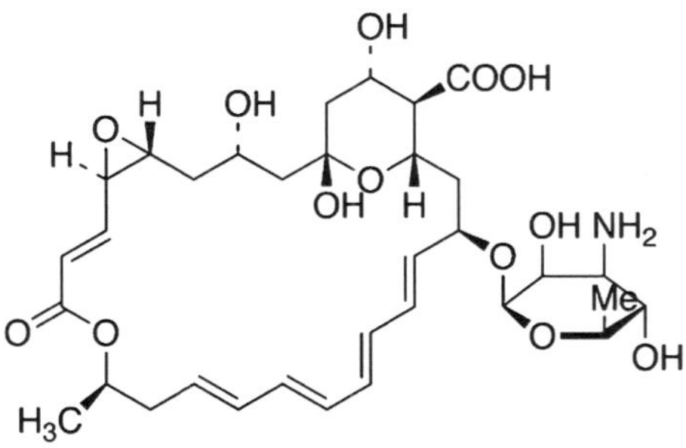

Fig. 2 The chemical structure of natamycin

Table 2 Sensitivity to natamycin of fungi occurring on sausages (Adapted from [83]

Molds and yeasts	Minimal inhibitory concentration (ppm)	Source of microorganisms
Aspergillus flavus	10–20	Air
Aspergillus niger	<5	Fruit
Cladosporium cladosporioides	<5	Meat stamp
Eurotium appendiculatum	<5	Smoked sausage
Mucor racemosus	<5	Sausage
Penicillium chrysogenum	<5	Meat
Penicillium nalgiovense	<5	Sausage
Rhizopus stolonifer	5–10	Bread
Candida zeylandoides	<5	Sausage
Debaryomyces hansenii	<5	Sausage
Rhodutorula mucilaginosa	<5	Air
Trichosporon pullulans	<5	Frozen beef

disrupts the cell membrane and increases the cell permeability, which finally leads to cell death. The fungicidal of natamycin is an "all-or-none" effect, which destroys the cell membrane of the target cells [57].

Due its interaction with ergosterol, which is a major constituent of fungal cells, it is unlikely that fungi will develop resistance. So far, after many decades of use, no development of resistance has been reported. Natamycin is mostly used for surface applications, particularly for treating surfaces of hard cheese and salami-type sausages. One of the advantages over sorbate is that even the dissolved fraction of natamycin hardly migrates into the food matrix. As shown in Table 3, natamycin can be applied by spraying the surface (e.g. of cheese), by dipping, by applying natamycin via coating emulsions or by direct addition.

The antifungal efficacy of natamycin has been extensively studied and a substantial amount of scientific papers have been published. Comprehensive overview articles are available [20, 24, 83]. However, due to its long history of use, no data on application studies have been published recently.

Natamycin does not have acute toxicity. In animal studies, the lowest median lethal dose found was 450 mg/kg. The long history of safe use in food products

Table 3 Applications of natamycin, with suggested dosage levels and methods of application

Food application	Suggested natamycin dosage levels (ppm)	Method
Hard/semihard cheese	1,250–2,000	Surface treatment by spray or immersion
	500	Direct addition to coating emulsion
Meat products: dry sausage	1,250–2,000	Surface treatment by spray or immersion
Yogurt	5–10	Direct addition to yogurt mix
Bakery products	1,250–2,000	Surface treatment by spray
Tomato purée/paste	7.5	Direct addition during mixing
Fruit juice	2.5–10	Direct addition
Wine	30–40	Direct addition to stop fermentation
	3–10	Added after bottling to prevent yeast/mold growth

Source [88]

confirms that natamycin is a safe antifungal preservative. As acceptable daily intake, 0.3 mg/kg body weight/day was suggested Smith and Moss [82]. In the scientific opinion on the use of natamycin (E 235) as a food additive from the European Food Safety Authority [26], "The Panel considered that the proposed use of levels of natamycin are not of safety concern if it is only used for the surface treatment of the rind of semi-hard and semi-soft cheese and on casings of certain sausages. The Panel concluded that there was no concern for the induction of antimicrobial resistance."

Natamycin is allowed as antifungal preservative in many countries, but details on authorization vary from country to country. In the European Union, natamycin is permitted for the surface treatment of hard, semihard, and semisoft cheese and dried cured sausages. According the EU Directive 1333/2008, the maximum permitted level is 1 mg/dm^2 surface.

Commercial preparations are produced by fermentation of sugar-based substrates by *Streptomyces natalensis*. Natamycin is then recovered by extraction, filtration, and spray drying. The dried powder can be stored for years without any activity loss.

4 Other Bacteriocins

Microorganisms produce a diverse range of microbial defense molecules, including the classical antibiotics, numerous types of protein exotoxins, lytic agents, metabolic byproducts, and bacteriocins [74]. The latter group has received particular attention due to a perceived high potential for exploitation for food preservation [13].

Bacteriocins are ribosomally synthesized, extracellularly released antimicrobial peptides that have a bacteriocidal or bacteriostatic effect on other microorganisms.

Bacteriocin production is a common feature of food-grade lactic acid bacteria (LAB). The first discovered bacteriocin was nisin, as described previously, and it was subsequently estimated that up to 99 % of all bacteria produce at least one bacteriocin [56]. Since then, the variety of both bacteriocin-producing LAB and bacteriocin molecules has proven to be very diverse. It has thus become evident that the natural presence of bacteriocins in microbial ecosystems (e.g. fermented foods) is quite common.

Parallel to the discovery of new bacteriocins, schemes for classifying the molecules have been proposed and modified. Table 1 shows the modification of Klaenhammer's original classification [56], which was suggested by Cotter et al. [14]. However, it is likely that the classification will still be modified as more knowledge on the LAB bacteriocins emerges (e.g. [47]. Class I, the lantibiotics (including nisin), and class II, the unmodified bacteriocins (including the class IIa pediocin-like antilisterial bacteriocins), constitute the most abundant, the best characterized, and presumably also the most useful of the food-grade bacteriocins.

Food-relevant class I bacteriocins include the lactococcal lantibiotics, lacticin 3147 and lacticin 481, which have shown good efficacy for both shelf-life and food safety purposes, particularly in dairy products [43]. The lantibiotics generally have a wide target range conferred by their interaction with lipid II, similarly to nisin, as described in detail in Sect. 2.

The class IIa bacteriocins are small, heat-stable unmodified peptides with a conserved YGNGV motif in their N-terminal domain. They are often referred to as pediocin-like, reflecting that some of the first characterized IIa bacteriocins were produced by *Pediococcus* [70]. It is now apparent that this is a really widespread type of family of peptides [14]. As listed in Table 4, IIA bacteriocins are produced by a variety of LAB in addition to *Pediococcus* including *Lactobacillus*, *Enterococcus*, *Carnobacterium*, *Leuconostoc*, *Streptococcus* and *Weissella*. The IIa-producing LAB have been isolated from various dairy, fermented sausage, and vegetable products and also from the mammalian gastrointestinal tract. Furthermore, the non-LAB species *Bifidobacterium bifidum*, *Bifidobacterium infantis*, *Bacillus coagulans* and *Listeria innocua* have also been reported to produce IIa bacteriocins.

In contrast to nisin, the class IIa bacteriocins have a relatively narrow target spectrum. They are generally active against *Listeria* species, and against some species of *Clostridium, Enterococcus, Carnobacterium, Lactobacillus, Pediococcus* and occasionally *Streptococcus* and *Leuconostoc*. These bacteriocins have thus primarily been tested in foods for their antilisterial properties. The class IIa specificity reflects their mode-of-action: the IIa bacteriocins bind to a target molecule, a sugar uptake system called the mannose phosphotransferase system (man-PTS), and subsequently form membrane pores and kill the sensitive cells. The man-PTS sequence is not completely conserved, and the variation in man-PTS sequences are reflected in varying sensitivities to IIa bacteriocins, from highly sensitive to completely insensitive [14]. Furthermore, elimination of the man-PTS expression can confer high-level resistance to class IIa bacteriocins in otherwise sensitive organisms [55], and this appears as a general mechanism for spontaneous IIa resistance

Table 4 Examples of class IIa bacteriocins

Bacteriocin[a]	Producing species and strains	Source
Pediocin AcH	*Lactobacillus plantarum* WHE92	Muenster cheese
	Pediococcus acidilactici H	Fermented meat
Pediocin PA-1+	*Pediococcus acidilactici* Pac 1.0	Culture collection
	Pediococcus parvulus AOT 77	Vegetables
Enterocin A	*Enterococcus faecium* CTC 492	Fermented sausage
Divercin 41	*Carnobacterium divergens* V41	Fish viscera
Leucocin A	*Leuconostoc gelidum* UAL 187	Processed meat
Sakacin A	*Lactobacillus sakei* Lb 706	Raw meat
Piscicocin V16	*Carnobacterium piscicola* V1	Fish
Mundticin ATO6	*Enterococcus mundtii* ATO6	Vegetables
Bavaricin MN	*Lactobacillus sakei* MN	Meat

[a] Several of the published class IIa bacteriocins have the same amino acid sequence, but were originally named differently
Source [73]

development in *Listeria monocytogenes* [40]. The risk arising from possible resistance development can best be circumvented by including class IIa bacteriocins, and in fact bacteriocins in general, by multifactorial approaches in which the bacteriocins are combined with plant extracts or suitable processes [48, 91].

In addition to the well-characterized bacteriocins of class I and IIa, two more recently discovered groups of peptides could potentially provide new opportunities for food applications. One group, which appears to be a subclass of class II peptides, was reported to have the unusual property for LAB bacteriocins of high efficacy against gram-negative bacteria, including *Campylobacter* [85]. However, the potential for commercial exploitation of these compounds for food preservation remains to be realized. The second group constitutes bacteriocins produced by propionic acid bacteria (PAB). The PAB peptides characterized so far display some unusual properties. However, they have a relatively narrow target spectrum with activity against other PAB and, in some cases, certain lactobacilli [30], which may limit their usefulness for biopreservation. The PAB bacteriocins are not yet as well characterized as their LAB counterparts, and it is conceivable that future investigations on PAB bacteriocins may disclose new opportunities for food applications.

In summary, the number of known LAB and PAB bacteriocins and the number of publications reporting their potential use for food preservation is steadily increasing [6, 22, 36, 43]. However, so far the industrial options for use in food production have not developed accordingly. As described in Fig. 1, there are in principle three different possible formats for applying bacteriocins in food: as a concentrated antimicrobial, as a fermentate, or as a live culture that produces the bacteriocins in the food product. With the fermentation technologies of today, supplying such products at an economically feasible cost does not represent a hurdle. However, the regulatory situation and the general perception have a strong impact on which solution may be viable in different regions. Promotion of other bacteriocins as antimicrobial additives similarly to nisin is unlikely, partly because of the high investment needed today for approval and partly because approval as an additive would in itself defeat the purpose: the main driver for the use of bacteriocins is the demand for "preservative-free" foods (i.e. to provide a natural alternative to chemical preservatives). If a bacteriocin was approved, it would no longer be perceived as natural as it would have become an "E-number". Therefore, the potential industrial food applications of bacteriocins constitute the fermentates and the cultures; these are described in detail in Sects. 5 and 6, respectively.

5 Fermented Food Ingredients: The Fermentates

As the name indicates, fermentates are fermented food ingredients. These products may be produced from a variety of raw materials (typically milk, sugar, or plant-derived material), and the fermentation is done using food-grade microorganisms such as lactic acid bacteria (LAB) or propionic acid bacteria (PAB). The fermentation is designed to provide a high yield of antimicrobial metabolites, which may comprise organic acids (lactic, acetic or propionic acid), diacetyl, bacteriocins, and other secondary metabolites, depending on the properties of the strain(s) used for the fermentation. Fermentates are thus complex products that inherently do not have a well-defined composition. Fermentates are usually supplied as a dry, cell-free powder.

The currently commercially available fermentates for use in foods are the MicroGARD range (DuPont), the DuraFresh range (Kerry), which includes the former Alta and Perlac products from Quest, and various other products that are promoted as shelf-life extenders—namely spray-dried vinegar or fermented wheat flour products. There are only limited scientific reports available on the functionality of fermentates in foods. The original Alta and Perlac were whey-based products for use as shelf-life extenders. The initial MicroGARD products, which were produced by fermenting skimmed milk or dextrose with *Propionibacterium shermanii* or specific lactococci, were demonstrated to inhibit the psychotropic spoilage flora and thereby enhance the shelf-life of cottage cheese [1]. Inhibition of *Pseudomonas*, *Salmonella*, *Yersinia,* and certain fungi was shown.

Conversely, the MicroGARD and Alta products had no significant effect on aerobic mesophilic counts, *Escherichia coli* or *Brochothrix thermosphacta* when tested in an acidified chicken meat model stored at 22 °C [58]. In hamburgers, addition of 1 % MicroGARD provided some initial reduction of *E.* coli O157:H7 and a bacteriostatic effect against *L. monocytogenes* during refrigerated storage [19]. In fresh salmon stored at 6 °C, a combination of nisin and MicroGARD reduced the total aerobic count by 2 log, which provided 3–4 days prolongation of shelf-life and furthermore reduced the outgrowth of *L. monocytogenes* [94]. Similarly, certain combinations of MicroGARD and nisin provided an adequate control of *Listeria innocua* in liquid cheese whey [92]. The anti-listerial effect of nisin in seasoned salmon roe was furthermore enhanced by combining with a *Pediococcus pentosaceus* fermentate and pectin [5].

Even though the scientific documentation for the fermentates is much sparser than for the bacteriocins, the industrial applications are much wider. The main markets for fermentates are in the United States, where such products are labeled as "cultured milk" or "cultured sugar", for example, according to the substrate used for producing the fermentate. Toxicity tests have been performed for a cultured dextrose version of MicroGARD and no detrimental effects were observed [9]. The MicroGARD products are used for a wide range of food applications including cottage cheese, yogurt, sour cream, dairy desserts, sauces, dressings, pasta, baked goods, and prepared meals. An estimated 30 % of the US cheese production is made with MicroGARD [78]. Durafresh was approved by the FDA in 2011 for use in cottage cheese to control *Listeria*, being labeled as "cultured grade A skim milk and skim milk powder".

Labeling as an undefined cultured raw material is not an option in the EU. In the EU, it would be required to label all active components in the fermentate, which presents two main problems: not all active components are known, and most of the known ones have E-numbers. Therefore, the use of fermentates as natural preservatives is so far quite limited in the EU.

6 Bioprotective Cultures

Food fermentation using microbial starter cultures is one of the oldest known uses of biotechnology. Fermentation of perishable food raw materials to provide more stable products has been used by man since approximately 10,000 BC [72]. Fermented food and beverages are still today an important part of the human diet and constitute an estimated 20–40 % of the global food supply [67].

The raw materials used for producing fermented foods are very diverse, covering the range from milk, meat, fish, vegetables, fruits, cereals, and honey. The main desired functionalities provided by the fermentation processes comprise the following: (1) enhanced durability through formation of antimicrobial metabolites (e.g. organic acids, bacteriocins, ethanol), often in conjunction with decreased water activity (drying and/or salting); (2) enhanced safety by reducing the level of either pathogenic microorganisms or their toxins at the time of consumption; (3) enhanced nutritional value; and (4) enhanced organoleptic quality [7].

In addition, there are various detrimental properties that are evidently unwanted and unacceptable in food cultures, including virulence, toxicity, and antibiotic resistance. In the US, acceptable food microorganisms are granted the GRAS status, and in EU they are included in the Qualified Presumption of Safety list. An inventory list of currently used microbial food cultures, comprising 195 bacterial species and 69 fungal species, has recently been compiled [7].

Microbial cultures used in food production are often referred to as either starter cultures (providing nutritional and organoleptic characteristics) or protective cultures (providing durability and safety). However, these properties are inherently linked, such that durability is enhanced by formation of organic acids, which also contribute to the characteristic taste and texture of many fermented foods. All starter cultures are per se also protective cultures, but not all protective cultures are also starter cultures. A clear distinction between starter cultures and bioprotective cultures is therefore neither possible nor meaningful.

In the last two or three decades, substantial research activities have aimed to develop cultures that (1) enhance food safety by directly killing or inhibiting the outgrowth of pathogenic bacteria or by suppressing toxin formation or (2) improve durability by reducing or inhibiting growth of spoilage microorganisms. The resulting scientific papers have been summarized in comprehensive reviews, including solutions for fish and seafood [10, 59], dairy products [6, 39], and antifungal cultures in general (Dalié et al. [17, 18]. Overall, most of the reports can be allocated to one of the following main categories: (1) use of bacteriocin-producing LAB cultures to control *L. monocytogenes* in various ready-to-eat foods, (2) use of antifungal LAB and/or PAB to delay spoilage of various types of food, or (3) use of nonbacteriocinogenic LAB with other competitive properties.

The mode of action of the bacteriocinogenic antilisterial cultures relies on the production of class I or IIa bacteriocins, as described in Sect. 4, and is relatively well characterized. The antifungal cultures, on the other hand, have been discovered more recently, and scientific evidence has been gathered during the last 10–15 years (Dalié et al. [17, 18, 27, 50, 77]. The antifungal cultures have been reported to produce a variety of different metabolites, and the current understanding indicates that they work by a complex antifungal mechanism obtained by the combined effects of the described and also as yet not elucidated metabolites [60, 79]. Finally, nonbacteriogenic cultures with antibacterial properties have been reported; these seemingly rely on a variety of competitive advantages over the unwanted microbiota.

In the following lists, recent examples of application studies within each category for various food segments are provided.

Anti-listerial bacteriocinogenic LAB cultures:

- In meat products: *Lactobacillus sakei* together with 50 % CO_2 prevented outgrowth of *L. monocytogenes* in bologna-type sausage without an unacceptable pH drop [54]. *Pediococcus acidilactici* was efficient in reducing *L. monocytogenes* in dry-fermented Spanish sausages [68]. *L. sakeii* prevented listerial growth in a pork meat system while enhancing protein hydrolysis [12].
- In fish and seafood: Listerial control was achieved in cold-smoked salmon using *Carnobacterium divergens* [86] or *Lb. sakei* [93].
- In dairy products: *Lactococcus lactis* used as starter culture in cottage cheese [15] or *Enterococcus faecium* in smear of soft cheese [53] controlled outgrowth of *L. monocytogenes.*
- In vegetable products: *Leuconostoc mesenteroides* was used for reduction of *L. monocytogenes* in apples and iceberg lettuce [90].

Antifungal LAB and/or PAB cultures:

- In bakery products: *Lactobacillus plantarum* was used for delaying *Penicillium* spoilage of bread [16, 37].
- In dairy products: *Lactobacillus rhamnosus* and *Propionibacterium freudenreichii* ssp. *shermanii* were used for inhibiting yeast in yogurt [61]. *Lactobacillus harbinensis* was used as an antifungal cultures in yogurt [23].

Non-bacteriocinogenic LAB with other competitive properties:

- In cooked peeled shrimp, *Lactococcus piscium* inhibited outgrowth of *L. monocytogenes* and delayed sensory spoilage [28, 29] and inhibited outgrowth of *Staphylococcus aureus* [64].
- *Leuconostoc gelidum* delayed spoilage of shrimp and cold-smoked salmon [64].
- *Staphylococcus xylosus* was used for inhibiting biogenic amine formation in anchovies [62].
- Plant-associated *Pseudomonas* was used for inhibiting *Salmonella enterica* on alfalfa sprouts [31].
- Commercial culture had a protective effect by depletion of oxygen [80].

Commercial products of protective cultures are produced in the same way as starter cultures: by batch fermentation, subsequent concentration by centrifugation, and final formulation as frozen pellets or freeze-dried powders. Approaches for continuous fermentation have also been described, such as cultivation of *Lactococcus lactis* in a fixed bed reactor [71]. Even though protective cultures were first introduced about 10–15 years ago, they are now well established in the food industry and recognized as an efficient tool to ensure the safety and durability of food products. Table 5 summarizes the main functionalities, species, and producers of protective cultures.

7 Bacteriophages

Phages are the most abundant living creatures on the planet: the estimated total number of phages is 10^{31}. As example, one milliliter of sea water contains about 1,000,000 bacteria but 10,000,000 phages. Phages are also widely spread in foods of various origins [8]. Today, it is recognized that the interaction between phages and bacteria plays an important part in maintaining the natural balance in our ecosystems.

The short name *phage* comes from ancient Greek, meaning "eat"; bacteriophage thus means "bacteria-eater". Bacteriophages are host-specific. The specificity is due to the fact that a phage can only propagate on a certain bacterial species; the phage recognizes its specific host cell. Thus, they are harmless to humans, animals, and plants.

Phages are renowned in the dairy industry for attacking starter cultures during fermentation and thereby spoiling the production of yogurt. On the contrary,

Table 5 Examples of commercially available protective cultures

Protective function	Microorganisms	Fields of application	Producer
Growth inhibition of *Listeria monocytogenes*	Lactic acid bacteria (e.g. *Lb. sakei, Lb. curvatus, Lb. plantarum*)	Fermented meat products Dairy products	Chr. Hansen (Denmark) DuPont (USA)
	Carnobacterium sp.	Fish and seafood	Sacco (Italy)
Inhibition of mold and yeasts	*Lactobacillus sp.*	Fresh dairy products	Chr. Hansen (Denmark)
	Lb. rhamnosus, Lb. paracasei, Propionobacterium sps.	Fresh dairy products	DuPont (USA)
Inhibition of *Clostridia tyrobutyricum*; prevention of late blowing	*Lactococcus lactis*	Cheese	CSK (Netherlands)

phages which can inactivate pathogenic bacteria can be useful in food processing. So to speak, "the enemy of my enemy is my useful friend".

In the food chain, bacteriophages were initially reported to be useful as interventions in the primary production. Drug-resistant bacteria have become a global problem, urging for the prompt development of alternative control strategies in order to avoid growth promoters while maintaining or enhancing food quality and safety. Oral treatment of broilers with phages reduced the carriage of *Salmonella* [32]. Phage therapy has also been explored in aquacultures; however, a recent review emphasizes the need of further research in the field of the application [69].

In unprocessed foods, phages have been tested for reducing *Campylobacter* and *Salmonella* on chicken skin [4, 38]. A reduction of 1–2 log of the pathogens was achieved. In various ready-to-eat foods (hot dogs, sliced turkey, smoked salmon, seafood, sliced cabbage, and lettuce), application of bacteriophages against *L. monocytogenes* provided up to a 5-log reduction of the pathogen [42].

However, a recent review of the use of bacteriophages against pathogens in food products concluded that the technology has so far had a variable success [35]. This could perhaps partially be due to testing of unsuitable applications for this relatively new technology in food production; bacteriophage products are already in use in agricultural, food safety, and diagnostic applications, demonstrating the utility and viability of such approaches [66].

Bacteriophages differ from many bacteria in the respect that phages are not motile. Therefore, the application method must ensure that the phages are well distributed in the product, so the target cells are brought into contact with a suitable number of phages. Furthermore, phages will typically become inactivated or bound in the food matrix—that is, they will have an initial effect in reducing their target organism, but will often presumably not be able to prevent regrowth of surviving cells. It is important to consider these factors for developing successful application of phages, such as by using them with other hurdles that provide growth inhibition or in multifactorial systems. As example, a combination of phages and a bacteriocinogenic culture of *Lb. sakei* was used to provide both initial reduction and suppression of outgrowth of *L. monocytogenes* in cooked ham [51]. Successful and plausible applications of bacteriophages in foods were recently reviewed by Garcia et al. [36].

Similarly to the scientific reports demonstrating successful use of phages in the food chain, the options for applying bacteriophage products in industrial food production is relatively new. Again, the regulatory situation varies in different regions. In the EU, the use of bacteriophages in the food chain is being reviewed by the European Food Safety Authority to assess its efficacy and safety for use with food producing animals and in food products [3].

In the USA, OmniLytics Inc. received FDA approval for an anti-*E. coli* and an anti-*Salmonella* phage-based product to treat live animals prior to slaughtering. Phage cocktails against *L. monocytogenes*—Listex (formerly EBI Food Safety, now Micreos Food Safety) and LMP 02 (Intralytics)—were approved by the FDA in ready-to-eat meat.

Other commercially available phage-based bioprotective products are Agri-Phage from Omnilytics (specific formulations for strains of *Xanthomonas campestris* or *Pseudomonas syringae*) and EcoShieldTM (targets *E. coli* O157), as well as ListShield (antilisterial phages) from Intralytix. Recently, the Korean CheilJedang Corporation has introduced BioTector, a bacteriophage product for reducing *Salmonella* in poultry [66].

A main limiting factor for the industrial application of bacteriophages in foods, in addition to developing functional applications and avoiding negative publicity (e.g. "they are putting viruses in our food"), has been the high production costs. This issue will presumably be solved as several companies are investing in development and production facilities, and it appears likely that we will see new bioprotective solutions based on bacteriophages as alternative measures for controlling detrimental bacteria in food production in the coming years.

References

1. al-ZorekyN, Ayres JW, SandineWE (1991) Antimicrobial activity of microgard against food spoilage and pathogenic microorganisms. J Dairy Sci 74:758–763
2. Alkhatib Z, Abts A, Mavaro A, Schmitt L, Smits SH (2012) Lantibiotics: how do producers become self-protected? J Biotechnol 159:145–154
3. Anon (2009) The use and mode of action of bacteriophages in food production; scientific opinion of the panel on biological hazards. EFSA J 1076:26
4. Atterbury RJ, Connerton PL, Dodd CE, Rees CE, Connerton IF (2003) Application of host-specific bacteriophages to the surface of chicken skin leads to a reduction in recovery of *Campylobacter jejuni*. Appl Environ Microbiol 69:6302–6306

5. Bagenda DK, Yamazaki K, Kobayashi T, Kawai Y (2010) Assessing and enhancing the antimicrobial effect of nisin in soy-seasoned salmon *Oncorynchus keta* roe using a *Pediococcus pentosaceus* fermentate and pectin. Fish Sci 76:395–401
6. Beshkova D, Frengova G (2012) Bacteriocins from lactic acid bacteria: microorganisms of potential biotechnological importance for the dairy industry. Eng Life Sci 12:419–432
7. Bourdichon F, Casaregola S, Farrokh C, Frisvad JC, Gerds ML, Hammes WP, Harnett J, Huys G, Laulund S, Ouwehand A, Powell IB, Prajapati JB, Seto Y, Ter SE, Van BA, Vankerckhoven V, Zgoda A, Tuijtelaars S, Hansen EB (2012) Food fermentations: microorganisms with technological beneficial use. Int J Food Microbiol 154:87–97
8. Brûssow H, Kutter E (2005) Phage ecology. In: Kutter E, Sulakvelidze A (eds) Bacteriophages: biology and applications. CRC Press, Boca Raton, pp 129–163
9. Buard A, Carlton BD, Floch F (2003) Subchronic toxicity, mutagenicity and allergenicity studies of a cultured dextrose food product. Food Chem Toxicol 41:689–694
10. Calo-Mata P, Arlindo S, Boehme K, de Miguel T, Pascoal A, Barros-Velazquez J (2008) Current applications and future trends of lactic acid bacteria and their bacteriocins for the biopreservation of aquatic food products. Food Bioprocess Technol 1:43–63
11. Campos CA, Gerschensin LN, Flores SK (2011) Development of edible films and coatings with antimicrobial activity. Food Bioprocess Technol 4:849–875
12. Castellano P, Aristoy MC, Sentandreu MA, Vignolo G, Toldra F (2012) *Lactobacillus sakei* CRL1862 improves safety and protein hydrolysis in meat systems. J Appl Microbiol 113:1407–1416
13. Cotter PD, Hill C, Ross RP (2005) Bacteriocins: developing innate immunity for food. Nat Rev Microbiol 3:777–788
14. Cui Y, Zhang C, Wang Y, Shi J, Zhang L, Ding Z, Qu X, Cui H (2012) Class IIa bacteriocins: diversity and new developments. Int J Mol Sci 13:16668–16707
15. Dal Bello B, Cocolin L, Zeppa G, Field D, Cotter PD, Hill C (2012) Technological characterization of bacteriocin producing *Lactococcus lactis* strains employed to control *Listeria monocytogenes* in Cottage cheese. Int J Food Microbiol 153:58–65
16. Dal Bello F, Clarke CI, Ryan LAM, Ulmer H, Schober TJ, Strom K, Sjogren J, van Sinderen D, Schnurer J, Arendt EK (2007) Improvement of the quality and shelf life of wheat bread by fermentation with the antifungal strain *Lactobacillus plantarum* FST 1.7. J Cereal Sci 45:309–318
17. Dalié DKD, Deschamps AM, Atanasova-Penichon V, Richard-Forget F (2010) Potential of *Pediococcus pentosaceus* (L006) isolated from maize leaf to suppress fumonisin-producing fungal growth. J Food Prot 73:1129–1137
18. Dalié DKD, Deschamps AM, Richard-Forget F (2010) Lactic acid bacteria—potential for control of mould growth and mycotoxins: a review. Food Control 21:370–380
19. Dave RI, Sharma P, Julson J, Muthukumarappan K, Henning DR (2003) Effectiveness of microgard (R) in controlling Escherichia coli O157: H7 and Listeria monocytogenes. J Food Sci Technol Mysore 40:262–266
20. Davidson PM, Doan CH (1993) Natamycin. In: Davidson PM, Branen LM (eds) Antimicrobials in foods. Marcel Dekker Inc., New York
21. de Arauz LJ, Jozala AF, Mazzola PG, Penna TCV (2009) Nisin biotechnological production and application: a review. Trends Food Sci Technol 20:146–154
22. Deegan LH, Cotter PD, Hill C, Ross P (2006) Bacterlocins: biological tools for bio-preservation and shelf-life extension. Int Dairy J 16:1058–1071
23. Delavenne E, Ismail R, Pawtowski A, Mounier J, Barbier G, Le Blay G (2013) Assessment of lactobacilli strains as yogurt bioprotective cultures. Food Control 30:206–213
24. Delves-Broughton J, Thomas LV, Doan CH, Davidson PM (2005) Natamycin. In: Davidson PM, Sofos JN, Branen AL (eds) Antimicrobials in food. CRC Press, Boca Raton, pp 275–289
25. Delves-Broughton J, Weber G (2010) Nisin, natamycin and other commercial fermentates used in food biopreservation. In: Lacroix C (ed) Protective cultures, antimicrobial metabolites and bacteriophages for food and beverage biopreservation. Woodhead Publishing Ltd, Cambridge

26. EFSA Panel on Food Additives and Nutrient Sources added to Food (2009) Scientific opinion on the use of natamycin (E235) as a food additive. EFSA J 7:1412
27. Erginkaya Z, Kavas C, Var I, Kabak B, Guven M (2004) Antifungal activity of several lactic acid bacteria and bifidobacteria. Archiv fur Lebensmittelhygiene 55:52–55
28. Fall PA, Leroi F, Cardinal M, Chevalier F, Pilet MF (2010) Inhibition of *Brochothrix thermosphacta* and sensory improvement of tropical peeled cooked shrimp by *Lactococcus piscium* CNCM I-4031. Lett Appl Microbiol 50:357–361
29. Fall PA, Leroi F, Chevalier F, Guerin C, Pilet MF (2010) Protective effect of a non-bacteriocinogenic *Lactococcus piscium* CNCM I-4031 strain against *Listeria monocytogenes* in sterilized tropical cooked peeled shrimp. J Aquat Food Prod Technol 19:84–92
30. Faye T, Holo H, Langsrud T, Nes IF, Brede DA (2011) The unconventional antimicrobial peptides of the classical propionibacteria. Appl Microbiol Biotechnol 89:549–554
31. Fett WF (2006) Inhibition of *Salmonella enterica* by plant-associated pseudomonads in vitro and on sprouting alfalfa seed. J Food Prot 69:719–728
32. Fiorentin L, Vieira ND, Barioni W (2005) Oral treatment with bacteriophages reduces the concentration of *Salmonella* Enteritidis PT4 in caecal contents of broilers. Avian Pathol 34:258–263
33. Frazer AC, Sharrat M, Hickman JR (1962) The biological effects of food additives. I. Nisin. J Sci Food Agr 13:32–42
34. Galvez A, Lopez RL, Abriouel H, Valdivia E, Omar NB (2008) Application of bacteriocins in the control of foodborne pathogenic and spoilage bacteria. Crit Rev Biotechnol 28:125–152
35. Garcia P, Martinez B, Obeso JM, Rodriguez A (2008) Bacteriophages and their application in food safety. Lett Appl Microbiol 47:479–485
36. Garcia P, Rodriguez L, Rodriguez A, Martinez B (2010) Food biopreservation: promising strategies using bacteriocins, bacteriophages and endolysins. Trends Food Sci Technol 21:373–382
37. Gerez CL, Torino MI, Obregozo MD, de Valdez GF (2010) A Ready-to-use antifungal starter culture improves the shelf life of packaged bread. J Food Prot 73:758–762
38. Goode D, Allen VM, Barrow PA (2003) Reduction of experimental *Salmonella* and *Campylobacter* contamination of chicken skin by application of lytic bacteriophages. Appl Environ Microbiol 69:5032–5036
39. Grattepanche F, Miescher-Schwenninger S, Meile L, Lacroix C (2008) Recent developments in cheese cultures with protective and probiotic functionalities. Dairy Sci Technol 88:421–444
40. Gravesen A, Ramnath M, Rechinger KB, Andersen N, Jansch L, Hechard Y, Hastings JW, Knochel S (2002) High-level resistance to class IIa bacteriocins is associated with one general mechanism in Listeria monocytogenes. Microbiology 148:2361–2369
41. Gravesen A, Sorensen K, Aarestrup FM, Knochel S (2001) Spontaneous nisin-resistant Listeria monocytogenes mutants with increased expression of a putative penicillin-binding protein and their sensitivity to various antibiotics. Microb Drug Resist 7:127–135
42. Guenther S, Huwyler D, Richard S, Loessner MJ (2009) Virulent bacteriophage for efficient biocontrol of *Listeria monocytogenes* in ready-to-eat foods. Appl Environ Microbiol 75:93–100
43. Guinane CM, Cotter PD, Hill C, Ross RP (2005) Microbial solutions to microbial problems; lactococcal bacteriocins for the control of undesirable biota in food. J Appl Microbiol 98:1316–1325
44. Gupta SM, Aranha CC, Reddy KV (2008) Evaluation of developmental toxicity of microbicide Nisin in rats. Food Chem Toxicol 46:598–603
45. Hagiwara A, Imai N, Nakashima H, Toda Y, Kawabe M, Furukawa F, Delves-Broughton J, Yasuhara K, Hayashi SM (2010) A 90-day oral toxicity study of nisin A, an anti-microbial peptide derived from *Lactococcus lactis* subsp. *lactis*, in F344 rats. Food Chem Toxicol 48:2421–2428

46. Hara S, Keido Y, Koji M, Tokako T, Tatsumori K, Masachiro S, Ziro I, Takeshi S (1962) An investigation of toxicity of nisin with a particular reference to experimental studies of its oral administration and influences on digestive enzymes. Tokyo Med Univ 20:175–207
47. Heng NCK, Tagg JR (2006) What's in a name? Class distinction for bacteriocins. Nat Rev Microbiol 4:160–165
48. Hereu A, Bover-Cid S, Garriga M, Aymerich T (2012) High hydrostatic pressure and biopreservation of dry-cured ham to meet the Food Safety Objectives for *Listeria monocytogenes*. Int J Food Microbiol 154:107–112
49. Hirsch A, Grinsted E, Ho PH, Chapman HR, Mattick AT (1951) A note on the inhibition of an anaerobic sporeformer in Swiss-type cheese by a nisin-producing *Streptococcus*. J Dairy Sci 18:205–206
50. Ho PH, Luo JB, Adams MC (2009) Lactobacilli and dairy propionibacterium with potential as biopreservatives against food fungi and yeast contamination. Appl Biochem Microbiol 45:414–418
51. Holck A, Berg J (2009) Inhibition of *Listeria monocytogenes* in cooked ham by virulent bacteriophages and protective cultures. Appl Environ Microbiol 75:6944–6946
52. Islam MR, Nagao J, Zendo T, Sonomoto K (2012) Antimicrobial mechanism of lantibiotics. Biochem Soc Trans 40:1528–1533
53. Izquierdo E, Marchioni E, Aoude-Werner D, Hasselmann C, Ennahar S (2009) Smearing of soft cheese with *Enterococcus faecium* WHE 81, a multi-bacteriocin producer, against *Listeria monocytogenes*. Food Microbiol 26:16–20
54. Kaban G, Kaya M, Lucke FK (2010) The effect of *Lactobacillus sakei* on the behavior of *Listeria monocytogenes* on sliced bologna-type sausages. J Food Saf 30:889–901
55. Kaur G, Malik RK, Mishra SK, Singh TP, Bhardwaj A, Singroha G, Vij S, Kumar N (2011) Nisin and class IIa bacteriocin resistance among *Listeria* and other foodborne pathogens and spoilage bacteria. Microb Drug Resist 17:197–205
56. Klaenhammer TR (1988) Bacteriocins of lactic acid bacteria. Biochimie 70:337–349
57. Kotler-Brajtburg J, Medoff G, Kobayashi GS, Boggs S, Schlessinger D, Pandey RC, Rinehart KL Jr (1979) Classification of polyene antibiotics according to chemical structure and biological effects. Antimicrob Agents Chemother 15:716–722
58. Lemay MJ, Choquette J, Delaquis PJ, Claude G, Rodrigue N, Saucier L (2002) Antimicrobial effect of natural preservatives in a cooked and acidified chicken meat model. Int J Food Microbiol 78:217–226
59. Leroi F (2010) Occurrence and role of lactic acid bacteria in seafood products. Food Microbiol 27:698–709
60. Lind H, Sjogren J, Gohil S, Kenne L, Schnurer J, Broberg A (2007) Antifungal compounds from cultures of dairy propionibacteria type strains. FEMS Microbiol Lett 271:310–315
61. Liptakova D, Valik L, Bajusova B (2006) Effect of protective culture on the growth of *Candida maltosa* YP1 in yoghurt. J Food Nutr Res 45:147–151
62. Mah JH, Hwang HJ (2009) Inhibition of biogenic amine formation in a salted and fermented anchovy by Staphylococcus xylosus as a protective culture. Food Control 20:796–801
63. Malheiros P, Daroit DJ, Brandelli A (2010) Food applications of liposome-encapsulated antimicrobial peptides. Trends Food Sci Technol 21:284–292
64. Matamoros S, Leroi F, Cardinal M, Gigout F, Kasbi CF, Cornet J, Prevost H, Pilett MF (2009) Psychrotrophic lactic acid bacteria used to improve the safety and quality of vacuum-packaged cooked and peeled tropical shrimp and cold-smoked salmon. J Food Prot 72:365–374
65. Mattick AT, Hirsch A (1947) Further observations on an inhibitory substance (nisin) from lactic streptococci. Lancet 2:5–8
66. Monk AB, Rees CD, Barrow P, Hagens S, Harper DR (2010) Bacteriophage applications: where are we now? Lett Appl Microbiol 51:363–369
67. Naidu AS (2000) Natural Food Antimicrobial Systems. CRC Press, Boca Raton
68. Nieto-Lozano JC, Reguera-Useros JI, Pelaez-Martinez MD, Sacristan-Perez-Minayo G, Gutierrez-Fernandez AJ, de la Torre AH (2010) The effect of the pediocin PA-1 produced by

Pediococcus acidilactici against *Listeria monocytogenes* and *Clostridium perfringens* in Spanish dry-fermented sausages and frankfurters. Food Control 21:679–685
69. Oliveira J, Castilho F, Cunha A, Pereira MJ (2012) Bacteriophage therapy as a bacterial control strategy in aquaculture. Aquac Int 20:879–910
70. Papagianni M, Anastasiadou S (2009) Pediocins: the bacteriocins of Pediococci. Sources, production, properties and applications. Microb Cell Fact 8:3
71. Portner R, Seemuk M, Schlothauer RC, Elsser D (2004) Anaerobic cultivation of *Lactococcus lactis* in a fixed bed reactor. Chemie Ingenieur Technik 76:1599–1602
72. Prajapati JB, Nair BM (2003) The history of fermented foods. In: Farnworth ER (ed) Fermented functional foods. CRC Press, Boca Raton, pp 1–25
73. Ray B, Miller KW (2003) Bacteriocins other than nisin: the pediocin-like cystobiotics of lactic acid bacteria. In: Roller S (ed) Natural antimicrobials for the minimal processing of foods. Woodhead Publishing Ltd, Cambridge, pp. 64–81
74. Riley MA, Wertz JE (2002) Bacteriocins: evolution, ecology, and application. Annu Rev Microbiol 56:117–137
75. Rogers LA (1928) The inhibiting effect of *Streptococcus lactis* on *Lactobacillus bulgaricus*. J Bacteriol 16:321–325
76. Roller S (2003) Introduction. In: Roller S (ed) Natural antimicrobials for the minimal processing of foods. Woodhead Publishing Ltd, Cambridge, pp 1–10
77. Rouse S, Harnett D, Vaughan A, van Sinderen D (2008) Lactic acid bacteria with potential to eliminate fungal spoilage in foods. J Appl Microbiol 104:915–923
78. Sani AM, Ehsani MR, Asadi MM (2005) Effect of *Propionibacterium shermanii* metabolites on sensory properties and shelf life of UF-Feta cheese. Nutr Food Sci 35:88–94
79. Schwenninger SM, Lacroix C, Truttmann S, Jans C, Sporndli C, Bigler L, Meile L (2008) Characterization of low-molecular-weight antiyeast metabolites produced by a food-protective *Lactobacillus-Propionibacterium* Coculture. J Food Prot 71:2481–2487
80. Seibert TM (2010) Protective culture eliminates residual oxygen. Fleischwirtschaft 90:59–61
81. Settanni L, Corsetti A (2008) Application of bacteriocins in vegetable food biopreservation. Int J Food Microbiol 121:123–138
82. Smith JE, Moss MO (1985) Mycotoxins: Formation, Analysis and Significance. Wiley, New York
83. Stark J (2003) Natamycin: an effective fungicide for food and beverages. In: Roller S (ed) Natrural antimicrobials for the minimal processing of foods. Woodhead Publishing Ltd, Cambridge
84. Struyk AP, Hoette I, Drost G, Waisvisz JM, Van Eek T, Hoogerheide JC (1957) Pimaricin, a new antifungal antibiotic. Antibiot Annu 5:878–885
85. Svetoch EA, Stern NJ (2010) Bacteriocins to control *Campylobacter* spp. in poultry—a review. Poult Sci 89:1763–1768
86. Tahiri I, Desbiens M, Kheadr E, Lacroix C, Fliss I (2009) Comparison of different application strategies of divergicin M35 for inactivation of *Listeria monocytogenes* in cold-smoked wild salmon. Food Microbiol 26:783–793
87. Thomas LV, Clarkson MR, Delves-Broughton J (2000) Nisin. In: Naidu AS (ed) Natural food antimicrobial systems. CRC Press, Boca Raton, pp 463–524
88. Thomas LV, Delves-Broughton J (2001) Applications of the natural food preservative natamycin. Res Adv Food Sci 2:1
89. Thomas LV, Delves-Broughton J (2005) Nisin. In: Davidson PM, Sofos JN, Branen AL (eds) Antimicrobials in food. CRC Press, Boca Raton, pp 237–274
90. Trias R, Badosa E, Montesinos E, Baneras L (2008) Bioprotective *Leuconostoc* strains against *Listeria monocytogenes* in fresh fruits and vegetables. Int J Food Microbiol 127:91–98
91. Turgis M, Vu KD, Dupont C, Lacroix M (2012) Combined antimicrobial effect of essential oils and bacteriocins against foodborne pathogens and food spoilage bacteria. Food Res Int 48:696–702

92. von Staszewski M, Jagus RJ (2008) Natural antimicrobials: Effect of Microgard (TM) and nisin against *Listeria innocua* in liquid cheese whey. Int Dairy J 18:255–259
93. Weiss A, Hammes WP (2006) Lactic acid bacteria as protective cultures against *Listeria* spp. on cold-smoked salmon. Eur Food Res Technol 222:343–346
94. Zuckerman H, Ben Avraham R (2002) Control of growth of *L. monocytogenes* in fresh salmon using Microgard (TM) and Nisin. Food Sci Technol 35:543–548

Adv Biochem Eng Biotechnol (2014) 143: 51–89
DOI: 10.1007/10_2013_241

Published Online: 15 September 2013

Biotechnological Production of Colorants

Lex de Boer

Abstract The color of food and drinks is important, as it is associated with freshness and taste. Despite that natural colorants are more expensive to produce, less stable to heat and light, and less consistent in color range, natural colorants have been gaining market share in recent years. The background is that artificial colorants are often associated with negative health aspects. Considerable progress has been made towards the fermentative production of some colorants. Because colorant biosynthesis is under close metabolic control, extensive strain and process development are needed in order to establish an economical production process. Another approach is the synthesis of colors by means of biotransformation of adequate precursors. Algae represent a promising group of microorganisms that have shown a high potential for the production of different colorants, and dedicated fermentation and downstream technologies have been developed. This chapter reviews the available information with respect to these approaches.

Keywords Algae · Colorants · Large-scale production · Microbial sources · Plants · Precursor bioconversion · Process development

Contents

L. de Boer (✉)
Department of Food and Technology, Groen Agro Control B.V, Distributieweg 1, 2645 EG, Delfgauw, The Netherlands
e-mail: deboerlex@gmail.com

1 Introduction

Colorants have been added to food and drinks for a long time. Important functions of food and beverage colors are the identification and recognition of spoilage that are associated with the food or beverage as such, but also with its taste and smell. Colorants are added in order to emphasize and to keep the original color after manufacturing for prolonged periods and for perception of freshness and shelf life: color makes the food more attractive for the consumer. For example, many people associate curry with its specific taste as the typical color and taste of curry are closely related. This is also the case for saffron [1]. Furthermore, it is hardly possible to obtain the bright colors of soft drinks without the addition of colorants. Most of the natural dyes are obtained from plants, and the most commonly applied are β-carotene, lycopene, and chlorophyll. Some of them may have favorable health properties, for example, antioxidant or even anticarcinogenic effects.

At least four carotenoids have a dietary function, as vitamin A can be formed thereof, namely α-carotene, β-carotene, γ-carotene, and β-cryptoxanthin [2]. Further examples of suspected beneficial health effects are the anti-inflammatory effect of curcumin (E100), a potential positive effect of lactoflavin against migraine, and of medicinal carbon during food poisoning (by adsorption of toxins) [3, 4]. Gist-brocades (DSM) has developed commercial biotechnological production processes for astaxanthin and β-carotene production.

A very old application of natural colors has been reported by Rymbai et al. [5]: in Japanese shosoin texts, the use of natural colorants for the colorization of soy beans and adzuki was described (eighth century). In the nineteenth century, many natural colorants were replaced by synthetic dyes because of lower production costs [5], higher stability [6], and a more consistent color range. The new colorants have passed an extensive trajectory of tests, and were registered in Europe under a E-numbers before they were allowed for use in food and drinks. Table 1 summarizes examples of synthetic and natural colorants of a wide range of colors with their corresponding E-numbers, their origins and applications.

In addition, there are a number of food additives that cause a color modification by a pH shift or by a chemical modification of the food itself, for example, potassium hydroxide (E525). Potassium hydroxide is used to give olives their black color. Further examples include sodium citrate (E331) and sodium thiosulphate (E539) which are used to prevent browning reactions of fruit and potatoes, respectively.

Despite the fact that natural colorants are typically more expensive to produce, less stable to heat and light, and less consistent in color range, natural pigments have been gaining market share as food and feed colorants and nutraceuticals in recent years. Consumers increasingly prefer natural colorants as synthetic colorants are associated with allergenic reactions, hyperactivity, and even bad taste [3, 7, 8]. Furthermore, more and more evidence appears in the scientific literature about undesired and potential toxicological effects of synthetic colorants. This prompts the regulatory authorities to shorten the list of permitted synthetic food

Table 1 Origin and application of selected colorants (http://www.food-info.net/)

Colorant	Color	E-Number	Origin	Application	Potential undesirable health effects
Food red 17	Red	E129	Synthetic azo-dye	Candies, meat	Hyperactivity, asthma
Food blue 2	Blue	E133	Synthetic triaryl methane dye	Ice	Promoting allergenic response
Food brown 1	Brown	E154	Synthetic azo-dye	Fish products	Hyperactivity, asthma, suspected mutagen
Food yellow 13	Yellow	E104	Synthetic nonazo-dye	Broad range of products	Not known
Food green 4 Food green S	Green	E142	Synthetic triaryl methane dye	Broad range food products	Not known
Anatto Bixin Norbixin	Red/(brown)	E160b	Isolated from seeds of the annatto tree	Many products	Promoting allergenic response
Carotenes	Yellow, orange	E160a	Extraction from plants	Many products	No side-effects, at high levels they may cause yellow skin
Anthocyanins	Dependent on the chemical structure/ mixture: red, blue, purple, orange	E163 (a t/m f, I, ii, iii)	Fruit, berries, flowers	Many products, unstable colorant	Not known
Natural green 3 Chlorophyll A	Green	E140	Grass, nettle	Many products	Not known
Gold	Gold	E175	Natural metal	Surface coloring of few products	Not known
Active carbon Charcoal Norit	Black	E153	Oxidation of vegetable material	Many food products	None
Calcium carbonate	White	E170	Surface coloring	Many products,	None
Titanium dioxide	White	E171	Surface coloring	Many products, cheese	None

colorants drastically [5]. Furthermore, the FAO/WHO Expert Committee on Food Additives advised the further removal of small chemical impurities from food colors.

In at least one case, the presence of an added colorant may become undesirable during a food production process, such as during cheese production. Because of seasonal variations, the β-carotene content of milk is variable, and therefore annatto is supplemented with the purpose of standardizing the color of cheese. However, the resulting whey is yellow to orange which lowers its market value. The whey may be bleached by application of a peroxidase (MSP1), which was originally isolated from a basidiomycete [9]. The product is currently sold by DSM under the trade name MaxiBright™.

2 Fermentative Production of Food Grade Colorants

In nature, many microorganisms are found that are able to synthesize colored metabolites, for example, chlorophylls, carotenoids, melanins, flavins, violacein, and indigotin. This opens up opportunities for their production by means of biotechnology (Table 2).

There are a number of examples of "biocolorants" that are economically interesting to produce by fermentation, for example, saffron. Saffron is present in the stigma of *Crocus sativa,* albeit in high levels, but only comprises a very small part of the total plant. In order to meet the demand, a significant part of the agricultural area is necessary (Iran: 80,000 hectares), and harvesting and processing of the stigmas are very laborious [10]. Therefore, studies were performed to grow saffron-producing tissues in vitro. It was technically possible, but the yields were far away from economic requirements.

Therefore, it is recommended to investigate microorganisms that are easy to cultivate for their ability to produce colorants. Until now, only some of the colorants produced by plants have been known to be produced in microorganisms. If a suitable microorganism is not available yet, a screening program is an alternative. With regard to this, algae represent an interesting group: the interest in these autotrophic or facultative heterotrophic microorganisms is strongly increasing as some species produce lipids up to very high levels (30 %), which may contribute to a more sustainable fuel supply. There is currently a strong focus on the development of fermentation technologies that allow for economical production. Autotrophic algae need light for their energy supply, and therefore specific fermentation equipment (e.g., transparent fermenters that automatically take a position for optimal light collection, CO_2-supply). Furthermore, harvest and downstream processing steps have to be optimized [11–13]. These cost-reduction programs for biofuel production may also be helpful for the production of colors by algae, for example, of astaxanthin, β-carotene, and lutein [14–17].

Preferably, wild-type microorganisms are used that produce the desired colorant up to commercial levels. In this case, for example, with some algae, a long and

Table 2 Origin of natural colorants based on E-number list (http://www.food-info.net/)

E-Number	Name	Color	Origin	Microbial alternative	Ref
E101	Riboflavin	Yellow	Milk, yeast	Ashbya gossypii, *Candida famata*	[18, 19]
E132	Indigotin	Blue	Synthetic, *Indigofera tinctoria*	*Nocardia globerula;* traditional Japanese indigo fermentation; recombinant *Escherichia coli strains;* tissue culture *Polygonum tictorium*; *Morchella* nov.ES-1	[20–25]
E140	Chlorophyll A	Green	Grass, nettle	*Dunaliella viridis*	[26]
E160a (ii)	β-Carotene	Yellow, orange	Flowers, carrots	*Blakeslea trispora*, *Dunaliella salina*	[27–29]
E160b	Bixin Norbixin (annatto)	Red/(brown)	Seeds from the annatto tree	Root cultures of *Bixa orellana*[a]	[30]
E160c	Capsaicin, capsarubin	Red	Fruit of *Capsicum annuum*	Suspension cultures of *Capsicum annuum*	[31, 32]
E160d	Lycopene	Red	Tomatoes	Recombinant *Candida utilis* *Streptomyces sp*	[33–35]
E160e	β-Apo-8′-carotenal	Red	Various plants	Human gene in *Escherichia coli*[b]	[36]
E161b	Lutein	Yellow	Grass, algae, nettles, or *Tagetes*	Spirulina	[37–40]
E161c	β-Cryptoxanthin	Yellow	*Physalis*	*Flavobacterium lutescens* ITCCB008	[41]
E161 g	Canthaxanthin	Orange	Chanterelle, flamingo feathers, synthetic	*Dietzia* natronolimnae,[b] *Halobacterium* *Haloferax alexandrinus*[c]	[42–45]
E161j	Zeaxanthin	Yellow	*Zea mays* (corn)	*Flavobacterium*	[46]
E161j	Astaxanthin	Salmon red	Shellfish	*Haematococcus pluvialis, Xanthophyllomyces dendrorhous, Paracoccus carotinifaciens, Agrobacterium aurantiacum*	[28, 47–52]
E162	Betanin	Red	*Beta vulgaris* (red beets)	Hairy root cultures of *Beta vulgaris*	[53–57]
E163 (a t/m f, I, ii, iii)	Anthocyanins	Red, blue, purple, orange	Fruit, berries, flowers	Callus cultures of *Ajuga pyramidalis, Begonia, Ocimum, Tradescantia; Perilla frutescens; Daucus carota*[c]	[58–60]
E181	Tannin	Yellowish	Gall-nuts	Root tissue cultures of *Rheum palmatum*	[61]

[a] Established process
[b] Established process
[c] Experimental process

costly strain improvement trajectory is not needed [14–16]. However, in order to obtain profitable production, strain improvement is often needed. Other serious challenges to get biocolorants on the market are a process development trajectory, high investment in the fermentation equipment, and long and costly registration trajectories to obtain approval.

2.1 Microbial Sources of Colorants

Although most of the production processes for biocolorants are still experimental, a number of industrial production processes are running with microorganisms that possess the complete biosynthetic pathway for one or more colorants: for example, astaxanthin from the yeast *Xanthophyllomyces dendrorhous* (formerly *Phaffia rhodozyma*) and the alga *Haematococcus pluvialis*; riboflavin and *β*-carotene from the fungi *Ashbya gossypii* and *Blakeslea trispora;* and chlorophyll and lutein from the algae *Dunalliella viridis* and *Chlorella protothecoides*. In Table 2, natural colorants are presented with their corresponding E-numbers, production micro-organism, or realized tissue cultures.

Carotenoids are broadly applied in food, feed, and pharmaceutical and cosmetic industries [17]. Therefore, these industries have made successful efforts to develop commercial processes. Industrial production processes are running, for example, for *β*-carotene, astaxanthin, and lutein [15, 17, 62]. Carotenoids are the main sources for the colors yellow, orange, and red. They are tetraterpenoid organic compounds that are found in plants, algae, and some bacterial and fungal species. Generally, carotenoids cannot be synthesized by animals. Therefore, they need to take them up with their diets, and incorporate carotenoids in their metabolism. Carotenoids are categorized in two classes. Pure hydrocarbons are called "carotenes," whereas oxygen-containing representatives are addressed as "xanthophylls." The light absorption is mainly in the blue region, and the energy is used in photosynthesis. Another function is protecting the chlorophylls from damage by UV light. In animals, carotenoids are responsible for the pink color of salmon, sea bream, shrimp, and lobster (astaxanthin or canthaxanthin), and for the pink feathers of flamingos. When these carotenoids are missing in the diet, these animals turn pale after a certain period. After boiling, lobsters and shrimp turn pink because of the denaturation of carotenoid–protein complexes. It is certain that the pink color is extremely important for perception: pale lobsters or white salmon and shrimp do not look attractive as food.

Carotenoids also act as regulators of membrane fluidity [63] which has been reviewed earlier [64]. Approximately 700 natural carotenoids have been described. The structures of important tetraterpenoid carotenes and xanthophylls are presented in Fig. 1. Apart from their provitamin A activity, further biological functions of carotenoids are directly linked to their antioxidant properties, which are related to their molecular structure [65, 66]. Supplementation of carotenoids to food

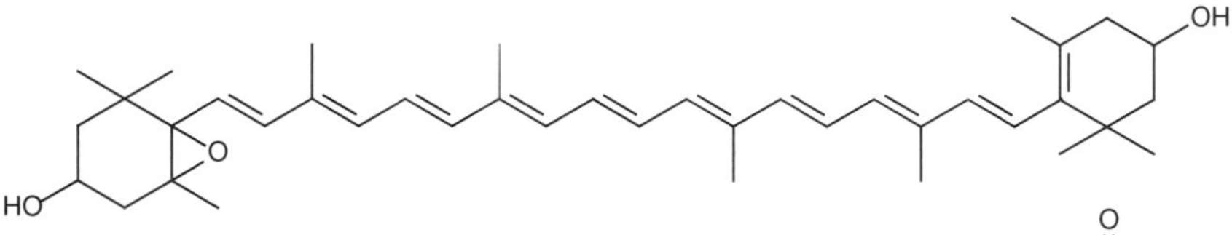

Antheraxanthin

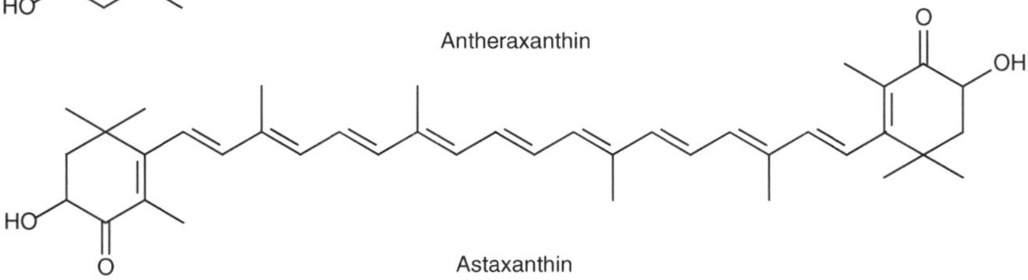

Astaxanthin

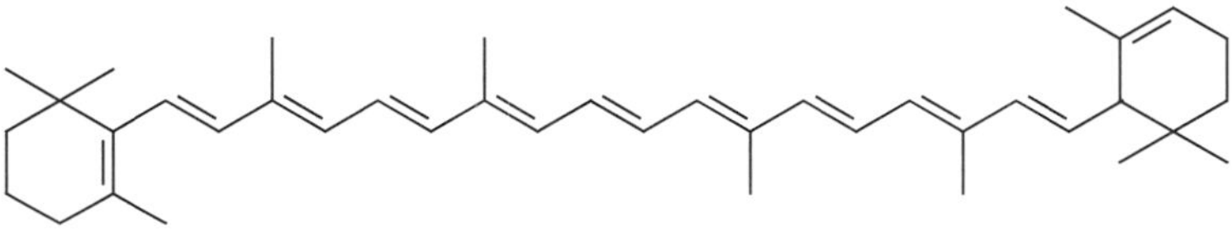

α-Carotene

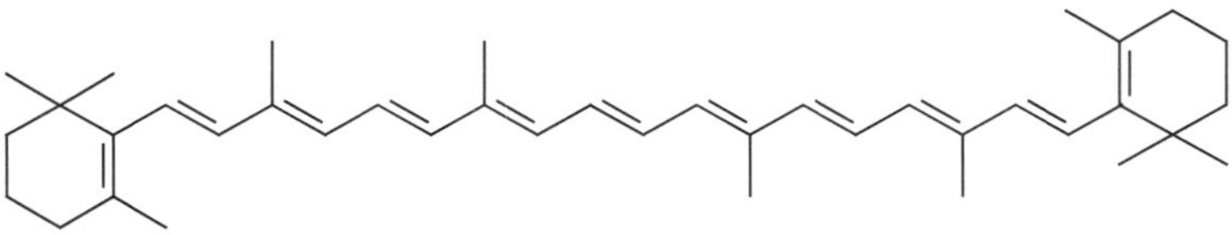

β-Carotene

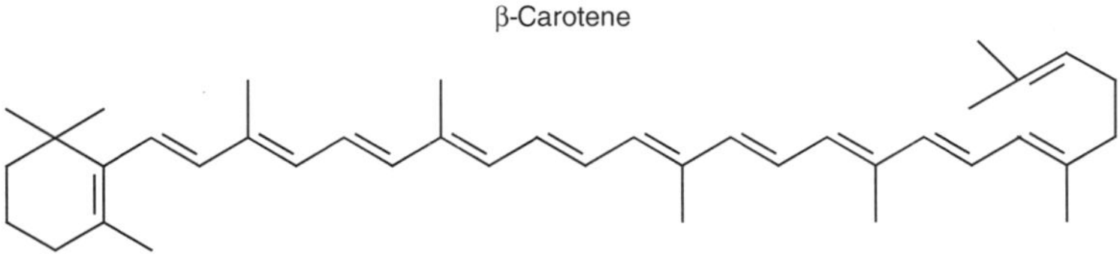

γ-Carotene

Fig. 1 Structures of tetraterpenoid carotenes and xanthophylls accessible by biotechnology

decreased the risk of suffering from degenerative diseases, and anti-inflammatory effects of carotenoids (*β*-carotene, lutein, lycopene) were also demonstrated [67].

There are a number of reasons to obtain colorants from biotechnological processes. First, biocolorants are not suspected to show undesired side effects which explain an increased consumer preference above synthetic dyes. Second, a number of natural colorants have potential positive health effects; for example, antioxidative/antitumor (lycopene, *β*-carotene, lutein, chlorophyll, anthocyanins) [4, 68], or provitamin A effects [2]. Third, being dependent on crops as sources of colorants has a number of disadvantages, namely agricultural availability (season and harvest) and downstream processing. Therefore, biotechnological production processes are attractive as production can be tuned on demand, and production is under tight control. The fourth reason is the price. In the eighties and the nineties of the last century, a number of companies started strain and process improvement on *X. dendrorhous*, because it was calculated that it was economically feasible to

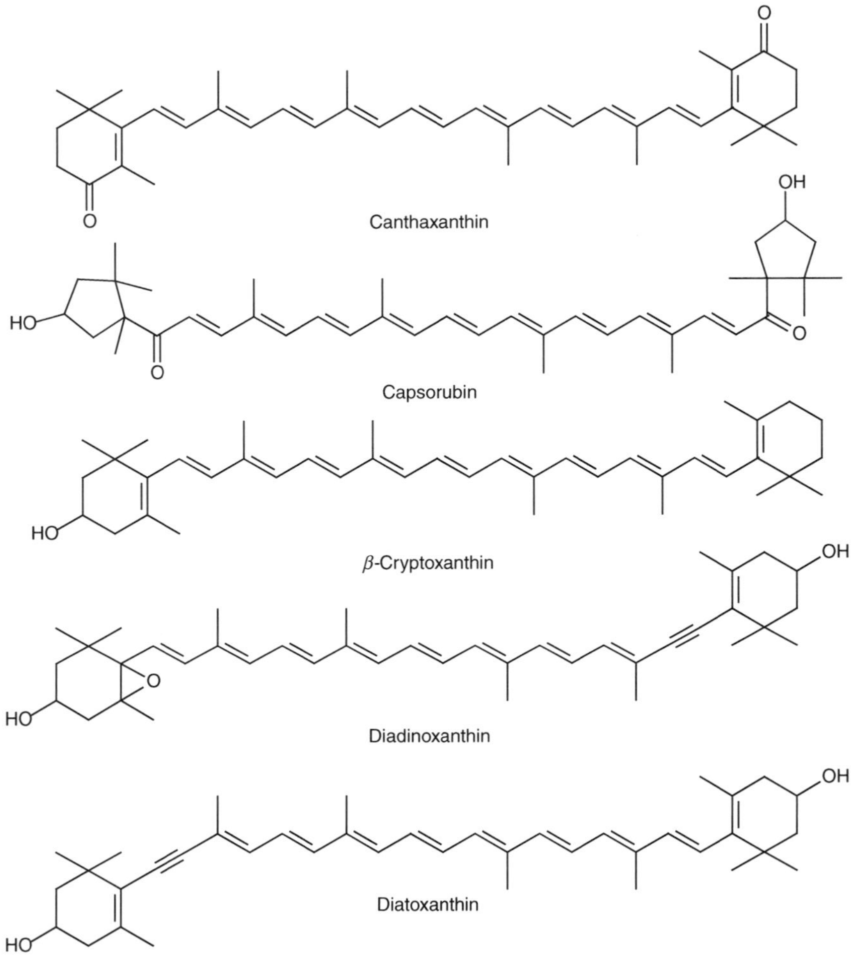

Fig. 1 (Continued)

develop a biotechnological production process that is cheaper than the Hoffman-la Roche (DSM Nutritional Products) chemical process. The fifth reason is the potential for discrimination based upon the existence of stereoisomers of some colorants. For example, in astaxanthin, two asymmetric carbon atoms are present and this results in four potential stereoisomers. The stereoisomeric composition of astaxanthin is dependent on the production process and organism. This allows for differentiation between salmon that were fed with synthetic astaxanthin, *X. dendrorhous* astaxanthin, and wild salmon by analyzing the composition of the stereoisomers of astaxanthin extracted from salmon (reviewed in [69]). Another example is *β*-carotene produced by *Dunaliella salina*. *β*-Carotene typically comprises 4 fractions, namely 9-(*Z*) (41 %), all-(*E*) (42 %), 15-(*Z*) (10 %), and other

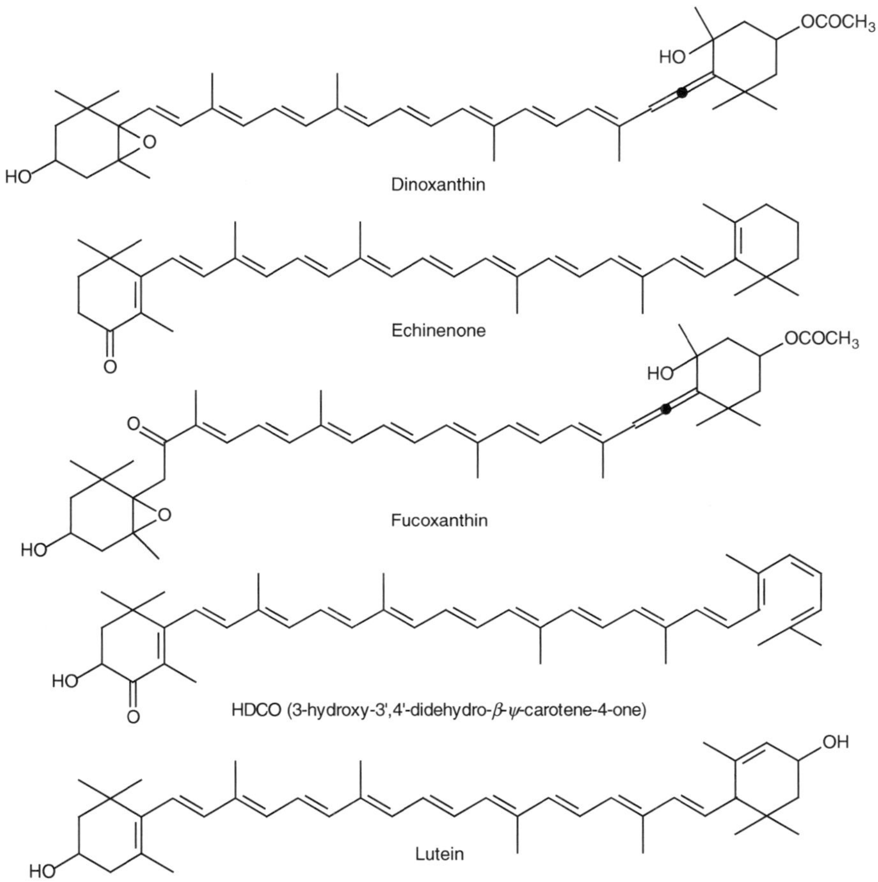

Fig. 1 (Continued)

isomers (6 %) [70]. Studies revealed no positive effect of synthetic *β*-carotene on cancer incidence, but even suggested carcinogenic activity at higher concentrations [71]. However, the 9-(Z) isomer is a better antioxidant compared to the all-(*E*) isomer, and the higher the share of the 9-(*Z*) isomer is, the higher are the antioxidant and antitumor effects [72, 73]. Therefore, the 9-(*Z*) isomer containing *β*-carotene obtained from *D. salina* is preferred by the health market [72]. An interesting note is that the 9-(*Z*) isomer shows a higher solubility in oil than the all-(*E*) isomer, and the latter is therefore easier to crystallize. This enables the selective enrichment of the 9-(*Z*) isomer in oil [74]. Taken together, natural *β*-carotene is preferred although the price of synthetic *β*-carotene is about half that of the natural product (€ ~ 700/kg).

However, fermentative production also shows a number of disadvantages. It is difficult to realize a stable production process for colorants with economic yields. Strain and process development are often tedious and expensive before routine

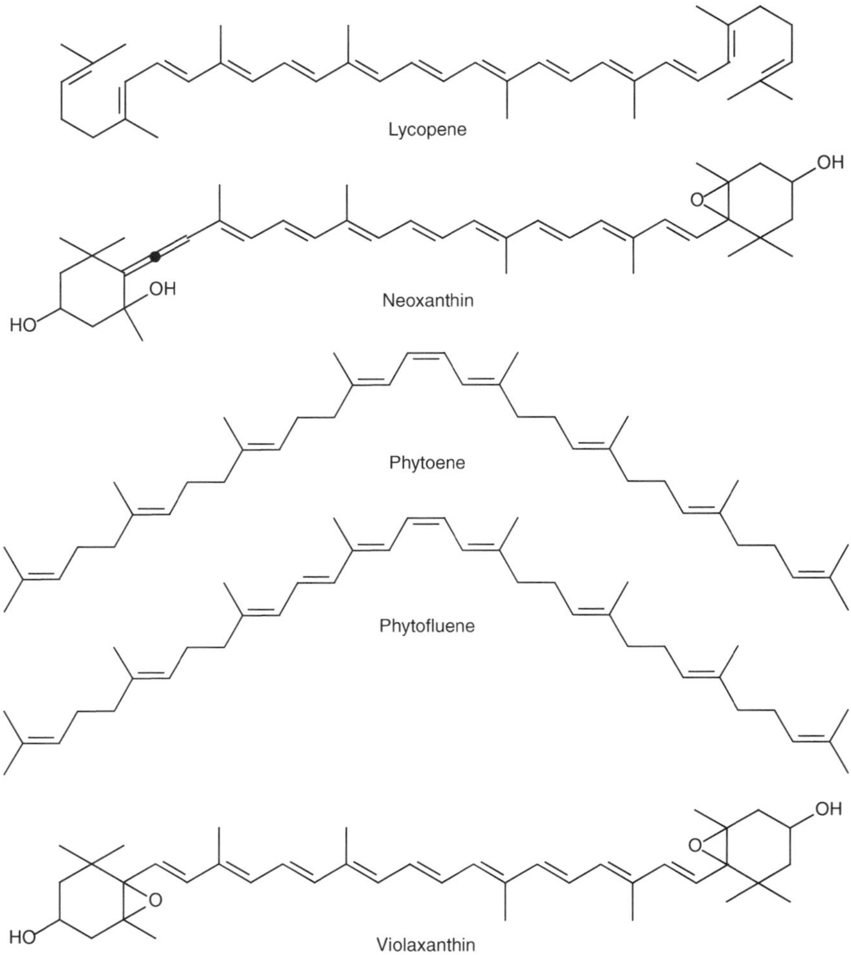

Fig. 1 (Continued)

production can start. Inasmuch as many colorants are typical plant metabolites, production may be realized by application of tissue cultures or metabolic engineering of crops and microorganisms [75, 76]. However, consumer acceptance for products obtained by genetic engineering is still low, and classical strain improvement is preferred. In the latter case, when a plant colorant is involved, a microorganism should be available that is able to synthesize the desired compound. If not, genetic engineering is inevitable to transfer the DNA involved in colorant synthesis into the production organism.

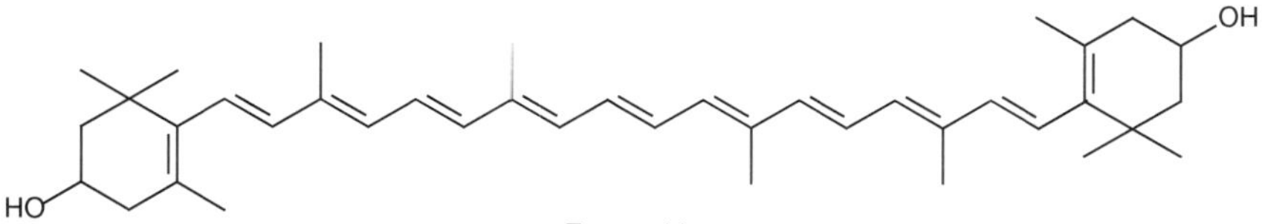

Zeaxanthin

Fig. 1 (Continued)

2.2 Colorant Production by Algae

Microalgae are a potential source of biochemicals with high added value including a high variability of pigments. Algae bring together properties of higher plants (photosynthesis and simple mineral needs) and the biotechnological advantages of microorganisms. Regarding metabolic aspects, it is a very versatile group, and algae are able to grow photo-auto-, hetero-, and mixotrophically, depending on the organism and the growth conditions. Despite that algae have already been commercially cultivated and utilized for more than 50 years, interest is currently increasing. This is mainly caused by their promising capability of producing next-generation biofuels production [77]. In this chapter, algae are discussed with respect to their economic interest for the production of the carotenoids *β*-carotene, astaxanthin, and lutein, and the blue noncarotenoid phycocyanin. Nevertheless, the prospects for the production of other carotenoids and biochemicals are promising. For example, green microalgae are able to synthesize xanthophylls that are synthesized by higher plants, such as neoxanthin, zeaxanthin, antheraxanthin, and violaxanthin. In addition, green algae are able to produce further xanthophylls, for example, canthaxanthin and loroxanthin, whereas fucoxanthin, diatoxanthin, and diadinoxanthin are synthesized by brown algae and diatoms [78, 79]. The distribution, biosynthesis, and functions of carotenoids in algae are reviewed in [80]. The pigments have various functions, and often mixtures of pigments are present in the cells [81]. For example, carotenoids protect chlorophyll from photooxidation and their presence is coupled with the presence of chlorophyll. Algae can also grow hetero- or mixotrophically, and the manner of carbon assimilation and energy generation highly affects growth yields and pigment composition, as reported for the green alga *Tetraselmis*. Under heterotrophic conditions, the total chlorophyll levels were less than 1 % compared to the levels obtained in the mixotrophically grown cultures, which indicates that chlorophyll synthesis is strongly subject to regulation [82].

In algae, carotenoids may have several functions in light-harvesting, scavenging reactive oxygen species and dissipating excess light, and structure stabilization of photosynthetic complexes [83]. Carotenoids are very intense dyes, even at ppm levels. Therefore, carotenoids are supplemented as a food colorant and feed additive for the pigmentation of the flesh of commercial fish (salmonids, sea bream; astaxanthin, canthaxanthin) and egg yolks (canthaxanthin). Other commercial applications are in nutraceuticals (for pharmaceutical and cosmetic purposes) because of their protective role against oxidative stress, and their use as nutrients in food and feed, cosmetics [84], and the removal of minerals (phosphate, nitrogen, sulfur) [11] or organic compounds from pollution streams as extensively reviewed in [12].

Generally, algae have simple growth requirements. They need water, small amounts of nutrients, such as nitrogen, phosphate, and sulfate, CO_2, and light. However, large-scale production with high cell densities is still problematic, because controlled illumination and harvest of biomass are still challenging [11]. In Table 3, the biological roles, origin, and some suspected beneficial health effects of the main colorants from algae are summarized.

Although carotenoids are still mainly produced by organic synthesis, production processes are running in which microorganisms (algae, fungi, yeast, bacteria) are used [85]. The estimated yearly production values of β-carotene, lutein, and astaxanthin are US$275, 250, and 230 million, respectively [86]; those for canthaxanthin, lycopene, and zeaxanthin amount to $80, 75, and 15 million. In Table 3, a number of interesting algal colorants are presented with their corresponding origin and potential health effects. Except for β-carotene, astaxanthin, and lutein, algae have the metabolic potential to synthesize many more carotenoids [80].

2.2.1 Cultivation Systems

Especially with the focus on carotenoid and biofuel production, fermentation and harvesting systems have been developed in order to maximize growth rates and product yields. Important challenges are the selection of the production organism, the development of the production process (medium composition, effect of growth (nitrogen) limitations), choice of fermenter materials (transparency for light, biomass adhesion), design of the fermenter (optimized light capture), CO_2 supply, and mode of fermentation (batch, fedbatch, in open [nonsterile] basins), harvesting, and cleaning. The developments are very fast and are characterized by fermentation equipment designs of high creativity.

The choice of the production alga is mainly based on the ability to synthesize the desired colorant under the given growth conditions, but also other strain-specific properties are taken into account. Strain properties crucial for economic production are the rate of biomass formation and product synthesis. Furthermore, nutrient costs must be low. A specific trait for algae grown in open basins is that they have to show high competitiveness towards other undesired algae species and that they are not too sensitive to predation (protozoan predators, snails) and various season-dependent growth conditions (variable light exposure, temperature) [117]. For production in open ponds, preferably production organisms are selected that are able to grow under very extreme conditions which minimizes the chance of development of contaminating organisms. Examples are the production of β-carotene by *D. salina* [17] and of canthaxanthin by *Haloferax alexandrinus* [115].

Table 3 Biological roles, origin, and potential beneficial health effect of the main colorants synthesized by algae

Colorant	Color	Production organism	Potential health effects	Ref
Chlorophyll a, b, and c	Green	*Chlorella vulgaris*	Potential anti-mutagenic activity	[87]
Phycoerythrobilin	Red	Cyanobacteria, red algae	No information available	[88]
Phycocyanobilin	Blue	Red algae	Potential against Parkinson's and Alzheimer's diseases	[88]
Phycocyanin	Blue	*Arthrospira (Spirulina) platensis, Galdieria sulphuraria*	Antioxidant, diagnostics, in food and medicine	[89, 90]
α-Carotene	Yellow/ orange	Cryptophyta, Chlorophyceae, *Dunaliella salina*	Antioxidant, against cardiovascular diseases and cancer	[80, 91]
β-Carotene	Yellow/ orange	*Dunaliella salina*	Provitamin A, antioxidant, preventive for several types of cancer and degenerative diseases, prevents progression of breast cancer and lung cancer	[92, 93]
Lycopene	Red	Algae, *Scenedesmus obliguus*	Antioxidant (most powerful carotenoid quencher of singlet oxygen), suggested prevention of cancers, particularly prostate cancer, cardiovascular disease	[86, 94–96]
Lutein	Yellow	*Muriellopsis* sp	Antioxidant, blue light absorption in retina and skin, age-related macular degeneration	[62, 97]
Echinenone	Orange/ red	Cyanophyta, *Thraustochitium Spirulina, Arthrospira*	Potential antioxidant	[80, 98]
Astaxanthin (diester)	Red	*Haematococcus pluvialis, Chlorococcum sp., Chlorella zofingiens*	Antioxidant, UV-light protection, anti-inflammatory, induces P450: protective against hepatocarcinogenesis, ocular and skin protective, fertility	[99–106]
Antheraxanthin	Yellow	*Chrysophaera magna*	Potential antioxidants	[80, 107–111]
Violaxanthin		Clorophyta *Mantoniella squamata*		
Neoxanthin		Chlorophyta, *Chlamidomonas reinhardtii,*		
Diadinoxanthin		Haptophyta, *Phaeodactilum tricornutum*		

(continued)

Table 3 (continued)

Colorant	Color	Production organism	Potential health effects	Ref
Diatoxanthin		Xanthophyceae, Dinophyta		
Dinoxanthin	Yellow	Dinophyta, *Protoceratium reticulatum*	Protection to ROS*, antioxidant	[112]
Fucoxanthin	Orange	Heterokontophyta, *Chrysophaera Magna*	Antioxidant	[107]
Peridinin	Orange	Dinophyta, Ceratium horridum		[113]
Canthaxanthin	Orange	*Chlorella zofingiensis*is	Induces P450: protective against hepatocarcinogenesis;	[86, 104, 114, 115]
		Haloferax alexandrinus	Protective against free radicals	
Zeaxanthin	Yellow/ orange	Rhodophyta, Glaucophyta, Cyanophyta, *Dunaliella salina*	Antioxidant	[86, 116]
Phytoene	Colorless	*Dunaliella salina*	Antioxidant	[12]

[a] *ROS* reactive oxygen species

Another selection criterion is the mode of harvesting by centrifugation or filtration, but these methods are relatively expensive. An alternative method is flocculation, which is also an algae species-dependent property.

With regard to the choice of cultivation system, the most straightforward designs are circular basins in the open air, eventually supplied with a pump or paddlewheel for water circulation. Continuous algae harvesting takes place by flocculation and deposition, followed by centrifugation. Afterwards, the algae are dried and processed for their definite application. The yields are often less than one g/l dry cells [118]. In order to realize an improved process control and higher production rates, basins are constructed in greenhouses, LED illumination is installed under the water surface, and combustion gas is flushed to elevate the CO_2 supply [119].

The designing of photobioreactors is proceeding rapidly. In order to optimize large-scale algae biomass production, much creativity has been put into the selection of appropriate materials to design the cultivation systems. With the objective to optimize light yields, the materials must have a high transparency. They should have a high resistance to fracture and they should not be too sensitive to scratches. The latter is especially important with respect to the adsorption of biomass to surfaces and cleaning. Biomass adhesion is also prevented by coating technologies. The design of the production systems is mainly determined by maximum utilization of light: ultrathin immobilized configurations, and flat plate and tubular bioreactors [120–124]. The design is based on an optimized homogeneous illumination of the biomass, keeping the light paths as minimal as possible, optimal mixing of the nutrients, the carbon dioxide supply, and the removal of oxygen. Together with the pH, the temperature, and the dissolved oxygen concentration, the mentioned growth parameters require accurate control.

An example of a tubular photobioreactor is a system composed of transparent parallel tubes that are closely placed on the ground and all tubes are connected [14]. After medium preparation and inoculation the culture is circulated in the tubes. Oxygen, a product of photosynthesis, is continuously removed during the exposure to light and may be an attractive by-product. During growth, light penetration in the tubes decreases because of the growing biomass, and a part of the captured light is used for maintenance rather than for growth. When the growth rate has become very low, harvesting is started to prevent deterioration of the biomass.

Despite significant progress in process development, the application of this technology for the production of biodiesel is also still hampered by high costs, and it is estimated that the price has to be reduced by a factor of 10–20 before large-scale production becomes economic [125].

After production of the pigment-containing biomass, the biomass can be separated from the medium, dried, extracted by vegetable oil or hexane, and utilized as a food or feed additive [12]. Because the pigments are sensitive to temperature and photo-oxidation, the extraction must be performed quickly and under controlled conditions. On the production scale, the application of zirconium ball mills is a proven technology and was applied for the extraction of astaxanthin from

X. dendrorhous in the presence of soybean oil. *Xanthophyllomyces dendrorhous* is a yeast with a very rigid cell wall, and the above method may also be effective for recovery of pigments from algae.

Furthermore, in [126], a production process is decribed in which *β*-carotene is extracted from the harvested biomass of *D. salina* using vegetable oil as the solvent.

2.2.2 *β*-Carotene

β-Carotene is applied as a food coloring additive, as provitamin A in food and feed, in multivitamin products, as an antioxidant, and as a colorant for cosmetics. In the literature, a number of microorganisms are described that are capable of accumulating *β*-carotene up to relatively high levels, for example, the fungi *B. trispora, Phycomyces blakesleanus*, the yeast *Rhodotorula glutinis,* and the above-mentioned alga *D. salina* [127]. Production processes were developed with mutants of *B. trispora* in the former Soviet Union (cf. Sect. 2.3) and for *D. salina*.

Among the *β*-carotene–producing algae, *D. salina* is currently the best-known and most efficient producer: *D. salina* cells are able to synthesize *β*-carotene in levels a thousand times higher than those of carrots [17]. *Dunaliella* is also able to synthesize α-carotene, violaxanthin, neoxanthin, zeaxanthin, and lutein. The unicellular green alga *D. salina* is grown in open ponds without (hardly any) process control [88, 93, 128]. It was shown that the highest production levels were obtained with high light intensities, nitrogen limitation, and stress-inducing temperatures. This has resulted in *β*-carotene levels up to 12 % of the dry mass [93]. The production facilities are located in places where optimum conditions are found, namely in Australia, Israel, and in the United States; these places benefit from a lot of sunshine, little cloudiness, the availability of saline water, and high average temperatures [85].

In Australia, *β*-carotene production is performed in an extensive manner. The ponds (~250 ha, 0.5 m deep) have a large surface, and there are no measures for active mixing or other ways of control. Therefore, mixing only occurs by wind, convection, and diffusion, and the biomass and *β*-carotene yields are low (~0.1 g *β*-carotene/m^3). The effect of the seasons on the production conditions is low, and this enables year-round production. The light intensities are optimal for maximum carotenoid synthesis. In order to maintain a stable production population without too much pressure from competitor and predator organisms, the salt concentration is kept under close control. At a salinity of 12 % salt (w/v), the formation of biomass is optimal, but the *β*-carotene synthesis is low. In addition, these conditions are favorable for the development of undesired organisms, whereas at salinities of 24–27 % most favorable production conditions are realized in terms of *β*-carotene production, biomass formation, and culture stability [126].

A more intensive two-stage cultivation process is performed in paddlewheel mixed circular basins of 3,000 m^3 (0.2 m deep) where CO_2 is flushed into the culture. In the first stage, the focus is on biomass production, whereas the conditions are optimized in order to realize a high β-carotene production [120, 128]. Optimization is also performed in semicontinuous outdoor cultures (20 m^2), using dilution circles of 3–4 days and nitrogen control [85]. It was found that the two-stage production process resulted in the highest β-carotene levels (up to 16 % of the dry weight, corresponding to 0.2 g β-carotene m^2/day) [93, 128]. The productivities of the culture systems are reviewed in [12]. In Australia, Whyalla, the Henkel-Cognis Company, uses about 800 ha for β-carotene production with *Dunaliella*, which is the largest surface with regard to β-carotene production in the world. Land is cheap, plenty of seawater is available, and the climate is excellent for growing *Dunaliella*. A modified natural salt lake of about 20 cm depth is used as a production pond. The salt lakes already contained dense natural *Dunaliella* populations. In order to control the salinity, seawater is used and growth nutrients are added for optimal culture development. After harvesting, the resulting medium is reused in order to prevent nutrient loss for cost reduction.

A potential advantage of a closed fermentation system is that it allows for an improved process control. This may be important for controlling the isomer composition of commercial carotenoid preparations. With respect to a commercial β-carotene product, a study suggested that the level of the 9-(Z) form may vary depending on the process control [129]. As discussed above, the 9-(*Z*) isomer is considered as the β-carotene isomer that shows the best antioxidative properties [72].

Generally, biomass harvest is performed by flocculation and sedimentation at the extensive producers, whereas centrifugation is applied at the high biomass production companies [93]. Inasmuch as *D. salina* is grown in a very saline and viscous medium, centrifugation causes significant cell damage and loss of β-carotene because of shear forces. The cells are very flexible and easily pass filters. Therefore, filtration efficiencies are rather low. The only successful filtration method was the application of diatomaceous earth [130]. Until now, the most effective technique to harvest the biomass has been the use of a continuous flow centrifuge [13]. An alternative technique is the application of a combined flocculation and floating method based on treatment of waste and drinking water [126]. Biomass flotation occurs because oxygen bubbles are formed that are entrapped in the biomass flakes. After biomass separation, extraction is performed with vegetable oil and in order to produce a powder with uniform color and shape. Finally, the β-carotene crystals are milled.

β-Carotene-containing products are available as gelatin capsules containing a 4 % β-carotene solution as food supplements, as food colorants (15–30 % β-carotene in soybean oil for margarine, 2 % emulsions for the coloration of beverages, spray-dried *Dunaliella* powder for coloration of prawns, as tablets, and as water-dispersible β-carotene containing powder and other formulations.

2.2.3 Astaxanthin

Commercial production processes are known in which astaxanthin production is performed in closed photobioreactors by Cyanotech Corporation (Hawaii) [128]. This mode of cultivation strongly improves process control, algae predation, and production, but the fermentation costs are significantly higher [131]. These higher costs have to be compensated by high productivity levels and more efficient downstream processing in order to make a competitive product.

Most of the astaxanthin available on the world market has been produced chemically since the 1950s, for example, by DSM and BASF, and the process is efficient and cost-effective [132]. There are also microbial sources, for example, *X. dendrorhous* (previously described as *P. rhodozyma)*, and the alga *H. pluvialis*. *X. dendrorhous* has been widely investigated as to its potential of astaxanthin production for salmonids (see below). This has resulted in a race between companies in the 1980s and 1990s. Natural astaxanthin is currently available as a spray-dried powder (5–10 mg astaxanthin/g), and is supplemented to fish feed to give salmonid flesh its pink color. Another astaxanthin application is in the nutraceutical market, as astaxanthin is regarded as a potential antioxidant [106].

With regard to production by algae, *H. pluvialis* is the only species commercially cultivated for astaxanthin production. This green alga is able to grow under autotrophic, heterotrophic, and mixotrophic conditions. An example of the latter is its ability to grow in the presence of acetate and light [15]. Astaxanthin accumulates in response to stress in lipid globules of the cells [133].

Studies were performed to define the optimal production conditions of *H. pluvialis* [134]. A cultivation design in which growth and astaxanthin production were separated was compared to production in a continuous culture under limited stress conditions at steady state. In the two-stage system, higher astaxanthin yields were obtained (4 % calculated on dry biomass) with a productivity of 13.9 mg/l per day compared to 5.6 mg/l per day in the continuous culture. However, biomass that was harvested from the two-stage culture was characterized by very rigid cell walls, and this made the downstream processing of astaxanthin more difficult. An advantage of the rigid cell walls is that the culture becomes less susceptible to predation. Probably an alternative intermediate high production cultivation regime can be developed that produces biomass more amenable for downstream processing.

Haematococcus pluvialis requires illumination and stress to induce carotenoid biosynthesis. It thus cannot be efficiently grown under heterotrophic conditions in the dark. For that reason, studies were performed with the green alga *Chlorella zofingiensis*, which is capable of accumulating astaxanthin and canthaxanthin and grows three times faster than *H. pluvialis*. This alga is able to synthesize substantial amounts of carotenoids in the dark under heterotrophic conditions, and may be used for large-scale production in high turbidity cultures [135, 136]. Under conditions of salt stress, low nitrogen availability, and low light, *C. zofingiensis* accumulated more canthaxanthin than astaxanthin. Experiments demonstrated that for canthaxanthin accumulation under salt stress and low availability of nitrogen, light is not the limiting factor. On the other hand, high light irradiance is mandatory for astaxanthin accumulation [114].

Irradiation causes oxidative stress by the formation of activated oxygen species, and induces a mechanism of cell protection. The formation of carotenoids is apparently part of this reaction and protects the cells from photo-oxidative damage [135]. With regard to the regulation of carotenoid biosynthesis, a study revealed that light quality was more important than light quantity by the use of flashing light compared to periods of continuous light exposure [137]. Other operating variables were culture turbidity, light path, cell sensitivity (dependent on cell stage), and a balanced medium composition [138]. With regard to the latter, nitrate dosage was crucial [139] and this was also the case for iron, which was crucial for the assimilation of nitrate and nitrite, nitrogen fixation, and chlorophyll synthesis [140, 141]. During illumination, the injection of CO_2 significantly stimulated the synthesis of astaxanthin in the production phase of *Haematococcus* [142], whereas intensive light exposure stimulated CO_2 fixation for astaxanthin synthesis [143], which can be optimized further by improving the carbon/nitrogen ratio supply [144].

Because *H. pluvialis* is very sensitive to microbial contamination, and no competitive advantageous conditions have been found to maintain *H. pluvialis* as a monoculture so far, mass cultivation generally takes place in closed bioreactors. Furthermore, growth is sensitive to deviations in temperature and light, and therefore accurate control of these growth parameters is necessary. Companies are not very open with regard to details of their production processes. However, a thorough review showed comprehensive information [79]. Resuming, astaxanthin commercial production is composed of two cultivation phases: in the first phase favorable growth conditions are set in order to obtain high biomass concentrations (high nutrient concentrations, optimum temperatures, and irradiation). The following growth phase is focused to induce hematocysts (aplanospores with high astaxanthin biosynthetic activity) and astaxanthin formation by realization of stress conditions: nutrient exhaustion (nitrogen and phosphorus), high temperature or high salt concentrations), and intensive illumination. Then the biomass is harvested by centrifugation, dried, and cracked. The cell wall of the cysts is very rigid, thus it is apparent that a special technology is needed for astaxanthin disclosure. However, companies consider this as strategic information and are not willing to communicate.

2.2.4 Lutein

Petals of marigold flowers (*Tagetes erecta* and *Tagetes patula*) currently represent the main source of commercial lutein. More than 95 % of the lutein is esterified, and about half of this fraction is esterified with fatty acid. Therefore, saponification is a part of the downstream processing [145]. Lutein is supplemented to food and feed for aquaculture and poultry farming [146]. Furthermore, lutein is suggested to be beneficial for health, for example, to prevent age-related macular degeneration

[97] and progression of early atherosclerosis [147]. In algae, lutein is accumulated in the nonesterified form. The alga *Muriellopsis* sp. is able to accumulate lutein up to high levels and is easy to cultivate photoautotrophically [62]. The effects of critical growth and production parameters in outdoor continuous cultures have been investigated. Under optimized conditions, 40 g dry cell mass/m^2 and 180 mg/m^2 lutein were produced per day, respectively. Further optimization was performed by introduction of agitation with a paddlewheel in a semicontinuous cultivation system and by CO_2 addition. Thus, the lutein content was increased to 0.4–0.6 % of the dry mass [85] at a productivity level comparable to that in a closed tubular photobioreactor [148]. Beneficial for lutein synthesis were high temperatures, high irradiance, an optimum pH value for biomass formation, and the addition of inducers such as H_2O_2 or NaClO in the presence of Fe^{2+} (for the generation of stress-inducing chemical species), especially under heterotrophic growth conditions where spontaneous oxidative stress is absent [16].

2.2.5 Phycocyanin

Phycocyanin is a blue dye formed by cyanobacteria, cryptophytes, and rhodophytes. It is an interesting nutraceutical because of its suspected antioxidative properties, and it is also applied for coloring and fluorescence in biochemical assays. Despite these interesting properties, the yearly market value is comparatively low. Currently, phycocyanin is produced in open basin cultures containing the cyanobacterium *Arthrospira* (*Spirulina*) *platensis*. However, productivities are low as the cultures are very sensitive to contaminating organisms. Recent studies addressing chemical modifications have enhanced the potential applications of phycocyanin in the fields of diagnostics and applications in food, nutraceuticals, and biotechnology [90]. Furthermore, optimization programs significantly increased productivity in heterotrophic and acidophilic cultures of the red alga *Galdieria sulphuraria* which are grown under well-controlled and axenic conditions [149]. This makes *G. sulphuraria* an attractive alternative for *Spirulina* [150]. Furthermore, improvement was achieved with regard to downstream processing: the application of two-phase aqueous extraction methods and optimized purification technologies have resulted in higher productivity and improved performance. Table 4 summarizes the main producers of natural β-carotene, astaxanthin, lutein, and phycoerythrin.

2.2.6 Prospects in Algae Colorants

The development of algae biotechnology is still in an early phase and a lot of work must be done to obtain a complete picture with respect to the full potential of algae applications. The biodiversity in nature is extremely high, just as the number of useful biochemicals that can be isolated from algae. In order to obtain the most appropriate algae with reference to the desired products and (related) production

Table 4 Producers of various colorants produced by various microalgae

Product	Production organism	Company
β-Carotene	*Dunaliella*	Cyanotech (Hawaii, USA)
		AquaCarotene (Washington,USA)
		Cognis Nutrition and Health (Australia)
		Nikken Sohonsha Corporation (Japan)
		Tianjin Lantai Biotechnology (China)
		Parry Pharmaceuticals (India)
	Blakeslea trispora	Combinat Verkhnedneprovsk'e Starch and Syrup Plant (Ukraine)
Astaxanthin	*Haematococcus*	Cyanotech (Hawaii, USA)
		Bioreal (Hawaii, USA)
		Alga Technologies (USA)
		Mera Pharmaceuticals (Hawai, USA)
		Parry Pharmaceuticals (India)
	Xanthophyllomyces dendrorhous	Fermic SA (Mexico)
Lutein	*Muriellopsis*	Sun Chlorella Co (Japan)
	Dunaliella salina	Nutrition & Health (Australia)
Phycoerythrin	*Arthrospira (Spirulina) platensis*	Cyanotech (Hawaii, USA)
		BlueBiotech International GmbH (Germany)

conditions, it is recommended to start a product development project with a screening program whose purpose is to find the most appropriate alga species. After its successful isolation, the next step is the selection of the fermentation system that enables economical production and the desired colorant quality (colorant by-products, isomers). Selection criteria are, for example, sensitivity to microbial contamination, fermentation parameter control, and the fermentation mode. This may influence the choice of, for example, out- or indoor cultivation and the use of photobioreactors. In Table 5, a selection of process and economic parameters are summarized that affect the choice of cultivation system. There is a lot of room for process and equipment optimization depending on the organism, product, and its economical added value. It is expected that each alga–product combination needs a specific fermentation medium, control and equipment, downstream processing, and infrastructure, among others.

If it is not possible to implement an economical production process by screening of algae species and process optimization, a strain improvement in order to enhance or to induce the overproduction of the desired metabolite may be necessary. *Dunaliella bardawil* mutants were isolated that were able to produce more β-carotene under a defined low-light regime than the parental strain [151]. However, the strength of mutant isolation by mutagenesis and selection to be implemented in running production processes has still to be demonstrated. Classical strain improvement is also applicable to create strains that accumulate interesting intermediates in a biosynthetic pathway. By mutagenesis and selection, strains were isolated that were unable to synthesize an active biosynthetic enzyme

Table 5 Comparison of selection criteria for open and closed photobioreactors

Parameter/variable	Closed system	Open system	Remarks
Microbial contamination/ predation	Easy	Difficult	Define selective advance production alga/ plaque control (snails, water fleas)
Biomass density	High	Low	
Fermentation system	Batch/ continuous	(Semi) Continuous	Batch: harvest by centrifuge (semi) continuous: by flocculation, adhesion, precipitation in coalescence/ precipitation drums
Process control	Easy	Difficult	Production control in ponds highly season-dependent
Production continuity	Under control	Season-dependent	Especially concerns light intensity and temperature, dependent on location
Needed area/ amount of product	Low	High	
Light utilization	High	Low	
Operation costs	High	Low	
Scale-up	Difficult	Easy	In closed systems much technical creativity and investments are needed
Cleaning intensity	Often	Seldom	Transparent tubes are sensitive for biomass adhesion

after the intermediate of interest. Target intermediates are, for example, echinone, diatoxanthin, diadinoxathin, dinoxanthin, antheraxanthin, violaxanthin, neoxanthin, and zeaxanthin (cf. Fig. 1). The added value compared to the carotenoids already commercially available has still to be evaluated for most of the mentioned intermediates. A successful example of this approach was the isolation of a mutant of *D. salina* after ethyl methyl sulfonate treatment. The mutant *zea1* was corrupted in the zeaxanthin epoxidation reaction and was thus not able to synthesize the β-branch xanthophylls neoxanthin, violaxanthin, and antheraxanthin, whereas zeaxanthin accumulated up to high levels [116]. Zeaxanthin may help to reduce the risk of diseases including cataracts, macular degradation, and arteriosclerosis [152], and has not been produced commercially by means of biotechnology yet. Furthermore, a mutant and stable strain of *Chlorococcum* has already been isolated. It showed two times higher carotenoid production than the wild-type and showed to be stable under the chosen production conditions [153].

The application of molecular genetics is another strategy for strain development. The heterologous expression of genes allows for the production of biochemicals specific to higher plants and animals in algae [154]. Currently, three alga species are routinely accessible for transformation, namely *Volvox carteri, Phaeodactylum tricornutum*, and *Chlamydomonas reinhardtii* [155]. An example is the successful elevation of lipid production by genetic engineering of diatoms [156]. The obligate phototrophic diatom *Phaeodactylum* was enabled to grow in the dark by heterologous expression of the functional glucose transporter [157].

Chlamydomonas is of particular interest, because many biochemical studies have been performed with regard to its pigments and photosynthesis [158]. Routine transformation is still not common practice and has only been achieved in a limited number of alga species. It is interesting to report that genetically modified transfer (T) DNA of *Agrobacter tumefaciens*, which causes tumor cells in plant tissues, was used to transform *C. reinhardtii* [159]. The current status of the developments in molecular engineering has been reviewed extensively in [12].

In order to prevent product losses, the biomass harvest has to be fast and efficient. Centrifugation and filtration are fast processes, but a higher risk for spoilage arises when containers are used for flocculation, precipitation, or coalescence. The most common way of processing is drying, either in the sun (which may result in significant carotenoid losses because of photo-oxidation), in drums, or by freeze-, and spray-drying. In many cases, an additional disruption of the biomass must be performed to make the carotenoids available, for example, for uptake by salmon. The biomass of algae contains very rigid cell walls and has to be disrupted in mechanical homogenizers or ball mills. Alternatively, chemical methods such as extraction with organic solvents are available. When the latter technology is applied, the carotenoid becomes sensitive to high temperatures and photo-oxidation and has to be stored refrigerated and in the dark, preferably under nitrogen. Furthermore, traces of organic solvents may still be left in the final product which is not desirable. These methods are applied for both β-carotene (*Dunaliella*) and astaxanthin production (*Haematococcus*).

In summary, biotechnological production of carotenoids by algae is a successful and proven technology. It is expected that the rapid developments in the field of biofuel production will also promote carotenoid production, especially in the fields of fermentation, downstream processing, and molecular engineering. New perspectives with regard to the production of unusual carotenoids with added value for nutritional and medicinal applications (phytoene, phytofluene) are emerging and bear promise for the times to come. Development and implementation of new fermenter designs, modes of cultivation (autotrophic, mixotrophic, and heterotrophic), and application of two-phase production systems will offer new chances with regard to product innovation and competitiveness.

2.3 Industrial Production Development Trajectories

Earlier in this chapter, examples were described with regard to the production of β-carotene, astaxanthin, lutein, and phycocyanin by algae. Commercial production was achieved by choosing the appropriate microorganism and optimum cultivation and production conditions. In these cases, no extensive strain improvement trajectory was required. However, there have been examples of production development trajectories for astaxanthin and β-carotene, respectively, in the 1990s. In this chapter, the development trajectories are discussed for the production of astaxanthin by the basidiomycetous pink yeast *X. dendrorhous* (formerly known as

P. rhodozyma [160]) and of *β*-carotene by the filamentous fungus *B. trispora* [161]. Despite the fact that the algae *D. salina* and *Haematacoccus pluvialis* were already known as *β* -carotene and astaxanthin producers, respectively, *B. trispora* and *X. dendrorhous* were considered as serious alternatives as both are heterotrophic organisms. This enables cultivation under well-controlled conditions in closed fermenters. However, production levels of both organisms were far too low, and inasmuch as knowledge with regard to the DNAomics and proteomics of carotenoid synthesis was unknown for the most part, a number of biotechnological companies have chosen to enhance *β*-carotene and astaxanthin production by repeated mutagenesis and selection.

2.3.1 Astaxanthin (Xanthophyllomyces dendrorhous)

Astaxanthin is generally recognized as safe (GRAS) by the FDA. In the United States it is, however, restricted to use in animal feed. The economical background of astaxanthin production is based on the aquaculture of salmon, sea bream, and rainbow trout [162, 163]. More than 90 % of salmon is grown in nurseries, and in order to obtain the nice pink color of salmon flesh, astaxanthin or canthaxanthin are necessary dietary supplements. The levels of these colorants in the feed determine the final color intensification of the fish. Canthaxanthin is used in farm-raised trout. The supplementation of these carotenoids contributes to a significant part of the cost price of the fish (4,000–5,000 €/kg pure astaxanthin, personal communication), whereas only a relatively small part of the supplemented astaxanthin is taken up by the salmon). Nearly all astaxanthin and canthaxanthin used for fish feed are chemically synthesized (by DSM and BASF).

In 1975 [64], the basidiomycetous yeast *P. rhodozyma* was isolated from exudates of deciduous trees from Alaska and Japan. Later, *Phaffia rhodyma* was classified as *X. dendrorhous*, but this classification is still under discussion [164]. The yeast is grown at relatively low temperatures (<25 °C), is Crabtree positive, and can be grown on complex and synthetic media. *Xanthophyllomyces dendrorhous* is economically interesting because it is able to synthesize astaxanthin [162, 163, 165], albeit in very low levels (<0.5 g/kg dry biomass). At first sight, the low levels of astaxanthin hamper the economic feasibility of the astaxanthin production process. However, the ease of discrimination between high and low astaxanthin levels by the color of agar-plate grown colonies and the ease of growing enables strain improvement by high-throughput screening. Strains with increased astaxanthin levels are reported frequently [166–168]. In the past, various approaches were tested in order to obtain high astaxanthin-producing strains, for example, the application of molecular genetics technology [169], protoplast fusion [170], stimulation of astaxanthin synthesis by medium nutrient supplementations, strain improvement by mutagenesis and selection [171, 172], and improved fermentation media and fermentation control. Usually, a fed batch fermentation process was performed [173].

To develop a commercial process, strain improvement by mutagenesis and selection was performed. Surprisingly, patents were filed by Danisco and other companies in this area [173–177]. This was surprising as strain improvement by mutagenesis and selection is a routine technology, generally applicable and accessible, and no specific measures or inventions are necessary to perform. Therefore, it is doubtful whether such a patent was granted because there were no relevant arguments to designate this work as an invention. The patent was indeed not granted [178] but these events indicated the importance of winning the race to the market and the economic importance of the astaxanthin strain improvement projects.

In order to obtain a profitable production process, astaxanthin levels had to be elevated at least by a factor of 20 compared to the wild-type [64]. The required technologies, designs, and strategies for strain improvement by mutagenesis and selection have been reviewed extensively [179–181]. During classical strain improvement, undesired mutations may be introduced that affect the growth rate and the yields. Therefore, a suspected high producer has to be screened as to whether it has not lost other production-related properties. This appeared a long and labor-intensive trajectory with many tricks and pitfalls. During strain improvement, the astaxanthin levels of the subsequent improved strains increased, but at the higher astaxanthin levels the human eye becomes too insensitive to discriminate between the higher astaxanthin producers on agar plates: all of the colonies were considered as very intensively colored. This was overcome by application of effectors that inhibit astaxanthin biosynthesis, for example, diphenylamine, nicotinamine, and β-ionone [166, 168]. The effectors were dosed in a minimal color inhibitory concentration: if a mutant with debottlenecked astaxanthin synthesis was present on the effector-containing agar plate, a high-producing colony was detectable between less-intensive colored colonies. Another approach is to find a selection tool directly correlated with survival and increased astaxanthin levels. Astaxanthin was thought to be important for radical scavenging, and this was later confirmed in the literature [67]. High-producing mutants were selected on growth-inhibiting concentrations of the radical-generating duroquinon: astaxanthin scavenges radicals which resulted in a survival advantage of astaxanthin-producing strains [182].

A strain that was isolated as a high producer on agar plates may show unaltered or even lower production levels in shake flask (batch) cultures. Even after confirmation of the increased production levels in shake flask cultures, productivity levels still have to be confirmed during scale-up. The results depend on the fermentation system used on an industrial scale (mostly batch or fedbatch, continuous cultures in some cases) and the scale-up factor. Productivity often decreases during scale-up, because production takes place under elevated hydrostatic pressures, mixing takes more time, and mass exchange rates are lower (oxygen, carbon dioxide, sugar and ammonia feed, etc.). Furthermore, it was often observed that higher astaxanthin levels went hand in hand with strain properties that are unfavorable for production: for example, low growth rates and yields or undesired growth requirements. This is mostly caused by mutations in the central metabolism that are not directly correlated with astaxanthin biosynthesis. Another reoccurring problem is that mutants may synthesize higher levels of additional carotenoids relative to astaxanthin, for example, HDCO (3-hydroxy-3′,4′-didehydro-β,ψ-carotene-4-one) which do not contribute to pigmentation of the fish or prawn and biosynthesis of which is mostly at the cost of astaxanthin synthesis. Therefore, the carotenoid composition of each suspected improved strain has to be analyzed accurately.

The designs of strain improvement at the various companies were generally performed as indicated below [173, 175, 176]. The first step was colony selection on agar plates. The most intensely colored colonies (1–2 %) were selected and tested in small liquid cultures (3–10 ml). From these cultures, the best 5–10 % were tested again in shake flasks, and the best cultures were transferred to 5–10 l fermenters. It is important that the same media are used for screening as for production in order to prevent false positives (strains that are improved under the selection conditions but do not show higher astaxanthin levels on the production medium) or false negatives. A statistical analysis is required in order to obtain an estimation of the numbers of colonies that have to be screened in order to realize the desired progress.

Frequently, false-positive strains were isolated that showed a more intense color than the parental strains, and it was often demonstrated that cultures, derived thereof, contained high levels of HDCO (3-hydroxy-3′,4′-didehydro-β,ψ-carotene-4-one), an undesired red-colored side-product of astaxanthin synthesis. The molecular structure of HDCO differs from astaxanthin with regard to one side of the molecule which is still linear because the ring is not formed and oxygenated. HDCO has no added value with respect to salmon pigmentation and a method was developed to isolate strains with low levels of HDCO relative to astaxanthin [183]. Since the HDCO biosynthetic pathway is a shunt of the biosynthetic pathway to astaxanthin, HDCO formation is at the cost of astaxanthin. HDCO is observed as red, whereas astaxanthin is observed as deep orange, and strains with increased HDCO levels were often selected as false positives.

It was also proven that carotenoids other than astaxanthin can be produced by *X. dendrorhous.* This was performed by the isolation of mutants that are blocked in specific steps of astaxanthin biosynthesis [164]. If an intermediate has a different color than astaxanthin, it is relatively simple to isolate mutants which accumulate that intermediate (e.g., β-carotene which is yellow at low levels and orange at high levels) [184].

Companies that are or have been active in the field of fermentative astaxanthin from *Xanthophyllomyces* are, among others, Universal Foods, DSM, Igene, and AmbroZea [173–178, 183]. These companies have put significant effort into the development of a profitable process, and great progress was made by Universal Foods and DSM. Commercial samples regularly appeared on the market, but to the best of the author's knowledge, Fermic SA in Mexico was the only company that produced *Xanthophyllomyces* astaxanthin on a regular base for Igene Inc. [185]. The natural astaxanthin derived from algae still dominates the market apparently using the cheaper mode of production. The rise of molecular genetics technology that enables the transfer of complete metabolic pathways into microorganisms which are easy to cultivate will probably change this situation in the future.

2.3.2 *β*-Carotene (Blakeslea trispora)

In 2008, the presence of *α*-, *β*-, *γ*-, *δ*-, and *ε*-carotenes was reported in 76 algae, 5 fungi, and 4 bacteria [186]. Among them, *β*-carotene is produced by a number of fungi, for example, by *Phycomyces, Mucor,* and *B. trispora*, and dry matter contents of 4–5 g /l are achieved [187].

Blakeslea trispora is a fungal plant pathogen and belongs to the order of Mucorales. The fungus is nonpathogenic and nontoxigenic for humans and animals. This order includes *Phycomyces blakesleeanus*, *Choanephora cucurbitarum*, and *B. trispora*. Two opposite mating types, namely (+) "plus" and (−) "minus" exist of the Phycomycete *B. trispora*, and both are able to form zygospores. When plus and minus mating types are cultivated on the same mating plate, they form progametangia. Then septation takes place in the progametangia, which results in the formation of gametangia, which fuse to form zygospores. On these spots high levels of *β*-carotene are formed [188]. When these types are separated, both types produce low levels of carotenoids, albeit the (−) mating type produces slightly more. However, when the two types are cultivated together in a specific ratio, the (−) type starts to produce 10–20 times more *β*-carotene and lycopene. The reason is that the (+) type produces the sex hormone trisporic acid, which serves as a stimulator of carotenogenesis [189]. The common denominator of the stimulating effect of trisporic acids are the ionone ring and the hydrocarbon side chain [190].

Production processes with *B. trispora* were developed by, among others, the Universal Food Coorporation [191]. In this patent, a representative description of an effective production process for *β*-carotene has been published. During process development, attention was paid to the composition of the culture medium that enables high production levels and strain improvement by mutagenesis and selection. For mutagenesis, spore solutions of the (−) mating type were used. The spores were plated out on agar plates and selection was performed on the most intense yellow, yellow/orange, or orange pigmented colonies. During selection, inhibitors of acetyl-CoA synthesis (acetate analogues), of isoprenoid biosynthesis (polyene antibiotics), and of carotenogenesis (diphenylamine, nicotinic acid, herbicides) were used. High-producing strains were obtained by 17 subsequent steps of mutagenesis and selection. In order to obtain a high variability of mutations, alternating UV (ultraviolet irradiation) and NTG (N-methyl-N′-nitro-N-nitrosoguanidine) mutagenesis were performed, and alternating selection was performed on lovastatine, acetoacetanilide, and *β*-ionone resistance. Strain improvement was performed by Panlabs in Taiwan.

Various production media were developed. All media contained well-defined medium components, but also complex medium supplements, such as corn steep liquor, soybean oil, or cottonseed oil. If a production culture was used that contained only a strain of the (–) mating type, the *β*-factor (trisporic acids) and carotene synthesis inducers had to be added approximately 50 hours after the fermentation start. The *β*-factor was prepared by coculturing the (+) and the (−) mating types, followed by centrifugation, acidification, and extraction of the culture by a suitable organic solvent. The *β*-factor was extracted from the organic

Table 6 Status of application of some promising colorants

Pigment	Color	Microorganism	Type	State of industrialization[a]	References
Ankaflavin	Yellow	*Monascus sp.*	Fungus	CP	[200]
Anthraquinone	Red	*Penicillium oxalicum*	Fungus	R	[201]
Astaxanthin	Orange/red	*Haematococcus pluvialis*	Alga	CP	[135, 136, 185, 202, 203]
		Xanthophyllomyces dendrorhous	Yeast	CP	
		Paracoccus sp. strain N-81106	Bacterium	R	
		Agrobacterium aurantiacum	Bacterium		
β-Carotene	Yellow/ orange	*Phycomyces blakesleanus*	Fungus	CP	[187, 204, 205]
		Blakeslea trispora	Fungus	R	
		Fusarium sporotrichioides	Fungus	R	
Canthaxanthin	Red	*Haloferax alexandricus*	Archae bacterium	R	[115]
Lutein	Yellow	*Muriellopsis*	Alga	CP	[16, 148]
		Dunaliella salina	Alga	CP	[129]
Lycopene	Red	*Blakeslea trispora*	Fungus	CP	[193, 196]
Melanin	Black	*Saccharomyces neoformans var. nigricans*	Yeast	R	[206]
Monascorubramin	Red	*Monascus pupureus*	Fungus	CP	[207]
Naphthoquinone	Red	*Cordyceps unilateralis*	Fungus	R	[208]
Phycocyanin	Blue	*Arthrospira (Spirulina) platensis*	Alga	R	[149, 150]
		Galdieria sulphuraria		R	
Riboflavin	Yellow	*Ahsbya gossypi*	Fungus	CP	[18]
Rubropunctatin	Red	*Monascus* sp.	Fungus	R	[209]
Torularhodin	Orange/red	*Rhodotorula* sp	Yeast	R	[210]
Zeaxanthin	Yellow/ orange	*Dunaliella salina mutant*	Alga	CP	[46, 116]
		Flavobacterium	Bacterium	R	

[a] *CP* commercial production; *R* R&D phase

phase by a buffer. Then, the solution was ready to stimulate carotenogenesis of a production culture. After 7 days of fermentation, β-carotene concentrations of 3 g/l were achieved. It is noteworthy that the carotenogenesis-stimulating β-factor was obtained from a coculture of (+) and (−) strains at the cost of fermentation capacity. This raised the question of whether it could be more efficient to perform β-carotene production in a mixed culture [192].

On the Internet website of Vitan Ltd, which has recently been acquired by LycoRed [193], an Israeli trade company, a process for β-carotene production with *B. trispora* in Ukraine was announced. Production of carotenoids by microbiological synthesis was started in Dneprovsky in August 1981. Because the process is also non-GMO [194, 195], most likely strain improvement was performed in a similar way as described for Universal Foods. Vitan's products are widely used as food colorants, feed additives, functional food ingredients, and dietary supplements. Vitan also produces lycopene and phytoene, thus it is apparent that *B. trispora* is also used for these carotenoids [196].

Another production process for β-carotene and lycopene with *B. trispora* has been performed at Léon, Spain, by the Vitatene SA company [197,198,199].

In Table 6, the state of industrialization of various other promising colorants is listed.

2.4 Microbial Synthesis from Precursors

In addition to *de novo* biosynthesis of colorants, there are examples of microorganisms that are able to synthesize the desired colorant from precursors. A traditional bacterial fermentation process is known from Japan, and population research and dynamics showed that *Halomonas, Alkalibacterium Amphibacillus* and *Oceanobacillus* play important roles in the biosynthesis and maintaining the reduced state of indigo [211].

An example is the biosynthesis of indigo from indole by *Pseudomonas sp.* HOB1 during cultivation on a naphtalene containing medium. It was demonstrated that the produced indigo could be used to dye cotton in a cotton factory [212].

A promising approach is the use of tissue cultures, because it is obvious that all the genes are present for colorant production, and no extensive screening program for biocolorant producing microorgansims is required. During this approach colorant biosynthesis regulation and cell instability are the challenges to overcome, for instance anthocyanin production suspension cultures of *Perilla frutescens* [213] and betacyanine production by hairy roots of *Beta vulgaris* L[214]. In another study annatto pigment production was stilumated by addition of plant growth regulators to the cultivation medium of Achiote (*Bixa orellana* L) [215]. To the best of the author's knowledge, there are still no large-scale commercial industrial processes. However, these conversion processes are scientifically interesting and challenging.

References

1. Abdullaev FI (2002) Cancer chemopreventive and tumoricidal properties of saffron (*Crocus sativus L.*). Exp Biol Med 227:20–25
2. Armstrong GA, Hearst JE (1996) Carotenoids 2: genetics and molecular biology of carotenoid pigment biosynthesis. Faseb J 10:228–237
3. Ben Mansour H, Latrach Tlemcani L (2009) Les colorants naturels sont-ils de bons additifs alimentaires? Phytotherapie 7:202–210
4. Counsell JN, Jeffries GS, Knewstubb CJ (1979) Some other natural colors and their applications. In: Counsell JN, Dunastable JA (eds) Natural colors for foods and other uses. Applied Science, London, 1979, pp 122–151
5. Rymbai H, Sharma RR, Srivastav Manish (2011) Biocolorants and its implications in health and food industry—A review. Int J Pharmtech Res 3:2228–2244
6. Kearsley MW, Rodriguez N (1981) The stability and use of natural colours in foods: anthocyanin, beta-carotene and riboflavin. J Food Technol 16:421–431
7. Apfelbaum M, Custot F, Manchon P, Souverain R, Thomas M (1977) Les colorants alimentaires: Toujours inutiles, parfois dangereux. Concours Med 99:236–246
8. Conners CK, Goyette CH, Newman EB (1980) Dose-time effect of artificial colors on hyperactive children. J Learn Disabil 13:48–52
9. Scheibner M, Hülsdau B, Zelena K, Nimtz M, de Boer L, Berger RG, Zorn H (2008) Novel peroxidases of *Marasmius scorodonius* degrade beta-carotene. Appl Microbiol Biotechnol 77:1241–1250
10. Ahmad M, Zaffar G, Mir SD, Razvi SM, Rather MA, Mir MR (2011) Saffron (*Crocus sativus* L). Strategies for enhancing productivity. Res J Med Plant 5:630–649
11. Barbosa MJ, Zijffers JW, Nisworo A, Vaes W, van Schoonhoven J, Wijffels RH (2005) Optimization of biomass, vitamins, and carotenoid yield on light energy in a flat-panel reactor using the A-stat technique. Biotechnol Bioeng 89:233–242
12. Hosseini Tafreshi A, Shariati M (2009) *Dunaliella* biotechnology: methods and applications. J Appl Microbiol 107:14–35
13. Ben-Amotz, A and Avron M (1989). The biotechnology of mass culturing of *Dunaliella* for products of commercial interest. In Algal and Cyanobacteria biotechnology ed. Creswell RC, Ress TAV, Shah N Page 90-114
14. NBC-14 Congress (2012) Biotechnology, delivering value: bridging Science and business. In: Lecture Prof.Dr.Ing. René Wijffels (Wageningen University. Microalgae, from Science to Business. 16–17 April 2012
15. Orosa M, Franqueira D, Cid A, Abalde J (2005) Analysis and enhancement of astaxanthin accumulation in *Haematococcus pluvialis*. Bioresour Technol 96:373–378
16. Fernández-Sevilla JM, Acién-Fernández FG, Molina-Grima E (2010) Biotechnological production of lutein and its applications. Appl Microbiol Biotechnol 86:27–40
17. Klausner A (1986) Algaculture: food for thought. Biotechnology 4:947–953
18. Schmidt G, Stahmann K-P, Kaesler B, Sahm H (1996) Correlation of isocitrate lyase activity and riboflavin formation of the riboflavin overproducer *Ashbya gossypii*. Microbiology 142:19–426
19. Heefner DL, Boyts A, Burdzinsky L, Yarus M (1993) Efficient riboflavin production with yeast. United States Patent US5231007. Zeagen Inc
20. Itoh K, Aoki s, Yatome C (1999) Production of indigo-related pigments by *Nocardia* globerula. Color Technol 115:233–235
21. Aino K, Narihiro T, Minamida K, Kamagata Y, Yoshimume K, Yumoto I (2010) Bacterial community characterization and dynamics of indigo fermentation. FEMS Microbiol Ecol 74:174–183
22. Oriel PJ, In Cheol K (1998) Method for production of indigo and indirubin dyes. Patent US 5834297

23. Kim SU, Song KS, Jung DS, Chae YA, Lee HJ (1996) Production of indoxyl derivatives in indole-supplemented tissue culture of *Polygonum tinctorium*. Planta Med 62:54–56
24. Berry A, Dodge TC, Pepsin M, Weyler (2002) Application of metabolic engineering to improve both the production and use of biotech indigo. J Ind Microbiol Biotechnol 28:127–133
25. Eyal J, Mabud MA, Walter JF (1991) Production of indigotin in submerged culture using *Morchella* nov. ES-1. Appl Biochem Biotechnol 30:303–312
26. Rosales-Loaiza N, Avendano D, Atero A, Morales E (2008) Crecimiento, producción de pigmentos y proteínas de la microalga Dunaliella viridis (Chlorophyta) en cultivos semicontinuos. Boletin del Centro de Investigaciones Biologicas Universidad del Zulia 42:323–334
27. Hadi MR, Shariati M, Afsharzadeh S (2008) Microalgal biotechnology: carotenoid and glycerol production by the green alga *Dunaliella* isolated from the gave-khooni salt marsch. Iran Biotechnol Bioprocess Eng 13:540–544
28. Jin E, Melis A (2003) Microalgal biotechnology: carotenoid production by the green algae *Dunaliella salina*. Biotechnol Process Eng 8:331–337
29. Jin E, Lee C-G, Polle JEW (2006) Secondary carotenoid accumulation in *Haematococcus* (Chlorophyceae): biosynthesis, regulation, and biotechnology. J Microbiol Biotechnol 16:821–831
30. Mahendranath G, Venugopalan A, Parimalan R, Giridhar P, Ravishankar GA (2011) Annatto pigment production in root cultures of Achiote (*Bixa orella* L.). Plant Cell Tissue Organ Cult 106:517–522
31. Bouvier F, Dogbo O, Camara B (2003) Biosynthesis of the food and cosmetic plant pigment bixin (annatto). Science 300:2089–2091
32. Sudhakar Johnson T, Ravishankar GA, Venkataraman LV (1991) Elicitation of capsaicin production in freely suspended cells and immobilized cell cultures of *Capsicum frutescens* Mill. Food Biotechnol 5:197–205
33. Miura Y, Keiji K, Shimada H, Saito T, Nakamura K, Misawa N (1998) Production of lycopene by the food yeast *Candida utilis* that naturally synthesize no carotenoids. Biotechnol Bioeng 58:306–308
34. Dharmaraj S, Ashokkumar B, Dhevendaran K (2009) Food-grade pigments from Streptomyces sp. isolated from the marine sponge Callyspongia diffusa. Food Res Int 42:487–492
35. Gavrilov AS, Kiselava AI, Matushkina SA, Kordyukova NP, Feofilova EP (1996) Industrial production of lycopene by a microbiological method. Appl Biochem Microbiol 32:492–494
36. Kim Y-S, Yeom S-J, Oh D-K (2011) Production of beta-apo-10′-carotenal from beta-carotene by human beta-carotene-9′,10′-oxygenase expressed in *E. coli*. Biotechnol Lett 33:1195–2000
37. Campo JA, Garcia-Gonzalez M, Guerrero MG (2007) Outdoor cultivation of microalgae for carotenoid production: current state and perspectives. Appl Microbiol Biotechnol 74:1163–1174
38. Leema JTM, Kirubagaran R, Vinithkumar NV, Dheenan PS, Karthikayulu S (2010) High value pigment production from Arthrospira (Spirulina) platensis cultured in seawater. Bioresour Technol 101:9221–9227
39. Zhang J-F, Hu Z-C, Xue Y-P, Zheng Y-G (2006) The production of lutein by microalga biotechnology. Food Sci Technol 12:106–109
40. Cordero BF, Obratztsova I, Couso I, Leon R, Angeles Vargas M, Rodriguez H (2011) Enhancement of lutein production in *Chlorella sorokiniana* (Chorophyta) by improvement of culture conditions and random mutagenesis. Marine Drugs 9:1607–1624
41. Serrato-Joya O, Jimenez-Islas H, Botello-Alvarez E, Rico-Martinez R, Navarette-Bolabos JL (2006) Production of β-Cryptoxanthin, a provitamin-a precursor, by *Flavobacterium Lutescens*. J Food Sci 71:E314–E319

42. Khodaiyan F, Razavi SH, Emam-Djomeh Z, Mousavi SMA (2007) Optimization of canthaxanthin production by *Dietzia natronolimnaea* HS-1 using response surface methodology. Pak J Biol Sci 10:2544–2552
43. Reza M, Nasrabadi N, Razavi SH (2010) Use of response surface methodology in a fed-batch process for optimization of tricarboxylic acid cycle intermediates to achieve high levels of canthaxanthin from Dietzia natronolimnaea HS-1. J Biosci Bioeng 109:361–368
44. Asker D, Ohta Y (1999) Production of canthaxanthin by extremely halophilic bacteria. J Biosci Bioeng 88:617–621
45. Asker D, Ohta Y (2002) Production of canthaxanthin by *Haloferax alexandrinus* under non-aseptic conditions and a simple, rapid method for its extraction. Appl Microbiol Biotechnol 58:743–750
46. Masetto A, Flores-Cotera LB, Diaz C, Langley E, Sanchez S (2001) Application of a complete factorial design for the production of zeaxanthin by *Flavobacterium* sp. J Biosci Bioeng 92:55–58
47. Fang TJ, Chiou TY (1996) Batch cultivation and astaxanthin production by a mutant of the red yeast, Phaffia rhodozyma NCHU-FS501. J Ind Microbiol 16:175–181
48. Parajo JC, Santos V, Vazquez M (1996) Producción biotecnológica de astaxanthina por *Phaffia rhodozyma*. Alimentacion Equipos y Tecnologia 15:153–160
49. Rodríguez-Sáiz M, de la Fuente JL, Barredo JL (2010) *Xanthophyllomyces dendrorhous* for the industrial production of astaxanthin. Appl Microbiol Biotechnol 88:645–658
50. Schmidt I, Schewe H, Gassel S, Jin C, Buckingham J, Huemblin M, Sandmann G, Schrader J (2011) Biotechnological production of astaxanthin with *Phaffia rhodozyma/Xanthophyllomyces dendrorhous*. Appl Microbiol Biotechnol 89:555–571
51. Martin AM, Acheampong E, Patel TR (1993) Production of astaxanthin by *Phaffia rhodozyma* using peat hydrolysates as substrate. J Chem Technol Biotechnol 58:223–230
52. Yokoyama A, Miki W (1995) Composition and presumed biosyntheticpathway of carotenoids in the astaxanthin-producing bacterium *Agrobacterium aurantiacum*. FEMS Microbiol Lett 128:139–144
53. Shin KS, Murthy HN, Ko JY, Paek KY (2002) Growth and betacyanin production by hairy roots of *Beta vulgaris* in airlift bioreactors. Biotechnol Lett 24:2067–2069
54. Pavokovic D, Krsnik-Rasol M (2011) Biotechnological production of betalains. Food Technol Biotechnol 49:145–155
55. Križnik B, Pavoković D (2010) Enhancement of betanin yield in transformed cells of sugar beet (*Beta vulgaris* L.). Acta Botanica Croatica 69:173–182
56. Moreno DA, Garcia-viguera C, Gil JL, Gil-Izquierdo A (2008) Betalains in the era of global agri-food science, technology and nutritional health. Phytochem Rev 7:61–280
57. Taya M, Mine K, Kino-oka M, Tone S, Ichi T (1992) Production and release of pigments by culture of transformed hairy root of red beet. J Ferm Bioeng 73:31–36
58. Juthankoon S (1995) Development of novel in vitro systems for natural anthocyanin production. www.ideals.illinois.edu/handle/2142/21306
59. Meng L, Lozano Y, Bombarda I, Gaydou E, Li B (2006) Anthocyanin and flavonoid production from *Perilla frutescens*: pilot plant scale processing including cross-flow microfiltration and reverse osmosis. J Agric Food Chem 54:4297–4303
60. Ravindra PV, Narayan MS (2003) Antioxidant activity of the anthocyanin from carrot (*Daucus carota*) callus culture. Int J Food Sci Nutr 54:349–355
61. Ishimaru K, Motoyoshi S, Shimomura K (1990) Production of (+)-catechin in root and cell suspension cultures of *Rheum palmatum* L. Plant Tissue Cult lett 7:159–163
62. Del Campo JA, Moreno J, Rodriquez H, Vargas MA, Rivas J, Guerro MG (2000) Carotenoid content of chlorophycean microalgae. Factors determining lutein accumulation in *Muriellopsis* sp. (Chlorophyta). J Biotechnol 76:51–59
63. Umeno D, Tobias AV, Arnold FH (2005) Diversifying carotenoid biosynthetic pathways by directed evolution. Microbiol Mol Biol Rev 69:51–78
64. Miller MW, Yoneyama M, Soneda M (1976) Phaffia, a new yeast genus in the Deuteromycotina (Blastomycetes). IJSEM 26:286–291

65. Okada T, Nakai M, Maeda H, Hosokawa M, Sashima T, Miyashita K (2008) Suppressive effect of neoxanthin on the differentiation of 3T3-L1 adipose cells. J Oleo Sci 57:345–351
66. Rao AV, Agarwal S (1999) Role of lycopene as antioxidant carotenoid in the prevention of chronic diseases: a review. Nutr Res 19:305–323
67. Vílchez C, Forján E, Cuaresma M, Bédmar F, Garbayo I, Vega JM (2011) Marine carotenoids: biological functions and commercial applications. Marine Drugs 9:319–333
68. Humphrey C (2004) Chlorophyll as a color and functional ingredient. J Food Sci 69:C422–C425
69. Schmidt I, Schewe H, Gassel S, Jin C, Buckingham J, Hümbelin M, Sandmann G, Schrader J (2011) Biotechnological production of astaxanthin with *Phaffia rhodozyma/Xanthophyllomyces dendrorhous*. J Appl Microbiol Biotechnol 89:555–571
70. Borowitzka LJ, Borowitzka MA (1989) Beta-carotene (provitamin A) production with algae. In: Vandamme EJ (ed) Biotechnology of vitamins, pigments and growth factors. Elsevier Applied Science, London, pp 15–26
71. Paolini M, cantelli-Forti G, Perocco P, Pedulli GF, Abdel-Rahman SZ, Legator MS (1999) Co-carcinogenic effect of beta-carotene. Nature 398:760–761
72. Chidambara Murthy KN, Vanitha A, Rajesha J, Mahadeva S, Swamy M, Sowmya PR, Ravishankar GA (2005) In vivo antioxidant activity of carotenoids from Dunaliella salina–a green microalga. Life Sci 76:1381–1390
73. Hu CC, Lin J, Lu F, Chou F, Yang D (2008) Determination of carotenoids in *Dunaliella salina* cultivated in Taiwan and antioxidant capacity of the algal carotenoid extract. Food Chem 109:439–446
74. Ben-Amotz A (1993) Production of beta-carotene and vitamin by the halotolerant algae *Dunaliella*. In: Ahaway A, Zabrosky O (eds) Marine biotechnology. Plenum Press, New York, pp 411–417
75. Namitha KK, Negi PS (2010) Chemistry and biotechnology of carotenoids. Crit Rev Food Sci Nutr 50:728–760
76. Khan MY, Aliabbas S, Kumar V, Rajkumar S (2009) Recent advances in medicinal plant biotechnology. Indian J Biotechnol 8:9–22
77. Sheehan J (1998) A look back at the U. S. department of energy's aquatic species program: biofuels from algae. National Renewable Energy Laboratory. Office of fuels development, US department of energy, forrestal building, 1000 Dependence Avenue SW, Washington, DC 20585-0121
78. Guedes AC, Amaro HM, Malcata FX (2011) Microalgae as sources of carotenoids. Marine Drugs 9:625–644
79. Olaizola M, Huntley ME (2003). Recent advances in commercial production of astaxanthin from microalgae. In: Fingerman M, Nagabhushanam R (eds) Biomaterials and bioprocessing. Enfield Science Publishers, Enfield, pp 143–164
80. Takaichi S (2011) Carotenoids in algae: distributions, biosyntheses and functions. Marine Drugs 9:1101–1118
81. Shi XM, Chen F (1999) Production and rapid extraction of lutein and the other lipid-soluble pigments. Nahrung 43:109–113
82. Day JG, Tsavalos AJ (1996) An investigation of the heterotrophic culture of the green alga *Tetraselmis*. J Appl Phycol 8:73–77
83. Demmig-Adams B, Adams WW (2002) Antioxidants in photosynthesis and human nutrition. Science 298:2149–2153
84. Gantar M, Svircev Z (2008) Microalgae and cyanobacteria: food for thought. J Phycol 44:260–268
85. Del Campo JA, García-González M, Guerrero MG (2007) Outdoor cultivation of microalgae for carotenoid production: current state and perspectives. Appl Microbiol Biotechnol 74:1163–1174
86. BCC Research Market forcasting (2011) The global market for carotenoids. http://www.bccresearch.com/report/carotenoids-global-market-fod025d.html
87. Dashwood R (1997) Chlorophylls as anticarcinogens. Int J Oncol 10:721–727

88. McCarty MF, Barroso-Aranda J, Contreras F (2010) Oral phycocyanobilin may diminish the pathogenicity of activated brain microglia in neurodegenerative disorders. Med Hypothesis 74:601–605
89. Mary Leema JT, Kirubagaran R, Vinithkumar NV, Dheenan PS, Karthikayulu S (2010) High value pigment production from Arthrospira (Spirulina) platensis cultured in seawater. Bioresour Technol 101:9221–9227
90. Eriksen N (2008) Production of phycocyanin–a pigment with applications in biology, biotechnology, foods and medicine. Appl Microbiol Biotechnol 80:1–14
91. Li C, Ford ES, Zhao G, Balluz LS, Giles WH, Liu S (2011) Serum α-carotene concentrations and risk of death among US Adults: the third national health and nutrition examination survey follow-up study. Arch Intern Med 171:507–515
92. García-González M, Moreno J, Manzano C, Florencio FJ, Guerrero MG (2005) Production of *Dunaliella salina* biomass rich in 9-cis-beta-carotene and lutein in a closed tubular photobioreactor. J Biotechnol 115:81–90
93. Ben-Amotz A (1999) Dunaliella beta-carotene: from science to commerce. In: Seckbach J (ed) Enigmatic microorganisms and life in extreme environments. Kluwer, Netherlands, pp 401–410
94. Venket Rao A, Agarwal S (2000) Role of antioxidant lycopene in cancer and heart disease. J Am Coll Nutr 19:563–569
95. Di Mascio P (1989) Lycopene as the most efficient biological carotenoid singlet oxygen quencher. Arch Biochem Biophys 274:532–538
96. Bishop NI, Urbig T, Senger H (1995) Complete separation of the beta, epsilon- and beta, beta-carotenoid biosynthetic pathways by a unique mutation of the lycopene cyclase in the green alga, Scenedesmus obliquus. FEBS Lett 367:158–162
97. Seddon JM, Ajani UA, Sperduto RD, Hiller R, Blair N, Burton TC, Farber MD, Gragoudas ES, Haller J, Miller DT et al (1994) Dietary carotenoids, vitamins A, C, and E, and advanced age-related macular degeneration. Eye disease case-control study group. JAMA 272:1413–1420
98. Carmona ML, Naganuma T, Yamaoka Y (2003) Identification by HPLC-MS of carotenoids of the *Thraustochytrium* CHN-1 strain isolated from the Seto Inland Sea. Biosci Biotechnol Biochem 67:884–888
99. Higuera-Ciapara I, Félix-Valenzuala L, Goycoolea FM (2006) Astaxanthin: a review of its chemistry and applications. Crit Rev Food Sci Nutr 46:185–196
100. Fabrégas J, Dominguez A, Maseda A, Otero A (2003) Interactions between irradiance and nutrient availability during astaxanthin accumulation and degradation in Haematococcus pluvialis. Appl Microbiol Biotechnol 61:545–551
101. Kobayashi M, Kakizono T, Nishio N, Nagai S (1992) Effect of light intensity, light quality, and illumination cycle on astaxanthin formation in a green alga, *Haematococcus pluvialis*. J ferment Bioeng 74:61–63
102. Olaizola M (2000) Commercial production of astaxanthin from *Haematococcus pluvialis* using 25,000 l outdoor photobioreactors. J Appl Phycol 12:499–506
103. Peng J, Xiang W, Tang Q, Sun N, Chen F, Yuan (2008) Comparative analysis of astaxanthin and its esters in the mutant E1 of Haematococcus pluvialis and other green algae by HPLC with a C30 column. J Sci China C Life Sci 51:1108–1115
104. Saleh ZA, El-Garawany GA, Assem F, El-Shibiny S (2007) Evaluation of the efficacy of whey protein to ameliorate the toxic effect of aflatoxin in rats. Int Dairy J 17:854–859
105. Yuan J-P, Peng J, Yin K, Wang J-H (2011) Potential health-promoting effects of astaxanthin: a high-value carotenoid mostly from microalgae. Mol Nutr Food Res 55:150–165
106. Guerin M, Huntley ME, Olaizola M (2003) *Haematococcus* astaxanthin: applications for human health and nutrition. Trends Biotechnol 21:210–216
107. Alberte RS, Andersen RA (1986) Antheraxanthin, a light harvesting carotenoid found in a chromophyte alga. Plant Physiol 80:583–587

108. Goss R, Böhme K, Wilhelm C (1998) The xanthophyll cycle of *Mantoniella squamata* converts violaxanthin into antheraxanthin but not to zeaxanthin. Planta 205:613–621
109. Niyogi KK, Björkman O, Grossman AR (1997) *Chlamidomonas* xanthophyll cycle mutants identified by video imaging of chlorophyll fluorescense quenching. Plant Cell 9:1369–1380
110. Dambek M, Eilers U, Breitenbach J, Steiger S, Büchel C, Sandmann G (2012) Biosynthesis of fucoxanthin and diadinoxanthin and function of initial pathway genes in *Phaeodactylum tricornutum*. J Exp Botany 63:5607–5612
111. Goss R, Ann Pinto E, Wilhelm C, Richter MJ (2006) The importance of a highly active and DeltapH-regulated diatoxanthin epoxidase for the regulation of the PS II antenna function in diadinoxanthin cycle containing algae. Plant Physiol 163:1008–1021
112. Gallardo Rodríguez JJ, Sánchez MA, Camacho FG, García MC, Belarbi EH, Chisti YM, Grima EM (2009) Causes of shear sensitivity of the toxic dinoflagellate *Protoceratium reticulatum*. Biotechnol Prog 25:792–800
113. Hackett JD, Anderson DM, Erdner DL, Debashish Bhattacharya D (2004) Dinoflagellates: a remarkable evolutionary experiment. Am J Bot 91:1523–1534
114. Pelah D, Sintov A, Cohen E (2004) The Effect of salt stress on the production of canthaxanthin and astaxanthin by *Clorella zofiengiensis* grown under limited light intensity. World J Microbiol Biotechnol 20:483–486
115. DiMascio P, Murphi ME, Sies H (1991) Antioxidant defense systems the role of carotenoids, tocopherols and thiols. Am J Clin Nutr 53:194S–200S
116. Jin ES, Feth B, Melis A (2003) A mutant of the green alga *Dunaliella salina* constitutively accumulates zeaxanthin under all growth conditions. Biotechnol Bioeng 81:115–124
117. Post FJ, Borowitzka LJ, Borowitzka MA, Mackay B, Moulton T (1983) The protozoa of a Western Australian hypersaline lagoon. Hydrobiologya 105:95–113
118. Borowitzka MA (1994) Large-scale algal culture systems: the next generation. Australasian Biotechnol 4:212–215
119. AF&F company, Hendrik Ido Ambacht, The Netherlands http://www.algaefoodfuel.com/nederlands/home/
120. Borowitzka MA (1999) Commercial production of microalgae: ponds, tanks, tubes and fermenters. J Biotechnol 70:313–321
121. Borowitzka MA (1996) Closed algal photobioreactors: design considerations for large-scale systems. J Mar Biotechnol 4:185–191
122. Bergmann P, Ripplinger P, Beyer L (2013) Disposable flat panel airlift photobioreactors. Chemie Ingenieur Technik 13(85):202–205
123. Tredici MR, Zitelli GC (1997). Cultivation of *Spirulina* (*Arthrospira*) *platensis* in flat plate reactors. In: Vonshak A (ed) *Spirulina platensis* (*Arthrospira*): Physiology, cell biology and biotechnology. Taylor and Francis, London, pp 117–130
124. Pulz O (2001) Photobioreactors: production system of phototrophic microorganisms. Appl Microbiol Biotechnol 57:287–293
125. Personal communication, NBC-14 Congress (2012) Biotechnology, delivering value: bridging Science and Business. 16–17 April 2012
126. Oswald JW (1988) In: Borowitzka MA (ed) Micro-algal biotechnology. University Press, Cambridge, pp 357–394
127. Johnson EA, Schroeder WA (1995) Microbial carotenoids. Adv Biochem Eng Biotechnol 53:119–178
128. Ben-Amotz A (2004) Industrial production of microalgal cell-mass and secondary products-major industrial species, *Dunaliella*. In: Richmond A (ed) Handbook of microalgal culture. Biotechnology and applied phycology. Blackwell Science, Oxford, UK, pp 273–280
129. García-González M, Moreno J, Manzano C, Florencio FJ, Guerrero MG (2005) Production of *Dunaliella salina* biomass rich in 9-cis-beta-carotene and lutein in a closed tubular photobioreactor. J Biotechnol 115:81–90
130. Ruane M (1977) Extraction of caroteniferous materials from algae. Australian Patent 7239574

131. Oilgae (2013) Cultivation of algae in photobioreactor. http://www.oilgae.com/algae/cult/pbr/pbr.html
132. BASF The Chemical Company (2013) http://www.basf.com/group/corporate/en/function:googlesearch:/submit?q=astaxanthin&as_epq=&as_oq=&as_eq=&lr=&as_ft=&as_filetype=&as_occt=&num=&wcms-form-submitted=true&startlocal=0&startglobal=&startproduct=&searchType=local&gs=globalsearch|global-europe|global-north-america|-global-south-america|global-asia|global-africa|global-australia&ps=zoa-products&pf=geo
133. Zhekisheva M, Zarka A, Khozin-Goldberg I, Cohen Z, Boussiba S (2005) Inhibition of astaxanthin synthesis under high irradiance does not abolish triacylglycerol accumulation in the green alga *Haematococcus pluvialis* (Chlorophyceae). J Phycol 41:819–826
134. Aflalo C, Meshulam Y, Zarka A, Boussiba S (2007) On the relative efficiency of two versus one- stage production of astaxanthin by the green alga *Haematococcus pluvialis*. Biotechnol Bioeng 98:300–305
135. Ip P-F, Chen F (2005) Production of astaxanthin by the green microalga *Chlorella zofingiensis* in the dark. Process Biochem 40:733–738
136. Del Campo JA, Rodriquez H, Moreno J, Vargas MA, Rivas J, Guerrero MG (2004) Accumulation of astaxanthin and lutein in *Chlorella zofingiensis* (Chlorophyta). Appl Microbiol Biotechnol 64:848–854
137. Kim ZH, Kim SH, Lee HS, Lee CG (2006) Enhanced production of astaxanthin by flashing light using *Haematococcus pluvialis*. Enzyme Microb Technol 39:414–419
138. Wang B, Zarka A, Trebest A, Boussiba S (2003) Astaxanthin accumulation in *Haematococcus pluvialis* (Chlorophyceae) as an active photoprotective process under high irradiance. J Phycol 39:1116–1124
139. Garcia-Malea MC, Acien FG, Rio E, Fernàndez JM, Ceron MC, Guerrero MG, Molina-Grima E (2009) Production of astaxanthin by *Haematococcus pluvialis*: taking the one-step system outdoors. Biotechnol Bioeng 102:651–657
140. Zhu MY, Mu XY, Li RX, Lü RH (2000) The effects of iron on the growth, the photosynthesis, and the biochemical composition of *Phaeodactylum triconutum*. Acta Oceanol Sinica 22:110–116
141. Liu CY, Zhang ZB, Chen XR (2005) Mutual effects of nitric oxide and iron on the growth of marine algae. Acta Oceanol Sinica 24:100–109
142. Boussiba S (2000) Carotenogenesis in the green alga. *Haematococcus pluvialis*: cellular physiology and stress response. Physiol Plant 108:111–117
143. Fábregas J, Dominguez A, Maseda A, Otera A (2003) Interactions between irradiance and nutrient availability during astaxanthin accumulation and degradation in *Haematococcus pluvialis*. Appl Microbiol Biotechnol 10:253–261
144. Kang CD, An JY, Park TH, Sim SJ (2006) Astaxanthin biosynthesis from simultaneous N and P uptake by the green alga *Haematococcus pluvialis* in primary-treated wastewater. Biochem Eng J 31:234–238
145. Zorn H, Breithaupt DE, Takenberg M, Schwack W, Berger RG (2003) Enzymatic hydrolysis of carotenoid esters of marigold flowers (*Tagetes erecta* L.) and red paprika (*Capsicum annuum* L.) by commercial lipases and *Pleurotus sapidus* extracellular lipase. Enzyme Microbiol Technol 32:623–628
146. Antos-Bocanegra E, Ospina-Osorio X, Oviedo-Rondón EO (2004) Evaluation of xanthophylls extracted from *Tagetes erectus* (Marigold Flower) and *Capsicum* Sp. (Red Literatur 123 Pepper Paprika) as a pigment for egg-yolks compare with synthetic pigments. Int J Poultry Sci 3:685–689
147. Dwyer JH, Navab M, Dwyer KM, Hasssan K, Sun P, Shircore A, Hama-Levy S, Hough G, Wang X, Drake T, Merz NB, Fogelman AM (2001) Oxygenated carotenoid lutein and the progression of early atherosclerosis. Circulation 103:2922–2927
148. Del Campo JA, Moreno J, Rodriquez H, Vargas MA, Rivas J, Guerro MG (2001) Lutein production by *Muriellopsis* sp. in an outdoor tubular photobioreactor. J Biotechnol 81:289–295

149. Sloth JK, Wiebe MG, Eriksen NT (2006) Accumulation of phycocyanin in heterotrophic and mixotrophic cultures of the acidophilic red alga *Galdieria sulphuraria*. Enzyme Microbiol Technol 38:168–175
150. Ankerstjerne K, Schmidt K, Marilyn G, Niels W, Eriksen T (2005) Heterotrophic high cell-density fed-batch cultures of the phycocyanin-producing red alga *Galdieria sulphuraria*. Biotechnol Bioeng 90:77–84
151. Shaish A, Ben-Amotz A, Avron M (1991) Production and selection of high b-carotene mutants of *Dunaliella bardawil* (Chlorophyta). J Phycol 27:652–656
152. Krinsky NI, Landrum JT, Bone RA (2003) Biologic mechanisms of the protective role of lutein and zeaxanthin in the eye. Ann Rev Nutr 23:171–201
153. Zhan DH, Lee YK (1997) Enhanced accumulation of secondary carotenoids in a mutant of the green alga, *Chlorococcum* sp. J Appl Phycol 9:459–463
154. León-Bañares R, González-Ballester D, Galván A, Fernández E (2004) Transgenic microalgae as green cell-factories. Trends Biotechnol 22:45–52
155. Walker TL, PurtonS Becker DK, Collet C (2005) Microalgae as bioreactors. Plant Cell Rep 24:629–641
156. Dunahay TG, Jarvis EE, Dais SS, Roesler PG (1996) Manipulation of microalgal lipid production using genetic engineering. Appl Biochem Biotechnol A 57(58):223–231
157. Zaslavskaia LA, Lippmeier JC, Shih C, Erhardt D, Grosman AR, Apt K (2001) Trophic conversion of an obligate photoautotrophic organism through metabolic engineering. Science 292:2073–2075
158. Grossman AR, Lohr M, Im CS (2004) *Chlamydomonas reinhardtii* in the landscape of pigments. Annu Rev Genet 38:119–173
159. Kumar SV, Misquitta RW, Reddy VS, Rao BJ, Rajam MV (2004) Genetic transformation of the green alga—*Chlamydomonas reinhardtii* by *Agrobacterium tumefaciens*. Plant Sci 166:731–738
160. Andrewes AG, Starr MP, Phaff HJ, Hope (1976) Carotenoids of *Phaffia rhodozyma,* a red-pigmented fermenting yeast. Phytochemistry 15:1003–1007
161. Jeong JC, Lee IY, Kim SW, Park YH (1999) Stimulation of -carotene synthesis by hydrogen peroxide in *Blakeslea trispora*. Biotechnol Lett 21:683–686
162. Johnson EA (1978) Astaxanthin production by the yeast *Phaffia rhodozyma* and its use as a pigment source in animal feeding. University of California, Davis, Master's Thesis
163. Johnson EA, Villa TG, Lewis MJ, Phaff HJ (1978) Simple method for isolation of astaxanthin from basidiomycetous yeast *Phaffia rhodozyma*. Appl Environ Microbiol 35:1155–1159
164. Libkind DA, Ruffin A, van Brook L, Alves L, Sampaio JP (2007) Biogeography, host specificity and molecular phylogeny of the biomycetous yeast *Phaffia rhodozyma* and its sexual form *Xanthophyllomyces dendrorhous*. Appl Environ Microbiol 73:1120–1125
165. Johnson EA, Lewis MJ (1979) Astaxanthin formation by the yeast *Phaffia rhodozyma*. J Gen Microbiol 115:173–183
166. An G-H, Schuman DB, Johnson EA (1989) Isolation of *Phaffia rhodozyma* mutants with increased astaxanthin content. Appl Environ Microbiol 55:116–124
167. Bon JA, Leathers TD, Jayaswal RK (1997) Isolation of astaxanthin-overproducing mutants of *Phaffia rhodozyma*. Biotechnol Lett 19:109–112
168. Lewis MJ, Ragot N, Berlant MC, Miranda M (1990) Selection of Astaxanthin-overproducing mutants of *Phaffia rhodozyma* with β-ionone. Appl Environ Microbiol 56:2944–2945
169. Fang TJ, Chiou TY (1996) Batch cultivation and astaxanthine production by a mutant of the red yeast *Phaffia rhodozyma*. J Ind Microbiol 16:175–181
170. Chung SB, Chin JE, Bai S, An G-H (1992) Strain improvement of *Phaffia rhodozyma* by protoplast fusion. FEMS Microbiol Lett 93:221–226
171. Calo P, de Miguel T, Velazquel JB, Villa TG (1995) Mevalonic acid increases transastaxanthin and carotenoid biosynthesis in *Phaffia rhodozyma*. Biotechnol Lett 17:575–578

172. Meyer PS, Preez JC (1994) Astaxanthin production by a *Phaffia rhodozyma* mutant on grape juice. World J Microbiol Biotechnol 10:178–183
173. Fleno B, Christensen I, Larsen, R (1988) Astaxanthine-producing yeast cells, Methods for their preparation and use. DSM IP Assets B.V. EPO Patent EP 0367765
174. Evans C T, Adams D, Wisdom R (1990) Yeasts and their use in astaxanthin production. Enzymatix Ltd. EPO Patent EP 0427405
175. Johnson EA, Schreiber D, Ho KP, Hall WT, Yang HH, Geldiay-Tuncer B (1989) Processes for in vivo production of astaxanthin and Phaffia rhodozyma yeast of enhanced astaxanthin content. Igene Biotechnology Ing. US Patents US5182208
176. Jacobson GK, Jolly SO, Sedmak JJ, Skatrud TJ, Wasileski JM (1993) Astaxanthin over-producing strains of *Phaffia rhodozyma*. Universal Foods Corporation US patents US5466599
177. David PD (2006) Compositions and methods for producing fermentation products and residuals. AmbroZea, Inc. US Patents US7309602
178. European Patent Office Decision EPO T 0737/96 (Astaxanthin/DSM) of 9.3.2000
179. Rowlands RT (1984) Industrial strain improvement: mutagenesis and random screening procedures. Enzyme Microbiol Technol 6:3–10
180. Rowlands RT (1984) Industrial strain improvement: rational screens and genetic recombination techniques. Enzyme Microbiol Technol 6:290–300
181. Parekh S, Vinci VA, Strobel RJ (2000) Improvement of microbial strains and fermentation processes. Appl Microbiol Biotechnol 54:287–301
182. Schroeder WA, Johnson EA (1995) Carotenoids protect *Phaffia rhodozyma* against singlet oxygen damage. J Ind Microbiol 14:502–507
183. De Boer L,Van Hell B, Krouwer AJJ (1992) Novel strains of *Phaffia rhodozyma*. Gist Brocades B.V. EPO Patent EP 0551676
184. An GH, Cho MH, Johnson EA (1999) Monocyclic carotenoid biosynthetic pathway in the yeast *Phaffia rhodozyma* (*Xanthophyllomyces dendrorhous*). J Biosci Bioeng 88:189–193
185. HighBeam Business (2003) Gene partners with Tate and Lyle for astaxanthin.(IGENE Biotechnology Inc.). http://business.highbeam.com/437101/article-1G1-168664530/igene-partners-tate-lyle-astaxanthin
186. Barbosa-Filho JM, Alencar AA, Nunes XP, de Andrade Tomaz AC, Sena-Filho JG, Athayde-Filho PF, Silva MS, Vanderlei de Souza MF, Leitão da-Cunha EV(2008) Sources of alpha-, beta-, gamma-, delta- and epsilon-carotenes: a twentieth century review. Rev Bras Farmacogn 18:135–154
187. Byng GS, Finkelstein M, Huang CC, Leach J, Tsau, BR J (1995) *Blakeslea trispora* mate culture capable of increased beta-carotene production. Patent Application US5422247
188. Mantzouridou F, Tsimidou MZ (2007) On the monitoring of caroteno- genesis by *Blakeslea trispora* using HPLC. Food Chem 104:439–444
189. Gooday GW (1978) Functions of trisporic acid. Phil Trans R Soc Lond B 284:509–520
190. Lampila LE, Wallen SE, Bullerman LB (1985) A review of factors affecting biosynthesis of carotenoids by the order mucorales. Mycopathologia 90:65–80
191. Finkelstein M, Huang CC, Byng GS, Tsau B-R, Leach J (1992) Fungal negative microorganism capable of producing high levels of beta-carotene. Universal Food Corporation US Patents US 5328845
192. Finkelstein M, Huang CC, Byng GS, Tsau B-R, Leach J(1993) *Blakeslea trispora* mated culture capable of increased beta-carotene production. Universal Food Corporation US Patents US5422247
193. Lycored buys Ukrainian carotenoid specialist (2012). http://www.nutraingredients.com/Industry/LycoRed-buys-Ukrainian-carotenoid-specialist
194. Vitan (2007) http://www.vitan.ru/manufacturesen/betacen/
195. www.foodmate.com
196. Lycored (2012) LycoRed group acquires vitan, strengthening its leadership position in the global natural carotenoids market. http://www.lycored.com/site/NewsEvents/News/NewsExpansion/tabid/95/ArticleId/24/-LycoRed-Group-Acquires-Vitan-strengthening-its-leadership-position-in-the-global-natural-carotenoid.aspx

197. Barredo FJ, Cabri W, Costa Perez J, Barredo FJ, De, La FuenteMoreno JL, Diez GB, Marcos Rodriquez AT, Peiro CE, Rodriquez Saiz M (2003) Method of production of beta-carotene by fermentation in mixed culture using (+) and (−) strains of *Blakeslea trispora*. Vitatene SA, USA Patents US 7,252,965
198. Bloomberg Business week (2011) http://investing.businessweek.com/research/stocks/private/snapshot.asp?privcapid=132259727
199. Costa Perez J, Marcos Rodriquez AT, FuenteMoreno JL, Rodriquez Saiz, Diez Garcia B, Peiro Cezon E, Cabri W, Barredo FJ, (2003) Method of production of beta-carotene by fermentation in mixed culture using (+) and (−) strains of *Blakeslea trispora*. Antibioticos SAU, USA Patents US 7252965
200. Jou PC, Ho BY, Hsu YW, Pan TM (2010) The effect of *Monascus* secondary polyketide metabolites, monascin and ankaflavin, on adipogenesis and lipolysis activity in 3T3-L1. J Agric Food Chem 58:12703–12709
201. Sardaryan E (2000) Strain of the microorganism *Penicillium oxalicum* var. armeniaca and its application. Council of Science and Industrial Research. USA Patent US 6340586
202. Ide T, Hoya M, Tanaka T, Harayama S (2012) Enhanced production of astaxanthin in *Paracoccus* sp. strain N-81106 by using random mutagenesis and genetic engineering. Biochem Eng J 65:37–43
203. Yokoyama A, Miki W (1995) Composition and presumed biosynthetic pathway of carotenoids in the astaxanthin-producing bacterium *Agrobacterium aurantiacum*. FEMS Microbiol Lett 128:139–144
204. Ootaki T, YamazakiY Noshita T, Takahashi S (1996) Excess carotenoids disturb prospective celltocell recognition system in mating responses of *Phycomyces blakesleanus*. Mycosci 37:427–435
205. Jones JD, Hohn TM, Leathers TD (2002) *Fusarium sporotrichioides* strains for production of beta-carotene. USA patent US 6372479
206. Morris-Jones RBL, Soraya Diez, Uran M, Morris-Jones SD, Casadevall A, Nosanchuk JD, Hamilton AJ (2005) Synthesis of melanin pigment by *Candida albicans* in vitro and during Infection. Infect Immun 73:6147–6150
207. Erdogrul Ö, Azirak S (2004) Review of the studies on the red yeast rice (*Monascus purpureus*). Turkish Electron J Biotechnol 2:37–39
208. Prathumpai W, Kocharin K, Phimmakong K, Wongsa P (2007) Effects of different carbon and nitrogen sources on naphthoquinone production of *Cordyceps unilateralis* BCC 1869. Appl Biochem Biotechnol 136:223–232
209. Zheng Y, Xin Y, Shi X, Guo Y (2010) Anti-cancer effect of rubropunctatin against human gastric carcinoma cells BGC-823. Appl Microbiol Biotechnol 88:1169–1177
210. Moliné M, Libkind D, van Broock M (2012) Production of torularhodin, torulene, and *β*-carotene by *Rhodotorula* yeasts. Methods Mol Biol 898:275–283
211. Aino K, Narihiro T, Minamida K, Kamagata Y, Yoshimune K, Yumoto I(2010) Bacterial community characterization and dynamics of indigo fermentation.FEMS Microbiol Ecol 74:174–183
212. Pathak H, Madamwar D (2010) Biosynthesis of indigo dye by newly isolatednaphthalene-degrading strain Pseudomonas sp. HOB1 and its application in dyingcotton fabric. Appl Biochem Biotechnol 160:1616–1626
213. Zhong JJ, Yoshida T (1993) Effects of temperature on cell growth and anthocyaninproduction in suspension cultures of Perilla frutescens. J Ferm Bioeng 76:530–531
214. Kim SH, Ahn SW, Bai DK, Kim KS Hwang B, Lee HY (1999) Continuousproduction of natural colorant, betacyanin, by Beta vulgaris L hairy root. J MicrobiolBiotechnol 9:716–721
215. Mahendranath G, Venugopalan A, Parimalan R, Giridhar P, Ravishankar GA(2011) Annatto pigment production in root cultures of Achiote (Bixa orellana L).Plant Cell Tiss Organ Cult 106:517–522

Adv Biochem Eng Biotechnol (2014) 143: 91–141
DOI: 10.1007/10_2013_262

Published Online: 26 November 2013

Acidic Organic Compounds in Beverage, Food, and Feed Production

Hendrich Quitmann, Rong Fan and Peter Czermak

Abstract Organic acids and their derivatives are frequently used in beverage, food, and feed production. Acidic additives may act as buffers to regulate acidity, antioxidants, preservatives, flavor enhancers, and sequestrants. Beneficial effects on animal health and growth performance have been observed when using acidic substances as feed additives. Organic acids could be classified in groups according to their chemical structure. Each group of organic acids has its own specific properties and is used for different applications. Organic acids with low molecular weight (e.g. acetic acid, lactic acid, and citric acid), which are part of the primary metabolism, are often produced by fermentation. Others are produced more economically by chemical synthesis based on petrochemical raw materials on an industrial scale (e.g. formic acid, propionic and benzoic acid). Biotechnology-based production is of interest due to legislation, consumer demand for natural ingredients, and increasing environmental awareness. In the United States, for example, biocatalytically produced esters for food applications can be labeled as "natural," whereas identical conventional acid catalyst-based molecules cannot. Natural esters command a price several times that of non-natural esters. Biotechnological routes need to be optimized regarding raw materials and yield, microorganisms, and recovery methods. New bioprocesses are being developed for organic acids, which are at this time commercially produced by chemical synthesis. Moreover, new organic acids that could be produced with biotechnological methods are under investigation for food applications.

Keywords Acidifier · Acidulant · Animal feed · Beverage · Biotechnological production · Food · Food acid · Food additive · Lactic acid production · Membrane bioreactor · Organic acid

H. Quitmann · R. Fan · P. Czermak (✉)
Institute of Bioprocess Engineering and Pharmaceutical Technology, University of Applied Science Mittelhessen, Wiesenstrasse 14, 35390 Giessen, Germany
e-mail: peter.czermak@kmub.thm.de; peter.czermak@lcb.chemie.uni-giessen.de

P. Czermak
Department of Chemical Engineering, Kansas State University, Durland Hall 1005, Manhattan KS, 66506-5102, USA

P. Czermak
Faculty of Biology and Chemistry, Justus Liebig University, Giessen, Germany

Contents

1 Introduction

Some of the most important and frequently used additives in beverage, food, and feed production are organic acids and their derivatives. Organic acids are acidic and contain carbon atoms. Often, they are products of metabolism. Therefore, many of these acids are advantageously produced via biotechnology.

Organic acids could be divided into several groups. The most common group comprises the carboxylic acids, which contain one or more carboxyl groups (–COOH). Important examples, especially for beverage, food, and feed applications, are acetic acid (one carboxyl group), malic acid (two carboxyl groups), and citric acid (three carboxyl groups). Organic acids are often weak acids that act as buffers in aqueous solutions. Buffering capacity is particularly interesting for beverage, food, and feed production. Organic compounds containing the functional group $-SO_2OH$ (sulfonic acids) are somewhat stronger acids than the carboxylic acids. One natural occurring example is taurine, which is added to dry food for cats [1]. Moreover, other functional groups such as alcohol-, thiol-, enol- and phenol groups can be responsible for a certain acidity of an organic compound. These substances are in general weak acids.

Organic acids can also be classified according to their occurrence in the metabolism of organisms. Some organic acids are part of the central metabolism that is essential for the energy supply of cells (see Fig. 1). These compounds have a low molecular weight and can often be produced at high titer by fermentation of microorganisms. Acidic additives produced in the primary metabolism are therefore frequently used in food, feed, and beverage manufacture.

Only a few organic acids belonging to secondary metabolic processes are used in beverage, food, and feed production. In general, these have special properties (e.g. ferulic acid and lactobionic acid) or could be easily produced by chemical synthesis (e.g. benzoic acid; see Fig. 2).

Most organic acids and their derivatives could in principle be produced with biotechnological methods. Unfortunately, not all of them are produced in an economically sustainable manner due to higher costs of such bioprocesses compared to synthetic production or lack of knowledge regarding biotechnological routes and downstream processing.

This review deals with organic acids and their derivatives (known as 'acidifiers,' 'acidulants,' or 'food acids') for the feed, food, and beverage industries. The compounds discussed here are involved in the primary and secondary metabolism, excluding special groups such as fatty acids, amino acids, and nucleotides.

After an overview of the properties of these substances regarding beverage and food production (Sect. 2) as well as animal nutrition (Sect. 3), individual organic acids and their derivatives with their production methods are discussed (Sect. 4). Three examples of current research in biotechnological production of organic acids for beverage, food, and feed applications are then discussed (Sect. 5). Future developments and applications of organic acids are discussed in Sect. 6.

2 Function of Acidic Compounds in Food and Beverage Production

2.1 Acidity Regulator

Acids and their derivatives are added as an acidity regulator in food and beverages to adjust and maintain pH at a desired level to stabilize beverages or foods. Additionally, in case of low final pH, unwanted growth of microorganisms is minimized due to unfavorable environmental conditions. A pH control is based on the establishment of a buffer system by using a weak acid and its salts. Which

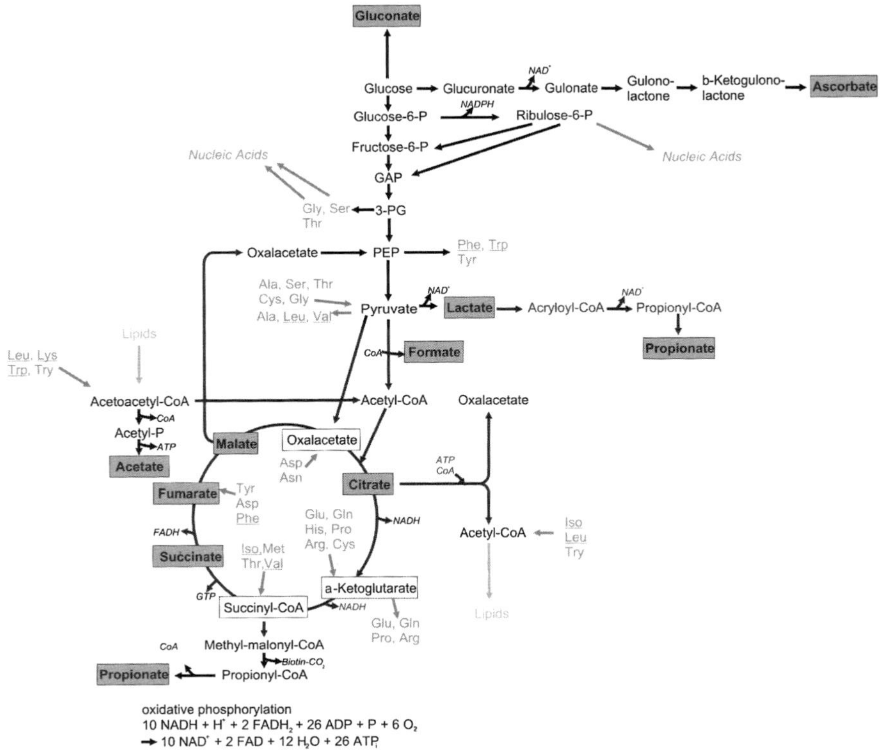

Fig. 1 Primary metabolic pathway according to KEGG database [2]

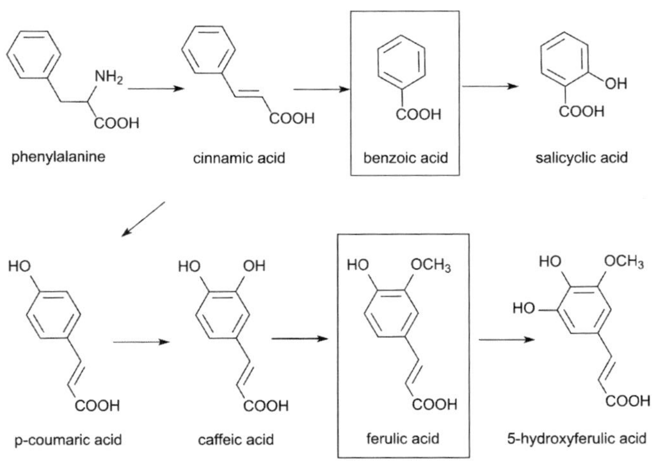

Fig. 2 Pathways of secondary metabolites using the example of benzoic and ferulic acid (adapted from [3])

acids or salts are chosen depends on the target pH of the product. The buffer capacity is high for pH values close to the pK_a value of the buffering agent. Therefore, acids and salts that have a pK_a value close to the target pH are used. A mixture of different acids and salts is often used to optimize the buffer system.

The acidity regulation is very important for gel-type foods such as jams, jellies, pectin gels, and products containing gelatin [4, 5] because the pH value impacts gelation. Gelatin has the lowest solubility at the isoelectric point [5]. Therefore, the pH has to be controlled to produce consistent batches.

Furthermore, sucrose inversion is influenced by the acidity [6]. Sucrose inversion is the hydrolysis to the monosaccharides fructose and glucose at elevated temperatures and it is catalyzed by acidic compounds. This has to be taken into account using acidity regulators in sucrose-containing products. For example, using citric acid as an acidity regulator minimizes sugar inversion [7].

2.2 *Antioxidant and Synergist*

Lipids as triglycerides are part of many foods and food raw materials. Lipids tend to deteriorate at high temperatures or over time (extended storage) [8, 9]. Fats and oils are hydrolyzed during high-temperature food processing (e.g. frying of foods with high water content). There are no food additives to protect against this unwanted reaction. To avoid this problem, procedures have to be chosen to minimize the water content if high-temperature food processing is used [8]. Secondly, the lipids deteriorate by oxidation over time (termed 'autoxidation'). The reasons are primarily spontaneous reactions with atmospheric oxygen. In a first reaction, unstable lipid radicals are generated, which then decompose to an unwanted volatile aroma compound. These 'off-flavors' influence the sensory quality of the product. The food turns rancid. Moreover, the free radicals may react with other substances such as fat-soluble vitamins and pigments (e.g. carotenoids), which results in losses of nutritional value or bleaching of the product [9]. Additionally, reactions with proteins and nucleic acids may occur, which also leads to unwanted off-flavors.

Substances that are capable of delaying, retarding, or preventing the oxidative deterioration of lipids are called antioxidants. Primary antioxidants are substances that are scavenging the free radicals directly. These are often phenolic compounds (e.g. vitamin E). Secondary antioxidants (also called synergists) avoid the primary oxidation reactions by binding metal ions (see Sect. 2.8), reacting with free oxygen, absorbing ultraviolet (UV) radiation, or converting unstable hydroperoxides to nonradical substances [9]. Synergists usually enhance the effect of primary antioxidants.

Organic acids and their derivatives can act as antioxidants or synergists. Phenolic acids, such as gallic acid [9–11] or ferulic acid [9, 12] and derivatives, act as primary antioxidants. Ascorbic acid and citric acid as well as their derivatives are examples of synergists [9, 11]. They improve the stability of primary antioxidants and lipids by acidifying the medium. Additionally, they deactivate pro-oxidant metal ions such as iron and copper by forming chelates. Moreover, ascorbic acid and their derivatives may scavenge oxygen and regenerate primary antioxidants such as tocopherols [8, 9].

2.3 Dough Conditioner

Dough conditioners, also known as flour treatment agents or improvement agents, are frequently used in bakeries. Organic acids have two functions in flour improvement. They are added for oxidation, leading to maturing of the flour and thus improving the baking quality [13, 14]. Secondly, organic acids are added for their reducing character in continuous dough mixing. Reduction reactions enhance the effectiveness of mixing. The mixing time to achieve proper dough development is decreased at a given mixing speed [14–16]. The needed energy input for dough mixing could be reduced.

2.4 Firming Agent

Firming agents are substances that are added in food processing to protect and retain the firmness (mechanical stability) of food [17, 18]. This is very important for products with pectin. The addition of firming agents causes a precipitation of the pectin, which leads to a strengthening of the structure. Salts of organic acid, such as gluconates and citrates, are frequently used as firming agents (see Sect. 4).

2.5 Flavor Enhancer

Sour taste is a very important flavor impression. The addition of organic acids gives a product the right tartness to balance high sweetness. Moreover, acidic compounds may act as flavor enhancers. Flavor enhancers are substances that intensify the flavor of the product without contributing a flavor of their own (e.g. succinic acid or tartaric acid) [17, 19, 20].

2.6 Functional Ingredient

Functional food is a term that has been used in recent decades for a food product that has health-promoting ingredients ('functional ingredients'). There are a few organic acids and their derivatives that have healthy effects. Ascorbic acid (vitamin C) is an important substance for humans. Because it cannot be produced by the human cells, it has to be taken up orally. Therefore, sometimes it is added to beverages and foods [21, 22]. Another example is lactobionic acid and its salts (lactobionates). Lactobionic acid is a novel ingredient. In recent years, several studies to analyze this compound's possibilities as a beverage and food additive have been carried out [23, 24]. There is some evidence for health-promoting activity. The acid is resistant to digestive enzymes, but it could be fermented by the intestinal flora. It may act as a prebiotic substance [25]. Furthermore, lactobionic acid and its derivatives have positive effects on the calcium absorption caused by their chelating properties [23]. Additionally, anti-inflammatory, anti-thrombotic, and anti-cancer activities have been suggested for ferulic acid, although this is not conclusive [26].

2.7 Humectant

In some foods, the control of moisture is important. To retain the moisture in food products, hygroscopic substances are useful. Some organic acids and their derivatives have this ability (e.g. lactic acid and its salts, [27]). By adding humectants, the thermodynamic water activity is reduced, which decreases microbial growth [28, 29].

2.8 Leavening Agent

Chemical leavening is very common in the production of bakery products. By producing gas through a chemical reaction, the dough rises and becomes fluffy, which is desired for most bakery products [30]. Usually, sodium bicarbonate (bakery soda) is used for this reaction. If bakery soda is heated, sodium carbonate is formed by releasing carbon dioxide and water. However, the sodium carbonate is unwanted in many applications because of its bitter taste and a tendency to produce yellowish color [31]. A possible solution is the addition of a weak acid. Then, in a first step, the sodium bicarbonate reacts with the acid. Carbonic acid and a sodium salt of the acid are produced. Afterwards, carbonic acid decomposes to water and carbon dioxide [13]. Figure 3 shows the chemical leavening reaction using the example of acetic acid as an additive.

Important organic acidic sources for chemical leavening with sodium carbonate are vinegar (acetic acid), lemon juice (citric acid), molasses or buttermilk (lactic acid) and cream of tartar (potassium bitartrate).

The correct mixture of soda and acid has to be used for a good product. If there is not enough acid, the product turns yellowish and bitter. Excess acid produces a sour taste. Commercial baking powder is an optimized mixture of sodium bicarbonate, an organic acid and a dry diluent (e.g. corn starch) [30].

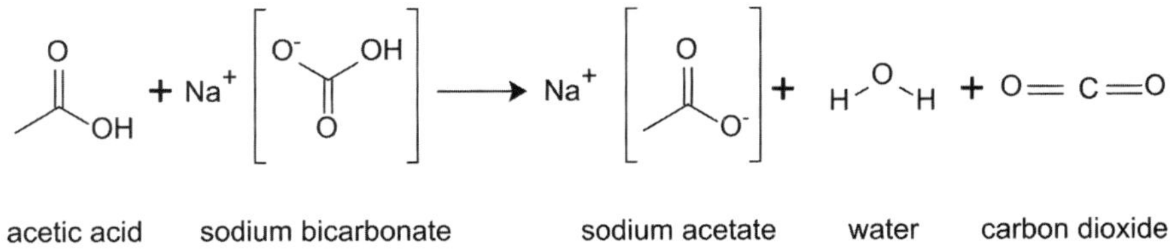

Fig. 3 Example for a chemical leavening reaction

2.9 Preservative

Preservation is essential in food production. Two types of preservation are distinguished. First, there are substances that prevent unwanted chemical reactions: antioxidants and sequestrants. Their functions are described elsewhere (see Sects. 2.2 and 2.8). The second type of preservatives prevent food spoilage by inhibiting unwanted microbial growth [29]. Requirements for spoilage by microbial growth are:

- The food has to be contaminated with an unwanted microorganism.
- Nutrients have to be bioavailable to the microorganisms.
- The environmental conditions (temperature, water activity, pH, presence or absence of oxygen) have to be favorable for microbial growth.

Preservatives either attack one of the conditions listed above or have a direct antimicrobial effect. For example, no microbial growth occurs at very high sugar concentrations. Therefore, sugar could act as a preservative. High salt concentrations or protective gases are other possibilities.

An antimicrobial action of a substance could be explained by different parts of the microorganism. The DNA, protein synthesis, enzyme activities, cell membrane or cell wall, and transport mechanisms of nutrients may be influenced by a preservative.

Organic acids and their derivatives are very often used as preservatives due to their properties. First, the environmental conditions become unfavorable for microbial growth by reducing the pH. Furthermore, the initial contamination could be minimized by adding acids during the food processing. Acidic conditions have a positive effect on the killing of microorganisms during heat treatment (e.g. sterilization) [29]. Moreover, organic acids and their derivatives have different direct antimicrobial effects. During germination of spore-forming bacteria, benzoates attack at the point of the spore coat breaking, whereas sorbates inhibit the formation of the vegetative cell [29].

2.10 Processing Aids

Processing aids are substances that are used in beverage and food processing, then removed later in the process [32]. For example, organic acids are used as extraction agents in the production of gelatin [33]. The cleavage of collagen cross-links can be achieved by acidifying the raw material followed by a treatment with warm water. Processing aids could be used for the cleaning of processing units. Acidic conditions are favorable for dissolving certain mineral salts. By treatment with weak organic acids (e.g. citric acid), the fouling of equipment with mineral salts can be minimized without impacting other materials of construction [7].

2.11 Sequestrant

Free metal ions are an issue in food production. They may react with food components or form insoluble or colored compounds. The results are unwanted precipitations, rancidity, or loss of nutritional quality. Sequestrants (also known as chelating agents) are a solution for this problem. They form stable and water-soluble complexes (chelates) with the free metal ions. Another application for chelating agents is the controlled release of metal ions. This could be used to control gelation [34]. Commonly used organic acid chelating agents are citric acid, gluconic acid, tartaric acid, and ethylenediaminetetraacetic acid (EDTA).

2.12 Thickener via Gelling or Coagulation

Thickeners are an important group of food additives. Often, organic compounds with high molecular weight and cross-linking, such as alginate, pectin, or gelatin, are used. As described elsewhere, organic acids could moderate the effect of this kind of thickener as firming agents (e.g. pectin, see Sect. 2.4) or they are involved in the production as processing aids (e.g. gelatin, see Sect. 2.10). Moreover, organic acids could be directly used as thickener in special cases. Proteins tend to precipitate and coagulate in acidic conditions. An example is the acid gelation of milk [35, 36]. The casein micelles are destabilized by adding an acid or its derivatives (e.g. glucono-δ-lactone at a pH around 4.9) [37]. Applications could be found in milk products, such as yogurts and fresh cheese. This process is carried out at ambient or elevated temperatures [37, 38]. The latter method is called heat-acid coagulation. In this procedure, milk is heated up to the boiling point and then cooled down to approximately 80 °C while stirring. Finally, an acid solution is added for coagulation. By using heat-acid coagulation, *chhana*, a traditional Indian counterpart of soft cottage cheese, is produced [39]. Other examples for acid thickening are tofu or other soybean protein gels [40, 41].

2.13 Other Applications

Organic acids and their derivatives can be used as base chemicals for the production of other food additives. Esters, for example, are the result of reactions between carboxylic acids and alcohols. Small ester molecules of organic acids are often used as flavors (e.g. butyl acetate or ethyl lactate). Parabens are esters formed from para-hydroxybenzoic acid with different alcohols. Although para-hydroxybenzoic acid is not used as food additive, parabens, such as butyl parahydroxybenzoate, are used as antimicrobial preservatives. Another group comprises diacetyl tartaric acid ester of mono- and diglycerides. These esters are used as emulsifiers. Emulsifiers are additives that enhance the stability of emulsions.

3 Acidifiers in Animal Nutrition

3.1 Function of Acidifiers in Feed

3.1.1 Preservation

Animal feed is produced, transported, and stored in large quantities. Therefore, a certain level of contamination with unwanted microorganisms is essentially unavoidable [42]. The level of microbes could rise rapidly under favorable temperatures and moisture conditions [43]. Contamination reduces the nutritional value of the feed. Additionally, certain pathogens (e.g. *Salmonella*) could be dangerous for the animals, even at low titers. By carryover, pathogens or their produced toxins may also enter the human food chain [44]. Hence, feed is often treated with heat and acidic substances are added. The preservation mechanism is the same as in food. The heat reduces the potential of initial contamination and microbial growth is inhibited by reducing the pH of the feed. Moreover, different acidic compounds act specifically against certain microorganism groups (see Sect. 2.9). Thus, the uptake of pathogens and toxins by farm animals with feed is minimized [45].

3.1.2 Beneficial Effects on Animal Health and Growth Performance

The addition of acidifiers into diets has positive effects for the health of the animals and the growth performance of farm animals can be enhanced. Several mechanisms describing the reasons of these beneficial effects have been proposed [46, 47]. First, certain organic acids and their derivatives have antimicrobial activities as described before (see Sect. 2.9). Pathogens may be reduced in animals fed using feed with acidifiers [47]. The gastric pH could be reduced by acidifiers. A low pH in the stomach promotes the activity of pepsin, which improves the protein digestion in swine, for example [46, 47]. The reduction of the gastric pH is somewhat controversial [48]. Further studies have shown that the influence of organic acids in this case depends on the chosen acidifier [46]. The chelating function of some organic acids and their derivatives has a positive effect on animal growth. An increased absorption and retention of different minerals has been shown (e.g. calcium, phosphorus, and zinc by addition of fumaric acid) [47]. Moreover, an improvement of enzyme secretion in the pancreas has been reported by adding organic acids to the feed [47].

Often mixtures of acidifiers are used to maximize the beneficial effects. Due to the positive effect on health of the animals, acidifiers are increasingly accepted as an alternative to antibiotics [42].

Additionally, acidifiers enhance the nutritional value of the feed, which leads to a better growth of the animals. Most organic acids and their derivatives have high energy contents (e.g. heat of consumption for propionic acid: 4,968 kcal/kg) [43]. This has to be taken into account when planning the feed rations to avoid over-feeding, if organic acids are added.

3.2 *Applications of Acidifiers in Animal Feedstuffs*

3.2.1 Effects of the Utilization of Acidifiers in Pig Diets

There are numerous studies of acidifiers used in pig farming. The influence in pork nutrition of all common organic acids and their salts, which are often used as preservatives in food, has been investigated in the last decade. Antimicrobial activities have been proven against *Escherichia coli*, *Salmonella spp.* and other microorganisms (e.g. yeasts and molds [46]). Moreover, the improvement of the growth performance has been shown using different organic acids and their salts [46, 47]. Very interesting results have been observed for the supplementation of diets for weaning pigs with acidifiers. In suckling pigs, the acid secretion in the stomach is reduced due to high production of lactic acid by fermentation of lactose from sow's milk. The lactose concentration and in consequence the lactic acid production by fermentation is reduced by switching from sow's milk to solid feed. The gastric pH can be elevated for days and may result in diarrhea. Additionally, an elevated stomach pH allows pathogens to survive, which may lead to an infection. By adding acidifiers to the solid feed, these risks could be minimized. The transition from milk to solid feed at weaning becomes more agreeable for the piglets [47, 49]. Many different organic acids (e.g. formic acid, fumaric acid, propionic acid, lactic acid, and benzoic acid [47]) are used as acidifiers for pig diets.

3.2.2 Applications of Acidifiers in Poultry Production

In poultry farming, the antimicrobial activity of certain organic acids and their derivatives is the most important aspect. Antibiotics are used frequently in poultry production. The reason is the high risk of an epidemic plague due to confined animal feeding operations. To reduce the amount of antibiotics in the food chain, the addition of acidifiers has been implemented in many cases [50, 51]. Different studies have shown the effective prevention of intestinal colonization with pathogens (e.g. *Salmonella*) [50, 52, 53]. Therefore, propionic acid [51] and mixtures of formic and propionic acids [50] are often used.

3.2.3 Organic Acids and their Derivatives in Aquaculture

Aquaculture is the fastest growing food-producing sector due to increased demand and overfishing of the oceans. As in poultry production, in aquacultures high numbers of fish or shrimp are living in a confined space. To reduce the risk of diseases, antibiotic growth promoters are used frequently. Acidifiers have been added in fish silage for decades as preservative agents to minimize the growth of unwanted microorganisms [43, 54, 55]. The application of organic acids and their derivatives in aquaculture diets to enhance health and to promote growth were investigated in recent years for fish [43, 55, 56] and shrimp [57].

3.2.4 Additional Reports on Acidifiers in Animal Nutrition

Organic acids and their derivatives have been tested in diets for rabbits. Improvements of the digestibility and final productivity of the nutrition with acidifiers added have been reported. However, the effects are not clear [46, 58]. Also, different studies regarding the health of rabbits using acidifiers instead of antibiotics have been carried out. A reduction of various unwanted microorganisms has been reported consistently. Furthermore, a few acidifiers, which are known to have antimicrobial activity from tests with other animals, showed no effect [58, 59]. More research in this field is clearly needed.

Organic acids and their derivatives have been added to feed for calves during weaning. Initial tests have been described using the acidification of milk, milk replacement, and post-weaning concentrates. Unfortunately, positive effects were not consistent [60].

4 Properties, Applications, and Production of Common Organic Acids

4.1 Monocarboxylic Acids

The four monocarboxylic acids—formic acid, acetic acid, propionic acid, and sorbic acid—are used in beverage, food, and animal nutrition (Fig. 4). Acetic acid ($C_2H_4O_2$; $pK_a = 4.75$) and its salts are used in various applications as emulsifiers, stabilizers, pH control agents, preservatives, flavor enhancers, and firming agents. Moreover, acetic acid is used as base chemical for the production of esters (e.g. amyl acetate, ethyl acetate and glyceryl diacetate), which can serve as solvents

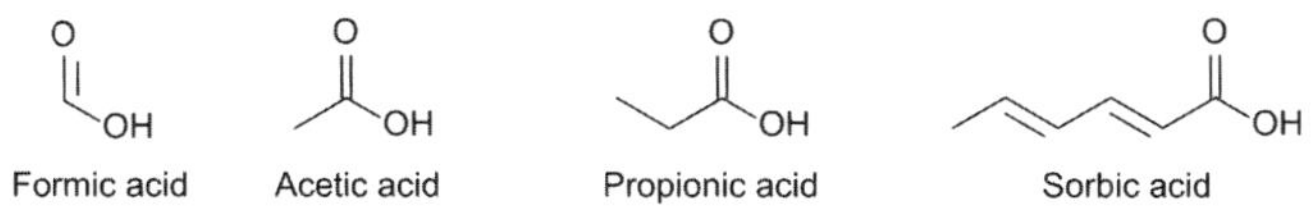

Fig. 4 Monocarboxylic acids

[19]. The annual world production of acetic acid was about 7,000,000 metric tons in 2007 (400–1,200 US$ per metric ton in 2013 [61]). Approximately 170,000 metric tons of this was produced by fermentation [62].

Formic acid (CH_2O_2; $pK_a = 3.77$) is the simplest carboxylic acid, with only one carbon atom. Formic acid and its salts, sodium formate and calcium formate, are used as preservatives and acidifiers in animal feed as well as flavoring agents at very low concentrations (between 0.5 ppm for gelatins and puddings and 6 ppm for hard candy) [63]. The acrid odor and taste limit the addition of formic acid and its derivatives in food at higher concentrations [19]. Nearly 49 % of formic acid annual word consumption is used for silage preservation and as animal feed additive. In 2009, the world production capacity was approximately 770,000 metric tons per year [64]. The price for formic acid was 550–950 US$ per metric ton in 2013 [61].

Propionic acid ($C_3H_6O_2$; $pK_a = 4.87$) and its derivatives can be used as synergists, pH control agents, preservatives, and flavor enhancers [19]. Propionic acid is often found as an acidifier in animal nutrition. About 45 % of the annual world production of 130,000 metric tons is used for animal feed and as grain preservative. Another 21 % is used as food preservatives [62, 65]. One metric ton of propionic acid in food-grade quality was offered for 800–2,000 US$ in 2013 [61].

Sorbic acid ($C_6H_8O_2$; $pK_a = 4.8$) and its salts are used as preservatives [19]. Approximately 30,000 tons of sorbic acid are produced annually [66]. The world price of sorbic acid was 3,000–5,500 US$ per metric ton in 2013 [61].

4.1.1 Biotechnological Production of Acetic Acid

Acetic acid is produced for beverage, food, and feed applications almost entirely using the traditional vinegar process [65]. First, ethanol is produced by fermentation with *Saccharomyces cerevisiae* in the absence of oxygen. Then, acetic acid is generated from ethanol by acetic acid bacteria, such as *Acetobacter aceti*, *Acetobacter pasteurianus,* or *Gluconacetobacter europaeus*, under aerobic conditions [65, 67]. Different substrates, such as malt, fruits, and sugarcane, are used for vinegar production [68]. Today, processes with two stages (e.g. two-tank cycle fermentation or two-stage submerged fermentation) are generally employed on an industrial scale. In a first step, biomass is produced in parallel to the acetic acid production. In the second part of the process, mainly acidification takes place. Acetic acid concentrations up to 200 $g{\cdot}L^{-1}$ can be achieved [65].

The vinegar process has been well studied over many decades [65]. However, there are still attempts to enhance vinegar production, especially regarding productivity and cost minimization through alternative substrates [69, 70], new process concepts (e.g. immobilized cells [68] or mixed cultures of yeasts and acetic acid bacteria [71]), and optimized acetic acid bacteria [72].

Acetic acid can be produced under anaerobic conditions by some microorganisms such as *Clostridium thermoaceticum* [65]. In free-cell batch fermentations, acetate concentrations of 50 $g \cdot L^{-1}$ were reached in less than 192 h. Acetic acid concentrations of 83–100 $g \cdot L^{-1}$, a yield of 0.74–0.80 g acetic acid per gram glucose, and a productivity of 0.60–0.85 $g \cdot L^{-1} \cdot h^{-1}$ were observed under optimized conditions in a cell-recovered fed-batch process with pH-control using glucose as substrate [65].

4.1.2 Biotechnological Production of Formic Acid

Formic acid is generally produced by chemical synthesis [73]. However, biotechnological routes are described in literature. First, formic acid could be produced from hydrogen and bicarbonate by whole-cell catalysis using a methanogen. Concentrations up to 1.02 $mol \cdot L^{-1}$ (47 $g \cdot L^{-1}$) have been reached within 50 h [74]. Another example is the formation of formic acid and ethanol as co-products by microbial fermentation of glycerol with genetically modified organisms. In small-scale experiments, 10 $g \cdot L^{-1}$ glycerol has been converted to 4.8 $g \cdot L^{-1}$ formate with a volumetric productivity of 3.18 $mmol \cdot L^{-1} \cdot h^{-1}$ and a yield of 0.92 mol formate per mole glycerol using an engineered *E. coli* strain [75, 76].

4.1.3 Biotechnological Production of Propionic Acid

Generally, propionic acid is produced via petrochemical routes. However, fermentative processes are interesting for food-grade production, although the price of biotechnologically produced propionic acid may be twice that of petrochemistry-based propionic acid. The microbial production of propionic acid is done with propionibacteria (e.g. *Propionibacterim freudenreichii*) [65, 77, 78]. Several fermentation methods have been studied. For example, an extractive fermentation is suggested to avoid low productivity and yields caused by product inhibition [77]. With this technique, a product concentration of 75 $g \cdot L^{-1}$ propionic acid, a yield of 0.66 g propionic acid per gram lactose, and a productivity of approximately 1 $g \cdot L^{-1} \cdot h^{-1}$ are reached [79].

Different substrates, such as glycerol [78], wheat flour [80], or mixtures of glycerol and glucose [81], have been analyzed to reduce costs. Also, techniques of cell immobilization show promising results [77]. Fibrous-bed reactor systems show the highest product concentrations: up to 106 $g \cdot L^{-1}$ propionic acid and a yield of 0.56 g propionic acid per gram glycerol [82]. In recent years, metabolic engineering has been used to improve the acid tolerance and to reduce byproduct formation [65].

For example, the acetate kinase gene has been inactivated by mutation of *Propionibacterium acidipropionici* [83]. Additionally, an adaptive evolution has been carried out. As result, the productivity was enhanced by approximately 50 %, up to 0.25 $g{\cdot}L^{-1}{\cdot}h^{-1}$ and a yield of 0.59 g propionic acid per gram glycerol, using immobilized cells adapted to high acid concentrations [82, 84].

4.1.4 Biotechnological Production of Sorbic Acid

Today, sorbic acid is produced solely by chemical synthesis [66]. However, fermentation and chemical synthesis might be combined to develop a new production route for sorbic acid [85]. In a first step, glucose would be converted to triacetic acid lactone by fermentation. It has been shown that triacetic acid lactone can be produced by genetically modified *E. coli* and *S. cerevisiae* strains [86, 87]. After a separation from the fermentation broth, triacetic acid lactone would be transformed into butyl sorbate in a multistage catalyst system (catalysis-hydrogenation and solid acid catalysis). Then, butyl sorbate would be purified and hydrolyzed to sorbic acid. Different scenarios are analyzed to evaluate the economic feasibility of such a production process [85].

4.2 Dicarboxylic Acids

Adipic acid, fumaric acid, and succinic acid are important dicarboxalic acids in beverage, feed, and food applications (Fig. 5). Adipic acid ($C_6H_{10}O_4$; $pK_{a1} = 4.43$ and $pK_{a2} = 5.41$) is known as raw material for nylon fabrication. However, adipic acid and its salts (e.g. calcium adipate and magnesium adipate) are used regularly in food production as sequestrants, acidity regulators, preservatives, flavor enhancers, and baking additives. The annual global production of adipic acid was 2,600,000 metric tons in 2010. The price for adipic acid was 1,500–2,000 US$ per metric ton in 2013 [61]. Less than 6 % is used for food applications [88, 89].

Fumaric acid ($C_4H_4O_4$; $pK_{a1} = 3.03$ and $pK_{a2} = 4.44$) and fumarates could be used as synergists, pH control agents, preservatives, flavor enhancers, and firming agents. Additionally, emulsifiers and dough conditioner could be produced by esterification (e.g. sodium stearyl fumarate). Currently, 22 % of the worldwide annual fumaric acid production of 90,000 metric tons is used as food and beverage additives [90, 91]. Fumaric acid in food-grade quality was offered for 700–2,500 US$ per metric ton in 2013 [61].

Succinic acid ($C_4H_6O_4$; $pK_{a1} = 4.16$ and $pK_{a2} = 5.61$) and its salts are used for pH control, preservation, flavor enhancement, and baking. In 2008, between 20,000 and 30,000 metric tons of succinic acid were produced worldwide [92]. Succinic acid was sold for 2,000–3,200 US$ per metric ton in 2013 [61].

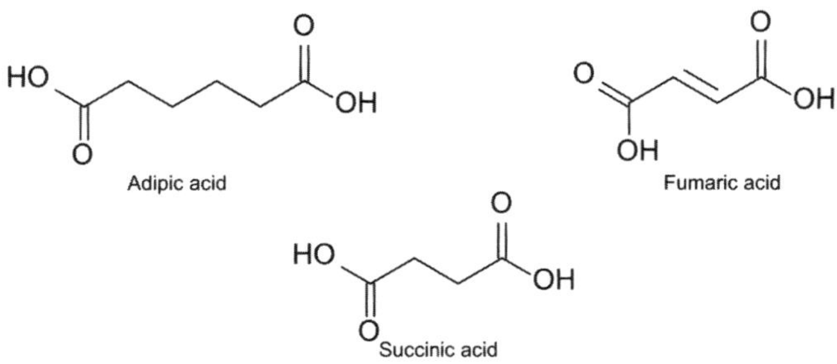

Fig. 5 Dicarboxylic acids

4.2.1 Biotechnological Production of Adipic Acid

Adipic acid is industrially produced by chemical synthesis [93]. However, there are new efforts to develop an adipic acid production process using biorenewable sources [88]. A direct biosynthesis route has not yet been reported. The possible precursors *Z,Z*-muconic acid and glucaric acid can be produced biotechnologically by fermentation. *Z,Z*-muconic acid can be made from benzoate with concentrations up to 130 mM with a yield of close to 100 % (mol/mol) by *Pseudomonas putida* KT2440-JD1 grown on glucose [94]. Alternatively, it can be produced by engineered *E. coli* directly from glucose at up to 260 mM with a yield of 0.2 mol *Z,Z*-muconic acid per mole glucose [95].

The production of the second possible precursor, glucaric acid, by engineered *E. coli* growing on glucose has been reported. However, the product titers were low (e.g. 4.8 [96] and 12 mM [97]). To overcome the problem of low product concentrations, an alternative synthetic pathway has been suggested but not yet demonstrated [98].

In a hydrogenation process, *Z,Z*-muconic acid and glucaric acid could be converted chemically into adipic acid. Therefore, bimetallic nanoparticles or platinum on activated carbon as catalysts have been studied [99]. In particular, nanoparticles of $Ru_{10}Pt_2$ anchored within pores of mesoporous silica showed high selectivity and conversion rates, greater than 0.90 mol adipic acid per mole *Z,Z*-muconic acid [99]. With platinum on activated carbon, conversion rates of 0.97 $mol{\cdot}mol^{-1}$ of *Z,Z*-muconic acid into adipic acid have been shown [95]. Another possibility would be the production of adipic acid from glucose via the α–aminoadipate pathway [88]. Finally, the production of adipic acid from long-chain carbon substrates has been suggested [88]. The conversion of fatty acids into dicarboxylic acids by engineered yeast strains has been reported [100, 101].

4.2.2 Biotechnological Production of Fumaric Acid

Currently, fumaric acid is mainly manufactured by chemical synthesis via the precursor maleic acid, which is produced using either benzene or n-butane via catalytic oxidation [90]. However, there are enzymatic and fermentative production routes for fumaric acid. Prior to the advent of inexpensive petroleum-based chemistry, fumaric acid was produced commercially by fermentation using organisms of the genus *Rhizopus* with an annual production of 4,000 metric tons [102]. Product concentrations from 30 to 130 $g{\cdot}L^{-1}$ with yields from 0.3 to 1.0 g of fumaric acid per gram of glucose and productivities of 0.46–2.0 $g{\cdot}L^{-1}{\cdot}h^{-1}$ have been reported growing on glucose [90].

In recent years, new approaches using metabolic engineering have been studied. For example, fumaric acid concentrations of 28.2 $g{\cdot}L^{-1}$ with a productivity of 0.448 $g{\cdot}L^{-1}{\cdot}h^{-1}$ have been reached in fed-batch cultivation of a genetic modified *E. coli* [103]. To achieve this result, eight modifications have been implemented.

Fumaric acid could be alternatively synthesized by an enzymatic process starting from maleic acid as in the chemical synthesis. By whole-cell biocatalysis of the *Pseudomonas alcaligenes* strain XD-1, a yield of 0.698 g of fumaric acid per gram of maleic acid and a production rate of 6.98 $g{\cdot}L^{-1}{\cdot}h^{-1}$ have been reached [104]. The process has been optimized. The formation of the byproduct malic acid was avoided due to an inactivation of fumarase by a heat treatment of the cells beforehand. Finally, a yield of 0.95 g fumaric acid per gram maleic acid and a production rate of 14.25 $g{\cdot}L^{-1}{\cdot}h^{-1}$ have been observed [105].

4.2.3 Biotechnological Production of Succinic Acid

Traditionally, succinic acid is produced by petrochemical synthesis using the precursor maleic acid [106]. However, there are some microorganisms that are able to produce succinic acid (e.g. *Actinobacillus succinogenes*, *Anaerobiospirillum succiniciproducens* and *Mannheimia succiniciproducens*). Maximum product concentrations of 106 $g{\cdot}L^{-1}$ with a yield of 1.25 mol of succinic acid per mole of glucose and a productivity of 1.36 $g{\cdot}L^{-1}{\cdot}h^{-1}$ have been achieved by growing *A. succinogenes* on glucose [106]. A high productivity of 10.40 $g{\cdot}L^{-1}{\cdot}h^{-1}$ has been reached with *A. succinogenes* growing on a complex medium with glucose in a continuous process with an integrated membrane bioreactor-electrodialysis process. In this process, the product concentration has been 83 $g{\cdot}L^{-1}$ [107]. Moreover, metabolic engineering methods were used to develop strains (e.g. *C. glutamicum*, *E. coli*, *S. cervisiae* and *Y. lipolytica*) with high productivity and titer as well as low byproduct formation [106, 108, 109]. For example, growing *C. glutamicum* strain ΔldhA-pCRA717 on a defined medium with glucose, a high productivity of 11.80 $g{\cdot}L^{-1}{\cdot}h^{-1}$ with a yield of 1.37 mol of succinic acid per mole of glucose and a titer of 83 $g{\cdot}L^{-1}$ has been reported after 7 h. An extended cultivation resulted in a product concentration of 146 $g{\cdot}L^{-1}$ after 46 h [110].

4.3 Alpha Hydroxy Acids

The alpha hydroxy acids citric acid, lactic acid, and malic acid are often used in beverage, food, and animal nutrition (Fig. 6). Citric acid ($C_6H_8O_7$; $pK_{a1} = 3.14$,

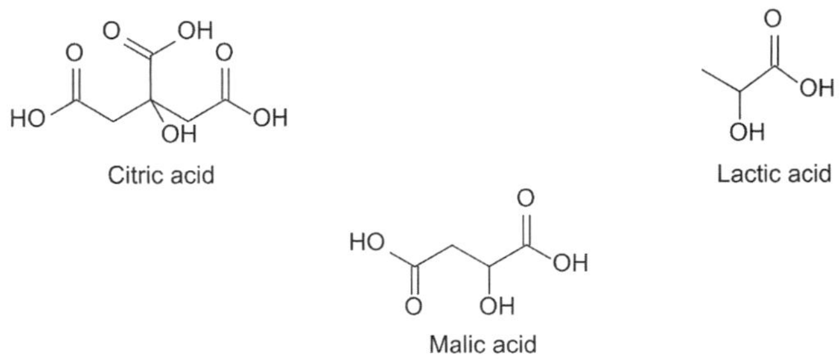

Fig. 6 Alpha hydroxy acids

pK_{a2} = 4.77 and pK_{a3} = 6.39) is frequently added as a synergist. Moreover, citric acid and its salts are used as sequestrants, pH regulators, preservatives, flavor enhancers, and firming agents. Additionally, its esters may be used as emulsifiers and solvents. The annual world production of citric acid was approximately 1,600,000 metric tons in 2009. Approximately 70 % has been used in the food sector [111, 112]. The price for citric acid was 400–1,300 US$ per metric ton in 2013 [61].

Lactic acid ($C_3H_6O_3$; pK_a = 3.08) and its derivatives have been used for a long time as acidity regulators, preservatives, baking additives, and flavor enhancers. Moreover, due to its hygroscopic activity, it also serves as humectant. The esters based on lactic acid serve as emulsifiers and solvents. More than 200,000 tons of lactic acid was used in 2012 for beverage, food, and animal feed applications [27, 113, 114]. Food-grade lactic acid was offered for 1,000–1,500 US$ per metric ton in 2013.

Malic acid ($C_4H_6O_5$; pK_{a1} = 3.40, pK_{a2} = 5.11) and its salts are used as synergists, acidity regulators, preservatives, and flavor enhancers. Approximately 55,000 tons of malic acid was consumed in 2006 (800–3,000 US$ per metric ton in 2013 [61]). The major applications have been in beverage (51 %), food (41 %), and other industrial applications [106].

4.3.1 Biotechnological Production of Citric Acid

Fermentation is the technology of choice for citric acid synthesis. Different bacteria (e.g. *Arthrobacter paraffinens* and *Bacillus licheniformis*), filamentous fungi (e.g. *Aspergilus niger* and *Penicillium citrinum*) and yeasts (e.g. *Candida tropicalis* and *Yarrowia lipolytica*) are able to produce citric acid [112]. Due to high productivity and easy handling, citric acid is usually produced by fermentation with *A. niger* [115]. For example, a product concentration of 114 $g \cdot L^{-1}$ within 168 h has been reached by cultivation of *A. niger* GCMC 7 on cane molasses [116]. On the industrial scale, submerged cultivation, surface fermentation and solid-state fermentation are used [111].

In general, molasses, starch hydrolyzate and starch are used as substrates. However, there are various studies for alternative raw materials [111]. Solid-state fermentation of inexpensive agricultural wastes is one possibility [112]. For example, high yields up to 88 % have been achieved using grape pomace as substrate [7]. Lowering the cost of product recovery is crucial. Different methods using precipitation, solvent extraction, adsorption, or in situ product recovery have been described [112]. One interesting process could be the in situ crystallization of citric acid during fermentation to improve the economics [112, 117].

4.3.2 Biotechnological Production of Lactic Acid

Lactic acid is produced biotechnologically in general by fermentation of lactic acid bacteria. More information about this process and new trends are described later in this chapter (see Sect. 5).

4.3.3 Biotechnological Production of Malic Acid

DL-malic acid as well as L-malic acid can be used in beverage, food, and animal nutrition. DL-malic acid is mainly derived from chemical synthesis, whereas L-malic acid is produced biotechnologically by enzymatic or fermentative processes [106].

Fumaric acid can be converted to L-malic acid using fumarases. Different microorganisms (e.g. *Brevibacterium flavum*, *Brevibacterum ammoniagenes,* and *Corynebacterium* species) are able to form naturally high amounts of fumarase intracellularly [106]. For example, *B. flavum* has been immobilized in κ-carrageenan and polyethyleneimine for whole-cell biocatalysis. A fumarase activity of 2.16 $mmol{\cdot}ml(gel)^{-1}{\cdot}h^{-1}$ at 55 °C has been reported [118]. This process has been used to produce 30 metric tons of L-malic acid per month in a continuous process with a 1,000 L column fed at a flow rate of 450 $L{\cdot}h^{-1}$ of 1 M sodium fumarate solution [118]. Genetic engineering has been used to improve productivity, by which *S. cerevisiae* is modified to overexpress fumarase. With this method, a conversion rate of 65 $mmol{\cdot}g^{-1}{\cdot}h^{-1}$ has been observed [119].

Another possibility is the cultivation of an L-malic acid forming microorganism (e.g. *Aspergillus flavus* or *Schizophyllum commune*) [106]. The best results have been achieved by cultivation of *A. flavus* on glucose. A final product concentration of 113 $g{\cdot}L^{-1}$ with a yield of 1.26 mol of malic acid per mole of glucose and a productivity of 0.59 $g{\cdot}L^{-1}{\cdot}h^{-1}$ has been measured. Moreover, new biotechnological routes have been described using metabolically engineered *S. cereviciae*. In batch cultivations, concentrations up to 59 $g{\cdot}L^{-1}$ with a yield of 0.42 mol of malic acid per mole of glucose and a productivity of 0.19 $g{\cdot}L^{-1}{\cdot}h^{-1}$ have been observed [120].

4.4 Sugar Acids

Ascorbic acid ($C_6H_8O_6$), its isomer erythorbic acid, and their salts are used as antioxidant synergists, sequestrants, and reducing agents [19]. The world

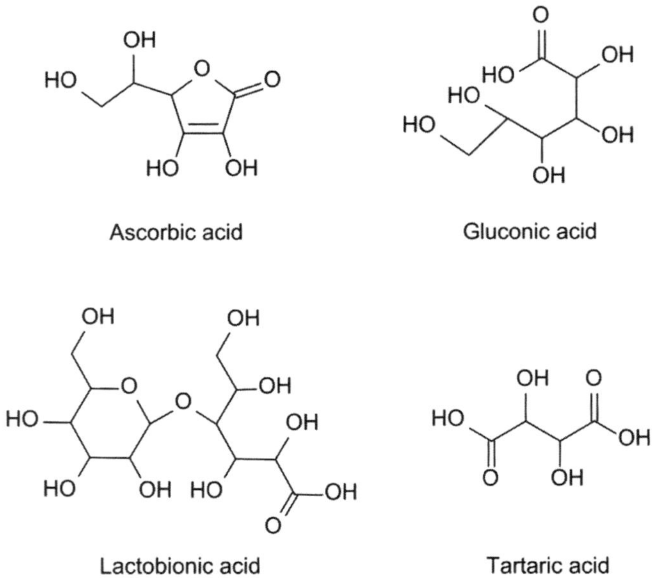

Fig. 7 Vitamin C and sugar acids

production of L-ascorbic acid has been estimated at 80,000 metric tons per annum in 2000 (2,650–4,900 US$ per metric ton in 2013 [61]). About 25 % is used as an antioxidant in food and another 15 % in beverage manufacturing. Moreover, 10 % of the global annual ascorbic acid production is used in animal nutrition (Fig. 7).

Gluconic acid ($C_6H_{12}O_7$; $pK_a = 3.7$) and its salts are used as processing aids (e.g. prevention of milkstone in the dairy industry) or in animal nutrition. The derivative glucono-δ-lactone is used as a chemical leavening agent, acidity regulator, sequestrant, preservative, and thickener by coagulation of proteins (e.g. tofu) [19]. The utilization of gluconic acids and its derivatives is limited due to high production costs [121]. In 2009, about 90,000 metric tons of gluconic acid were produced worldwide [122]. Approximately 34 % is used by the food industry [121]. The price for food-grade gluconic acid was 300–1,640 US$ per metric ton in 2013 [61].

Lactobionic acid ($C_{12}H_{22}O_{12}$; $pK_a = 3.6$) and its salts could be used as acidity regulators, antioxidants, gelling agents, flavor enhancers, firming agents, and baking additives. Additionally, the utilization of lactobionic acid as carrier for calcium supplementation has been reported due to its chelating ability [23, 24]. In 2007, the market was estimated between 15,000 and 17,000 metric tons per year worldwide. The annual growth is expected to be about 5 % [24]. Sodium lactobionate was offered for 11,850–30,000 US$ per metric ton in 2013 [61].

Tartaric acid ($C_4H_6O_6$; $pK_{a1} = 2.98$ and $pK_{a2} = 4.34$) and its salts are used for pH regulation, preservation, flavor enhancement, and chelating. Furthermore, they can be used as humectants, firming agents, baking additives, and emulsifiers [19]. The global consumption of tartaric acid for beverage and food applications has been estimated at 28,000 metric tons in 2010 [123]. An average of 3,000 US$ per metric ton was paid for tartaric acid in 2013 [61].

4.4.1 Biotechnological Production of Ascorbic Acid

Traditionally, ascorbic acid is produced via the Reichstein process, which is a chemical synthesis route with several process steps [22, 124]. In the last 20 years, the development of biotechnological processes for ascorbic acid has been in focus. More information about ascorbic acid and its utilization in beverage, food, and animal nutrition as well as its biotechnological production are offered in Industrial Production of L-Ascorbic Acid (Vitamin C) and D-Isoascorbic Acid.

4.4.2 Biotechnological Production of Gluconic Acid

Currently, gluconic acid is commercially produced by submerged fed-batch cultivations of *Aspergillus niger* using glucose as substrate [122]. *A. niger* produces citric acid and gluconic acid growing on glucose. The product concentration and yields of the product depend on the fermentation conditions. For optimal gluconic acid production, high glucose concentrations (110–250 $g{\cdot}L^{-1}$), low concentrations of nitrogen and phosphorus in the medium, a limitation of metal ion concentrations, a pH value in the range of 4.5–6.5, and high aeration rates for the oxygen supply are needed [122].

Much research has been carried out to find new ways for cheaper production. Different microorganisms have been studied (e.g. *G. oxydans*, *Z. mobilis*, *A. methanolicous,* and *P. fluorescence* [121, 125]). Moreover, new microbial strains have been developed by mutagenesis or genetic engineering [121]. Additionally, the fermentation process and recovery have been optimized [122]. New inexpensive substrates (e.g. cornstarch, grape or banana must, figs, and cheese whey) have been tested [121].

One example of a new and efficient production process of gluconic acid is the cultivation of *Aureobasidium pullulans* growing on glucose [126, 127]. Using a continuous process with biomass retention by crossover filtration, a product concentration of 375 $g{\cdot}L^{-1}$, a yield of 0.83 g of gluconic acid per gram of glucose, and a productivity of 17 $g{\cdot}L^{-1}{\cdot}h^{-1}$ could be achieved at a residence time of 22 h. In this process, 100 % of the glucose is converted [127]. This process might be interesting for industrial applications. In continuous gluconic acid production with immobilized mycelia of *A. niger,* product concentrations of 120–140 $g{\cdot}L^{-1}$ have been achieved [128].

4.4.3 Biotechnological Production of Lactobionic Acid

Currently, lactobionic acid is produced by chemical synthesis using refined lactose as feedstock. This process is expensive due to the energy demand. Alternatively, enzymatic processes have been suggested. For example, lactose could be reacted to lactobionic acid using an enzymatic system with co-factor regeneration. First, lactose is converted to lactobionolactone by a cellobiose dehydrogenase. This reaction requires an electron acceptor, which is regenerated by a laccase reducing oxygen to water. Finally, lactobionolactone spontaneously hydrolyzes to lactobionic acid [129].

Moreover, microbial production of lactobionic acid has been described [23]. In a fed-batch cultivation of *Burkolderia cepacia* growing in a complex medium (lactose, salts, peptone, and yeast extract), a final titer of 400 $g{\cdot}L^{-1}$, a yield of approximately 1.0 g of lactobionic acid per gram of lactose, and a productivity of 1.67 $g{\cdot}L^{-1}{\cdot}h^{-1}$ have been achieved [130].

Another promising strategy for an inexpensive biotechnological process is the utilization of cheap raw materials. For example, lactobionic acid could be obtained from concentrated cheese whey by fermentation with *Pseudomonas taetrolens*. In a fed-batch process, a product concentration of 164 $g{\cdot}L^{-1}$ with a productivity of 2.05 $g{\cdot}L^{-1}{\cdot}h^{-1}$ and a yield of 0.82 g of lactobionic acid per gram of lactose have been observed [131]. Furthermore, whole-cell biocatalysis using permeabilized *Zymonmonas mobilis* cells and an equimolar mixture of lactose and fructose has been tested. In a batch process, a maximum lactobionic acid concentration of 268 $g{\cdot}L^{-1}$ and a conversion rate of 72 % within 24 h have been measured. The productivity of lactobionic acid was 11.2 $g{\cdot}L^{-1}{\cdot}h^{-1}$ [132].

4.4.4 Biotechnological Production of Tartaric Acid

Tartaric acid is generally produced from crude tartar and lees, which are byproducts of wine production [133]. However, there are a few reports of fermentative production of tartaric acid by *Gluconobacter suboxydans* growing on glucose [134] or sorbitol [135, 136]. Vanadate plays a central role in this process. The microorganism forms 5-keto-D-gluconic acid, which is oxidized to tartaric acid. The vanadium catalyzes this reaction [134, 137]. Product concentrations up to 2.96 $g{\cdot}L^{-1}$ have been observed after 3 days of fermentation [136].

4.5 Aromatic and Phenolic Acids

Benzoic acid ($C_7H_6O_2$; $pK_a = 4.2$) and its derivatives are often used as antimicrobial preservatives [19, 138]. The world production capacity was estimated to be 680,000 metric tons per year in 1998 [139]. About 32,000 metric tons (5 % of the world production capacity) were consumed as benzoic acid or benzoates in 1998 [139]. Approximately 41 % of the benzoic acid world consumption is used for beverage and food applications as sodium and potassium benzoate [140]. The price for benzoic acid was 1,000–2,300 US$ per metric ton in 2013 [61] (Fig. 8).

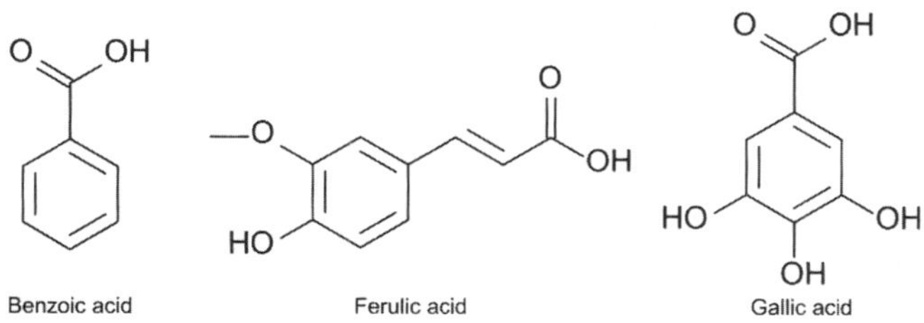

Fig. 8 Aromatic and phenolic acids

Ferulic acid ($C_{10}H_{10}O_4$) could be an interesting organic acid for food applications. However, ferulic acid is allowed as food additive only in a few countries (e.g. Japan) [12, 26]. In the United States and most European countries, natural extracts (e.g. natural extracts of herbs, coffee or vanilla beans) with high content of ferulic acid are allowed to be added to foods [26]. Ferulic acid has two isomers. However, only trans-ferulic acid occurs naturally. Ferulic acid is used as natural antioxidant and preservative in food and beverages [26, 141]. Furthermore, ferulic acid could be used as thickener due to the ability to cross-link with polysaccharides [142]. Another interesting application is the utilization of ferulic acid as precursor for the production of the high-value flavor compound vanillin [143]. High-quality ferulic acid for food applications was offered for 11.85–365 US$ per kilogram in 2013 [61].

Gallic acid ($C_7H_6O_5$) itself is currently not used in food. However, esters of gallic acid (e.g. propyl gallate) serve as antioxidants. Different functional activities (e.g. antibacterial and antiviral) have been described, which could lead to new application fields in the future. The global production of gallic acid was estimated at 8,000 metric tons per year [144]. The price for food-grade gallic acid was 26,000–29,000 US$ per metric ton in 2013 [61].

4.5.1 Biotechnological Production of Benzoic Acid

Benzoic acid is exclusively chemically synthesized on an industrial scale. Toluene from petrochemical routes is oxidized in the presence of the catalyst potassium permanganate to benzoic acid [140]. However, a recent study described for the first time a benzoic acid production process by fermentation using *Streptomyces maritimus* [145]. The production of benzoic acid during cultivation on glucose, starch, and cellobiose has been investigated. The best results have been achieved with product concentrations of 460 $mg{\cdot}L^{-1}$ in 6 days using starch as substrate. Additionally, a genetically modified *S. maritimus* optimized for endo-glucanase-secretion has been tested on phosphoric acid swollen cellulose. A final product concentration of 125 $mg{\cdot}L^{-1}$ was observed after 4 days of cultivation [145].

4.5.2 Biotechnological Production of Ferulic Acid

There are three different natural sources for ferulic acid. It could be produced from low-molecular-weight ferulic conjugates. For example, ferulic acid has been isolated from the waste material of rice bran oil production by hydrolyzing with sodium hydroxide or potassium hydroxide at 90–100 °C. Ferulic acid with a purity of 70–90 % was produced within 8 h under atmospheric pressure [146].

Another possibility is a direct extraction of ferulic acid from plant cell walls by using feruloyl esterases. Various microorganism are able to secrete feruloyl esterases (e.g. *A. niger*, *Bacillus species* and *Clostridium thermocellum*) [26]. The enzymatic hydrolysis of sugar-beet pulp has been analyzed using a mixture of carbohydrases from *Aspergillus aculeatus* with a final ferulic acid concentration of 200 $mg{\cdot}L^{-1}$ in the hydrolyzate [147]. Moreover, a purification method to isolate ferulic acid from sugar-beet pulp after enzymatic hydrolysis using a fixed-bed adsorption with activated carbon has been developed. With this process, a purity of 50 % has been achieved [148]. Finally, ferulic acid could be produced by cell culture fermentations [26]. For example, free ferulic acid (up to 50 $mg{\cdot}L^{-1}$) and also conjugated to anthocyanins (up to 150 $mg{\cdot}L^{-1}$) has been accumulated in cell cultures of *Ajuga pyramidalis* [149].

4.5.3 Biotechnological Production of Gallic Acid

The production of gallic acid is challenging. Conventionally, it has been produced by acid hydrolysis of tannic acid. However, this process is expensive due to low yields and high impurities [10, 11]. To overcome this problem, microbial production of gallic acid has been suggested. For example, in a solid-state fermentation of Teri pod cover powder containing tannin using *Rhizopus oryzae*, a yield of 90.9 % based on the tannin content of 58 % of the substrate was observed [150]. In a submerged culture of *Aspergillus aceleatus* DBF9 growing on a medium with 3 % tannin, a maximal product concentration of 6.8 $g{\cdot}L^{-1}$ was reported [151]. With tannic acid, even higher product concentrations of up to 25 $g{\cdot}L^{-1}$, a yield of 0.83 g of gallic acid per gram of tannic acid, and a productivity of 0,56 $g{\cdot}L^{-1}{\cdot}h^{-1}$ were shown using *Apergillus fischeri* MTCC 150 in submerged cultivation [144]. An alternative is the enzymatic hydrolysis of tannic acids using tannase produced by microorganisms (e.g. *Aspergillus fischeri* or *R. oryzae*) [152]. For example, propyl gallate could be produced using a tannase from *Emericela nidulans* immobilized on ionic and covalent supports [153].

5 Lactic Acid as an Example for Bioprocess Development

5.1 Lactic Acid and its Applications

Lactic acid (2-hydroxypropionic acid, CH_3-CHOH-COOH) is the most widely occurring organic acid in nature. Due to its chiral α-carbon atom, lactic acid (LA) has two enantiomeric forms (Fig. 9). Of these, L-(+)-lactic acid is more important

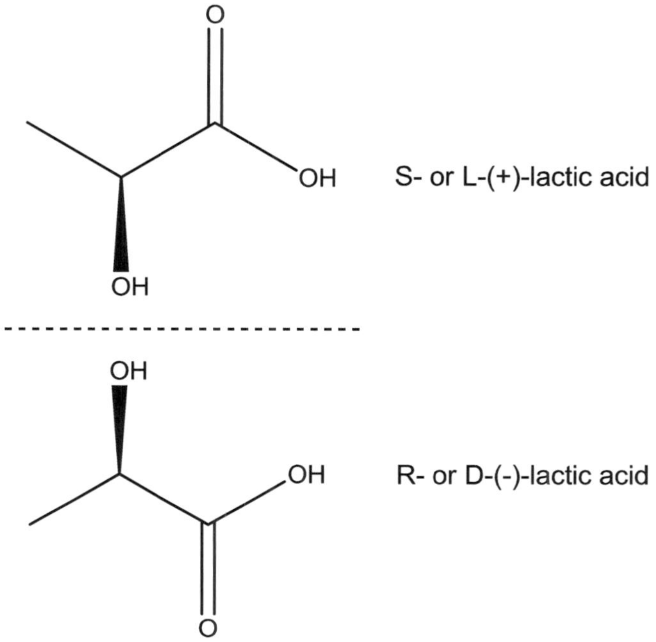

Fig. 9 Structure of lactic acid

in food and pharmaceutical industries because humans have only L-lactate dehydrogenase [27, 154]. The chemical behavior of lactic acid is mostly determined by the two functional groups. Besides the acidic character in aqueous medium, the bifunctionality (a terminal carboxylic acid and a hydroxyl group) allows lactic acid molecules to form "interesters" such as the cyclic dimers, the trimers, or longer lactic acid oligomers [155, 156].

After its first isolation by the Swedish chemist Scheel in 1780 from sour milk, lactic acid has been produced commercially since the 1880s in the United States and later in Europe [156, 157]. Worldwide, lactic acid production was approximately 250,000 metric tons per year in 2012 and is expected to reach 330,000 metric tons by the year 2015 [27, 113, 114], with an average price of 1.25 US$ per kilogram in 2013 (food grade, 80–85 % purity) [61].

Approximately 85 % of the demand for LA is from the food industry. The primary use of lactic acid is as a pH-adjusting agent in the beverage sector and as a preservative in the food industry. It is included in the Generally Recognized as Safe (GRAS) by the U.S. Food and Drug Administration [158] as a food ingredient and was deemed safe by the European Food Safety Authority as well [159]. The acceptable daily intake for LA was defined by the Joint FAO/WHO Expert Committee on Food Additives as "not limited," and it is also supported by the Scientific Committee of Food [160, 161].

In recent decades, the consumption of lactic acid due to its novel applications has grown quite rapidly, by 19 % per year [27]. Nonfood use of lactic acid for polymer production contributes to this growth. Biodegradable polylactic acid is considered to be an environmentally friendly alternative to other plastics from petroleum [162]. It is used in various fields, including drug delivery systems, medical devices, fibers, and packaging materials [27, 155, 162, 163].

Lactic acid can be produced via chemical synthesis or carbohydrate fermentation. The chemical route has various issues, including toxic raw materials, low conversion rates, and especially the inability to produce the optically pure isomer. Therefore, approximately 90 % of lactic acid worldwide is produced by biotechnological processes, namely fermentations using renewable resources [163], which is relatively fast, economical, and able to supply selectively one or two stereoisomers of lactic acid [164].

5.2 Biotechnological Production of LA

Biotechnological LA production consists of the following steps [27, 155, 163]:

- Pretreatment of the substrate: hydrolysis of carbohydrate to mono- or disaccharides.
- Fermentation of the substrate to LA by suitable microorganisms.
- Removal of biomass and other solids from the fermentation broth.
- Recovery and purification of LA.

Since the 1990s, a 2- to 4-day batch or fed-batch fermentation has been employed industrially. Under anaerobic conditions, carbohydrates are first transformed into pyruvic acid and then converted to LA by microorganisms while energy is supplied for cell growth and metabolism. Because the pK_a of lactic acid is 3.86 at 25 °C, the fermentation broth must be neutralized to hold the pH within the range of the organism's tolerance [156]. In industry, this is usually done by adding calcium hydroxide or calcium carbonate so that the LA converts to calcium lactate. LA is then recovered from the calcium lactate using sulfuric acid.

The byproduct calcium sulfate (gypsum) has very little value and must be dealt with. Calcium lactate may form a thick and hard layer on the liquid surface at high concentrations, resulting in problems for product recovery. The LA concentration is therefore generally controlled at 10 $g{\cdot}L^{-1}$, with a yield of approximately 0.9 g of LA per gram of carbon source, in order to avoid a precipitation of the calcium lactate salt during the separation steps.

The crude product is then purified using active carbon adsorption and ion exchange. Afterwards, it is concentrated through evaporation to produce technical-grade LA. The technical-grade LA has to be refined through esterification with alcohol, distillation, hydrolysis of the ester, and finally evaporation to obtain highly pure LA (>98 wt %), which is heat stable and suitable for polymerization, synthesis of solvents, and the other value-added applications.

Over the past decades, various strains of microorganisms with high productivity (>5 $g{\cdot}L^{-1}{\cdot}h^{-1}$) were isolated and are available for industrial applications.

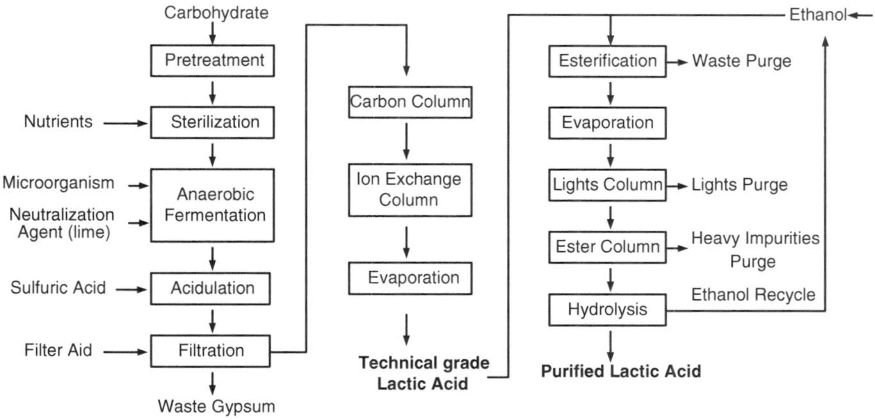

Fig. 10 Conventional process for lactic acid production using lime as neutralization agent

The product concentrations could reach >180 $g{\cdot}L^{-1}$, with yields higher than 0.9 g of LA per gram of substrate provided [154, 165–167].

The highest values of LA production reported in the literature are:

- LA concentration of 225 $g{\cdot}L^{-1}$ in fermentation broth using fed-batch fermentation [166].
- Maximum LA productivity in the range of 52–144 $g{\cdot}L^{-1}{\cdot}h^{-1}$ through high-cell density culture using cell retention.
- A yield of 1.01 g of LA per gram of glucose growing in a complex medium with glucose as the main carbon source [168].

Figure 10 shows schematically a conventional carbohydrate fermentation process. As discussed, the disadvantages of the conventional process (either economically or ecologically) are obvious. In this process, nearly 1 metric ton of calcium sulfate (gypsum) is formed per metric ton of LA as a waste byproduct, which has to be deposited and/or treated [169]. A large amount of wastewater is also produced, requiring costly water treatment.

The cost of lactic acid production is determined by rate, titer, yield, and disposal of waste water and other wastes from both fermentation and product separation process [170]. To optimize the productivity of lactic acid fermentation, microbial strains, culture medium, fermentation condition, fermentation mode, cell retention, waste minimization, and product recovery must be considered:

- *Economical feedstock sources:* Commercial prices of food-grade LA are modest, at approximately US$ 1.50 per kilogram, depending on the purity [171]. The feedstock needs to be low cost and easy to obtain. Therefore, raw materials should be selected according to the location. In addition, rapid production rate, high utilization rate, low level of contaminants, and minimum pretreatment are also important to choose optimum feedstock materials [27].

- *Selection of suitable microorganisms:* The microorganisms that are most widely employed in industrial LA production include some filamentous fungi, lactic acid bacteria (LAB), and genetically modified yeasts [155, 156, 169, 171, 172]. Because the LA productivity by microorganisms is influenced by the feedstock, a suitable strain should be able to deal with the locally available substrate rapidly and efficiently. It is important to select either a homo- or a heterofermentative organism with regard to the desired LA isomer. Furthermore, resistance against acidic pH is a preferred/positive attribute for LA production because the cost of neutralization and the resulting separation processes could be reduced.
- *Fermentation mode*: Besides the batch fermentation, which is employed most commonly for LA production, fed-batch, continuous, and simultaneous fermentation have also been investigated or applied to achieve higher LA concentration or productivity [154, 166, 167].
- *Cell recycling or immobilization*: LA is a primary metabolite, and it is well known that LA production is coupled to cell growth [173]. Thus, the fermentation efficiency could be increased in high-cell-density cultures, which is promoted by cell recirculation or immobilization [174].
- *pH control*: Limited by the tolerance of the organism to an acidic environment, fermentations in which the pH value is maintained as constant show better results compared to those without control [156]. Due to the generation of a great deal of gypsum, calcium hydroxide and calcium carbonate as neutralization agents tend to be replaced by ammonium carbonate. Instead of the low-value solid-waste gypsum, ammonium sulfate is generated, which is an important chemical fertilizer. Another alternative is continuous LA removal from the fermentation medium by filtration, extraction, or electrodialysis [175].
- *Product recovery*: Approximately half of the LA manufacturing cost is consumed in downstream processes involving precipitation, filtration, acidification, electrodialysis, carbon adsorption, evaporation, and crystallization [176].

5.3 Fermentation Process

5.3.1 Raw Materials

Theoretically, pentose and hexose could be used by certain microorganism as carbon sources and fermented to LA [155]. However, costly and industrially unattractive pure sugars, such as glucose or sucrose, are unfortunately preferred in laboratory work. Therefore, renewable resources are becoming more interesting to academia and industry. Recently, some waste products from agriculture, stockbreeding, and forestry were successfully tested for industrial LA production. These renewable resources can be divided into two groups: simple sugar containing and polymeric substrates [155].

The substrates that contain mono- or disaccharides can be converted directly into LA without complex pretreatments, such as liquefaction and saccharification. Whey (with lactose), molasses (mainly with sucrose), sugarcane, and sugar beet juice (with sucrose) are most widely used because they are cheap and easily obtained in certain locations.

Different from simple sugars, polysaccharides such as starch and lignocellulosic biomass must be hydrolyzed to dimeric or monomeric sugars to become fermentable. Currently, the simultaneous fermentation system is one of the most interesting areas in bioprocess engineering and is developing rapidly. The fermentation is coupled with the substrate hydrolysis with the hope of improved efficiency. Starch could be easily obtained from crops, potato, or cassava bagasse, for example, whereas waste paper, wood, straw, or cottonseed hulls could supply plenty of lignocellulosic materials. Nonfood feedstocks have the advantage of avoiding the food-versus-chemicals or fuels debate.

Apart from carbon sources, complex nutrients, such as amino acids and vitamins, are required for maintaining the growth and reproduction of microorganisms [27]. Many nitrogenous substrates, such as whey permeate, yeast extract, beef extract, grass extract, peptones, corn steep liquor, and soybean hydrolyzate, have been investigated. Among all these substrates, yeast extract led to the highest LA productivity because it is rich in B vitamins, purines, and pyrimidines [155, 163].

5.3.2 Microorganisms

Since the role of microorganisms in milk acidification was discovered by Pasteur in the 1860s, thousands of species and strains were investigated for LA production. In modern biotechnology, candidates for LA production are mainly divided into three groups: filamentous fungi, lactic acid bacteria, and genetically engineered yeasts. The selection of an organism depends not only on the fermented carbohydrate, the produced LA concentration (titer), and the yield of LA per mass of substrate, but also on the productivity (LA production rate, related to capital cost) and its pH tolerance [27, 156, 163, 171]. Some organisms that were investigated for their potential for industrial application are listed in Table 1.

It can be seen from Table 1 that work with industrially meaningful feedstock is very rare in the literature. The sources of micronutrients are usually elaborate and cost prohibitive for industry. This should be kept in mind for the discussion below.

Rhizopus is the best-known genus of fungi for the production of LA in industry, including strains such as *R. oryzae, R. nigricans, R. chinenis,* and *R. stolonifer*. Because they can release extracellular amylases, starchy materials from different sources (e.g. rice, corn, potato, wheat) can be hydrolyzed and used directly. Thus, the saccharification stage is not necessary. The nutrient requirements of fungi are simple and cheap because urea and ammonium may be used as nitrogen sources instead of expensive organic supplements, such as peptone and yeast extract. In addition, the biomass can be separated easily due to its mycelium formation during the fermentation. However, mass transfer limitations are commonly encountered

Table 1 Comparison of different strains for lactic acid production

Organisms		C_{LA} ($g \cdot L^{-1}$)	Yield ($g \cdot g^{-1}$)	Productivity ($g \cdot L^{-1} \cdot h^{-1}$)	Substrate	References
Fungi	*Rhizopus oryzae As3.819*	100.8	0.84	1.4	Glucose	[177]
	Rhizopus oryzae OX-1	73.1	0.75	2.1	Glucose	[178]
	Rhizopus oryzae NRRL 395	104.6	0.87	1.8	Glucose	[179]
	Rhizopus oryzae NRRL 395	127	1.00	1.7	Starch	[172]
Lactic acid bacteria	*Lactobacillus casei* NRRL B-441	82	0.91	5.6	Glucose	[180]
	Lactobacillus rhamnosus ATCC 10863	68	0.76	3.5	Glucose	[180]
	Streptococcus salivarius spp. *thermophilus*	18	0.50	5.9	Whey permeate	[181]
	Lactococcus lactis sp. *lactis* 2432	8.3	0.21	2.1	Whey permeate	[181]
	Lactobacillus delbrueckii sp. *Bulgaricus* AU	20	0.45	–	Molasses	[182]
	Lactobacillus rhamnosus ATCC 7469	18	0.40	–	Molasses	[182]
Yeasts	*Kluyveromyces lactis* PMI/C1	29	0.70	0.4	Glucose	[183]
	Saccharomyces cerevisiae	70	0.93	1.0	Glucose	[184]
	Candida boidinii	86	1.01	1.8	Glucose	[168]
	Pichia stipitis CBS6054	41	0.44	0.9	Glucose	[185]
Others	*Bacillus coagulans* SIM-7 DSM 14043	93.7	0.98	3.9	Glucose	[186]
	Bacillus subtilis MUR1	183.2	0.99	3.5	Glucose	[154]

by the pellet-like mycelium and cause the reduction of LA productivity [179]. Moreover, the yields are not as high as with the use of other organisms due to the formation of byproducts, such as fumaric acid and ethanol [172].

Therefore, LAB are still the main fermentation agents in LA production. LAB belong to the Gram-positive genera and include *Carnobacterium*, *Enterococcus* (*Ent*), *Lactobacillus* (*Lb*), *Lactococcus* (*Lc*), *Leuconostoc* (*Leu*), *Oenococcus*, *Pediococcus* (*Ped*), *Streptococcus* (*Str*), *Tetragenococcus*, *Vagococcus*, and *Weissella* [187]. LAB are mainly classified as homofermentative or heterofermentative according to the different pathways of sugar metabolism. In theory, 1 mol of glucose can be converted by homofermentative LAB via the Embden-Meyerhof-Parnas pathway to 2 mol of LA without other products, whereas heterofermentative LAB would produce other substances besides LA, such as CO_2

and ethanol. Because the product yield is up to >0.90 g LA/g substrate and much higher than in heterofermentation, only homofermentative LAB are appropriate for the industrial production of L-LA. In contrast to fungi, LAB have more complex requirements for nutrients because certain enzymes for the synthesis of B vitamins and amino acids are missing [188].

The optimal temperature for mesophilic LAB is in the range from 28 to 45 °C; for thermophilic LAB, it is between 45 and 62 °C. Thermophilic LAB, such as *Lb. delbrueckii*, of which the optimal activity is at 50 °C, are preferred because the contamination risks may be reduced as other microorganisms cannot survive at this temperature. Besides the temperature, the fermentation is also strongly influenced by the pH (i.e. product inhibition by LA). The optimal pH value for LA formation varies between 5 and 7 according to the strain, and the fermentation comes to a standstill when the pH drops below 4.5. A pH control during fermentation is therefore required, either through neutralization with lime or in situ removal of the product.

Metabolic engineering has been recently applied to develop improved LA-producing organisms. It has been reported that some recombinant yeasts show higher resistance against low pH in comparison to conventional LAB and could produce LA at industrially relevant yields and productivity below pH 3.0 [156]. They have been reported to grow on simple chemically defined media and use different sugars [168, 183, 184]. It remains to be seen if fermentation with industrially relevant feedstock and micronutrient sources will confirm this work.

5.3.3 Process Operation

Batch, fed-batch, repeated-batch, and continuous fermentations are most commonly employed for LA production. Although higher LA concentrations could be achieved in batch or fed-batch fermentations, cell-recycle systems, repeated-batch, or continuous processes are available to provide higher LA productivity (Table 2). In addition, there has also been a series of studies on cell immobilization, which allows convenient biomass removal downstream [173, 178].

5.4 Recent Developments of Product Treatment

Downstream (from the fermenter) processing is considered to be one of the most costly parts in LA production [155]. Various methods for LA recovery and purification are patented or reported in the literature. However, there is no universally recognized optimum method. The LA recovery from the fermentation broth, which is (partly) neutralized by lime during the fermentation, could be generally performed via two routes [156, 175, 195, 196]:

Table 2 Efficiency of biotechnological lactic acid production by different fermentation strategies

Organism	Fermentation mode	C_{LA} ($g{\cdot}L^{-1}$)	Productivity ($g{\cdot}L^{-1}{\cdot}h^{-1}$)	Ref.
Bacillus coagulans SIM-7 DSM 14043	Batch	91.5	2.0	[186]
	Fed-batch	91.6	4.0	[186]
Rhizopus oryzae OX-1	Batch, coimmobilization	73.1	2.1	[178]
Lactobacillus rhamnosus ATCC 10863	Batch	120.0	2.1	[189]
	Continuous, cell-recycle via membrane	92.0	57.0	[189]
Lactobacillus rhamnosus IfO 3863	Batch	98.0	1.9	[190]
	Continuous, in situ removal via electrodialysis	20.0	8.2	[190]
Lactobacillus casei ssp. *Rhamnosus* ATCC 11443	Continuous, cell-recycle via immobilization	22.4	9.0	[191]
Lactobacillus delbrueckii NRRL b445	Fed-batch, in situ removal via solvent extraction	23.1	0.2	[192]
Lactobacillus delbrueckii CECT 286	Continuous, in situ removal via ion-exchange resin	26.1	10.4	[193]
Lactobacillus delbruckii NCIM-2025	Continuous, in situ removal via membrane	82.7	12.4	[194]

1. The fermentation broth is first concentrated to 20–30 % lactate and then acidified by sulfuric acid to release the LA. The generated gypsum is removed with the biomass simultaneously by filtration (conventional belt or drum) or centrifugation.
2. The fermentation broth is first clarified through removal of biomass and solid impurities. The solution is then concentrated by heating and evaporation and results in crystallization of calcium lactate. The crystals are collected, washed, and then redissolved in water, in which sulfuric acid is added to release the LA and form gypsum.

Apart from classical filtration, modern filtration processes (e.g. micro-, ultra-, or nanofiltration) could be an economical alternative for the separation of biomass from the fermentation broth.

The crude LA has to be purified to 99 wt % to be marketable. Purification options include the following:

- Extraction of LA with simultaneous generation of lactate salts using alkaline extractants and back extraction of LA with water [197].
- Simultaneous acidification and esterification with alcohol and subsequent hydrolysis of the ester in water to recover LA and alcohol for recycling.
- Direct purification of the LA using adsorption, reactive distillation, ion exchange, membrane processes, etc.

Reactive extraction: Instead of lime or other neutralization agents, the alternative for stabilization of the pH value in the fermentation broth is the continuous extraction of LA from the fermentation system [175]. Due to its hydrophilic nature, LA may not be extracted by organic solvents. Thus, a number of alkali extractants, such as trimethylamine, are used; the LA reacts with the extractant and the product complex dissolves into the organic diluent. The LA can be re-extracted into the aqueous solution and then recovered. Meanwhile, the extractant containing diluent is recycled to the fermentation process. Hence, it is ideally suitable for continuous fermentation and enhances the process efficiency. However, cell growth inhibition was observed in many studies when the organic solvent and the extractant were added to the fermentation system because they are more or less toxic to the microorganism [197]. This is not surprising because liquid–liquid extraction always establishes equilibrium between the liquids involved. Therefore, the choice of suitable solvent and extractant is the key point of reactive extraction, where the distribution coefficient, low toxicity, viscosity, density, solubility, and stability in water must be considered.

Adsorption: In this process, the LA anion is primarily combined with the sorbent and the LA is eluted by a suitable eluent from the saturated sorbent [175]. Due to the affinity between LA and anion ion exchange resins, the LA could be selectively separated from the other compounds in the fermentation broth. Competition with other organic or inorganic anions is an issue with this approach. Improved overall productivity could be expected when the solid sorbent is coupled online with the fermentation process for in situ product recovery [198]. The process efficiency could be raised by using ion exchange resin with high capacity and integration of automation techniques. However, ion exchange resins must be regenerated by inorganic strong acids after saturation. A large quantity of wastewater with high salinity is thus generated [156]. The treatment and disposal of large amounts of salts and effluents is a great challenge when this process is used along with competing anions besides LA.

Electrodialysis with bipolar membranes (EDBM): In conventional electrodialysis, the space between cathode and anode is divided by alternating anion- and cation-selective membranes into different compartments [199]. The charged species migrate along the electric field and are trapped in specific compartments by ion-selective membranes. This results in the concentration and the separation of the charged species from uncharged components in the broth. EDBM is one of the most important variants of electrodialysis, in which bipolar membranes are utilized [200].

The operating principle of EDBM for LA production with two-compartment-configuration is schematically shown in Fig. 11. In the electric field, water is split into a proton and a hydroxide ion by a bipolar membrane. Protons are driven into the acid cell by the electric field. EDBM essentially supplies protons and hydroxyl ions using electrical current without addition of chemicals. In the meantime, the lactate anion migrates from the broth across the anion-selective membrane towards the anode and converts into lactic acid when joining the proton from the BM. The alkali is regenerated in another cell and returned to the bioreactor for neutralization. The generation of gypsum could be thus markedly reduced. In addition, the costs of neutralization and acidification agents are also reduced [198, 200].

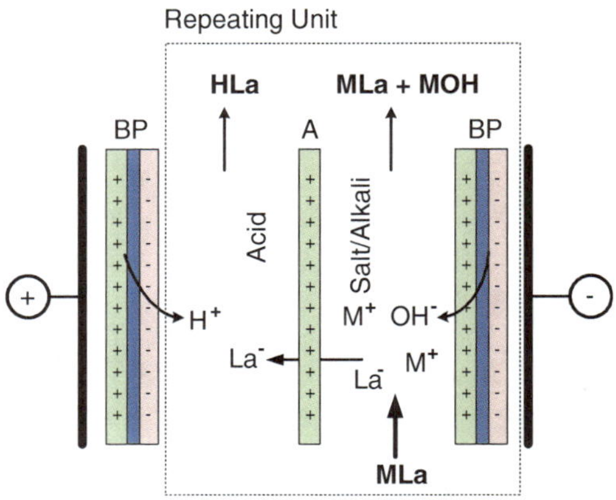

Fig. 11 Two-compartment electrodialysis with bipolar membranes for lactic acid (LA) production; A: anion selective membrane; BP: bipolar membrane; La^{-}: lactate anion; M^{+}: cations; HLa: lactic acid; MOH: alkali

EDBM plants for industrial LA production, rated at 2,600 metric tons LA per year, are being operated successfully in Europe [201]. The cost of the bipolar membrane and the consumption of electric energy are the main issues for the industrial application of EDBM [156]. Furthermore, undesirable parasitic ion transport through membranes can result in the contamination of products or an increased demand for electric energy [200].

5.5 Membrane-Bioreactor Systems for LA Production

As discussed above, cell growth and LA productivity are inhibited by the accumulation of LA during fermentation, and neutralization is therefore desirable. New methods to remedy the product inhibition could improve the productivity, which should result in reduced capital investment, less space requirements, lower energy consumption, and improved stability. Numerous investigations focus on applying in situ product recovery (ISPR) in the LA production in order to remove the LA from the reaction system during the fermentation. The negative effects such as product inhibition, degradation, or transformation of the product into unexpected compounds could be relieved. ISPR is most commonly associated with the combination of membrane technology with fermentation—that is, membrane-bioreactor systems (MBR) [202].

A membrane bioreactor refers to the coupling of membrane devices (micro-, ultra- or nanofiltration membrane) with a traditional bioreactor (e.g. stirred tank reactor) or to the function of the membrane itself (e.g. membrane module directly as a catalytic unit) [202, 203]. Two different configurations have tended to be established during the development of MBR in recent years: immersed and bypass membrane bioreactors (Fig. 12). With the bypass setup, the membrane module is located in a separated circulation outside the reactor (external loop, bypass). The typical operation conditions for this configuration are: transmembrane pressure (TMP): 0.5–5 bar, flux: 40–120 $L{\cdot}m^{-2}$ (depending on biomass concentration and

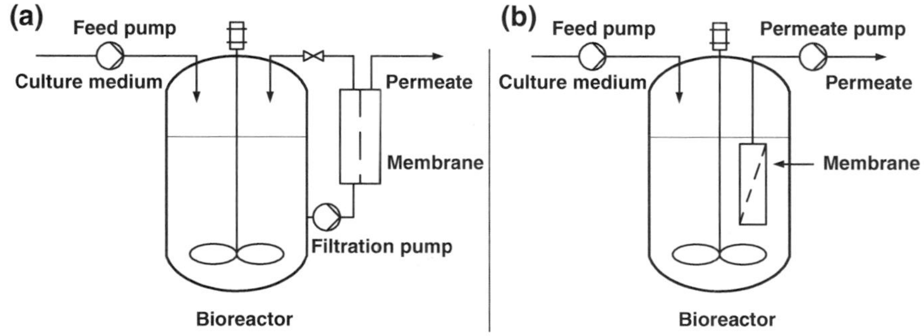

Fig. 12 Configurations of membrane bioreactor systems: **a** Membrane-bioreactor system with an external separation unit, **b** Membrane-bioreactor system with a submerged separation unit

medium complexity), energy consumption: 2–10 kWh·m^{-3} (permeate). In the immersed setup, the membrane module is located inside the reactor (submerged). Typical operation conditions for this configuration are: TMP: 0.05–0.5 bar, flux: 10–60 L·m^{-2} (depending on biomass concentration and medium complexity), energy consumption: 0.02–0.8 kWh·m^{-3} (permeate) [198].

Various organic or inorganic membranes were investigated for the integration with bioreactors. In comparison with organic membranes, ceramic membranes are more favorable for MBRs due to the following advantages [204]:

- Higher mechanical strength and thermal resistance (sterilizable by autoclaving).
- Better chemical resistance (stable to acid, alkali, or oxidants).
- Permanent operational lifetime.
- Possibility of back-flushing for membrane cleaning.
- Easier control of the separation limit and the selectivity.

Aluminum oxide (α- or γ-Al_2O_3) is generally used as a supporting material for inorganic membranes [205]. The active separation layer, which is made of aluminum oxide (α-Al_2O_3), titanium dioxide (TiO_2), or zirconium dioxide (ZrO_2), is coated on the inner surface of the tubular membrane (outer surface on the rotating or flat membrane). According to the producer, the membranes are generally classified with the pore size between several micrometers and nanometers. Figure 13 shows some examples of the ceramic membranes with different physical shapes and operation modes.

The permeate flux in liters per area of membrane is within limits as governed by the TMP at a certain temperature. The pressure-dependent flux is an important factor for evaluation of the membrane performance because it impacts the capital expenses and operating costs. Figure 14 depicts the permeate flux of some ceramic membranes at different pressures. The slopes represent the permeability of membranes with different pore sizes for pure water at 53 °C (Table 3).

However, due to the biomass and other nondissolved compounds in the fermentation broth, a covering layer forms on the membrane surface, or even in some cases inside the membrane pores [206]. This causes an attenuation of the

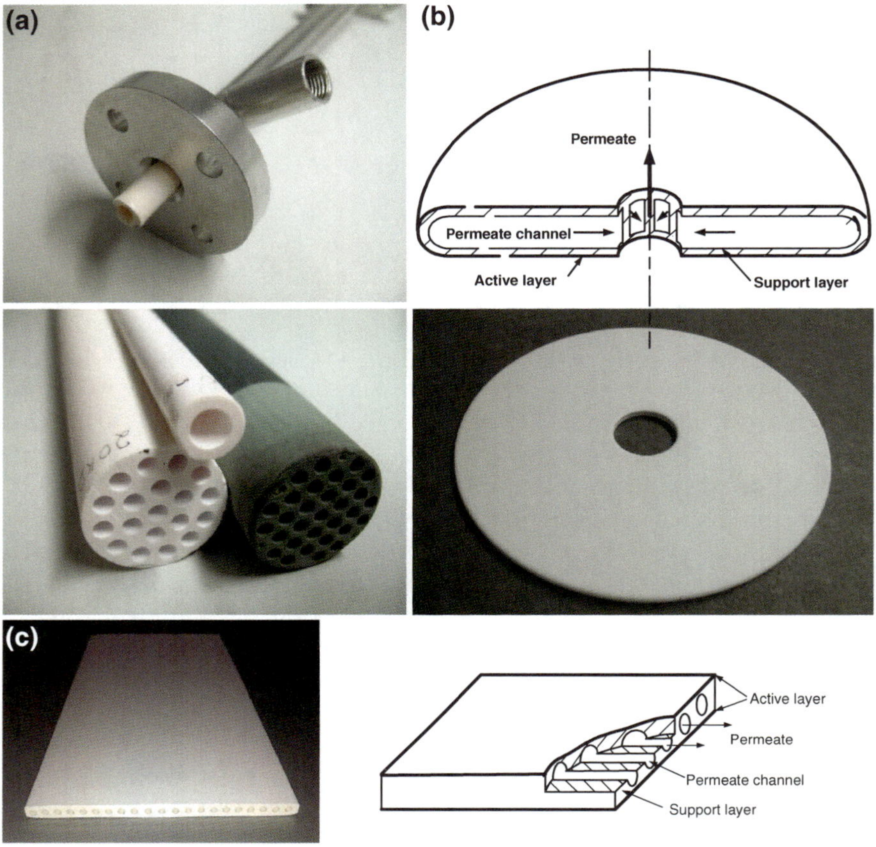

Fig. 13 **a** Tubular membranes with metal housing for external loop bypass systems; **b** Rotating ceramic membrane discs; **c** Flat membranes with an outer coated separation layer for submerged systems

membrane flux shortly after the filtration starts. Cross-flow filtration is used to partially remove the solids retained at the membrane surface by establishing feed fluid flow tangentially to the membrane surface. Thereby, the flux is influenced not only by TMP but also by the tangential flow velocity.

A filtration of LA fermentation broth is shown in Fig. 15. The tubular ceramic membrane device with monochannel (Fig. 13a) is configured as a bypass, as shown in Fig. 12a. After the operation begins, the flux strongly decreases within 30 min due to the development of the covering layer. Subsequently, when the equilibrium between the formation and the washing away of the covering layer was established, the permeability settled at a steady state. When the filtration is carried out at the same TMP (1.6 bar), a higher pressure-normalized flux of 30–40 $L{\cdot}h^{-1}{\cdot}m^{-2}{\cdot}bar^{-1}$ is obtained with a higher flow velocity (1.6 $m{\cdot}s^{-1}$) because the covering layer is reduced by a better flush effect and is not as thick as at a lower flow velocity. However, at this flow velocity, an even higher pressure-normalized

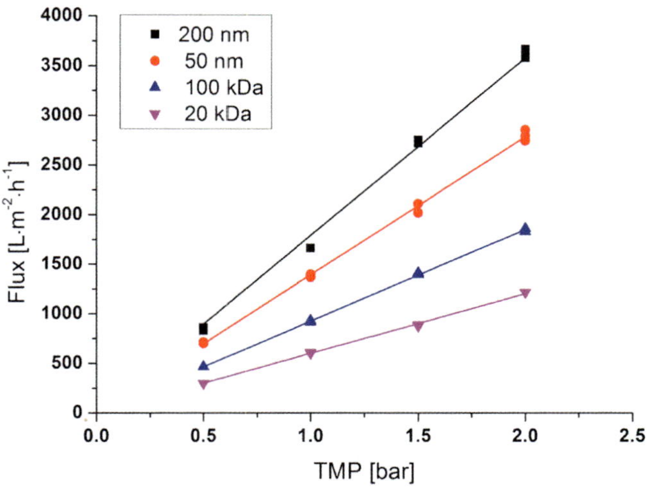

Fig. 14 Filtration profile in pure water using ceramic membranes of different pore sizes and membrane shapes: tubular, monochannel, $\Phi_{in} = 6$ mm, $\Phi_{out} = 10$ mm; membrane material: Al_2O_3, temperature: 53 °C. TMP, transmembrane pressure

Table 3 Permeability of ceramic membranes with different pore sizes (data from Fig. 14)

Pore size/MWCO	200 nm	50 nm	100 kDa	20 kDa
Permeability [$L{\cdot}h^{-1}{\cdot}m^{-2}{\cdot}bar^{-1}$]	1790.8	1394.4	926.5	601.4

flux of 150–175 $L{\cdot}h^{-1}{\cdot}m^{-2}{\cdot}bar^{-1}$ is observed at a much lower TPM (0.4 bar), because a more compact filtration cake at higher pressure is preferred to be formed on the membrane surface and causes a stronger mass transference resistance. An economic optimum must be established between energy cost (fluid velocity and pressure), capital cost (membrane area), and frequency and type (backflush, chemicals) of membrane cleaning. In the case of live organisms in the feed stream to the membrane, the viability of the organisms relative to fluid flow (shear) must also be taken in account.

One anti-fouling measure is backflushing, sometimes also called backwashing/backshocking. Because the cross-flow mode is generally applied in external loop systems, backflushing plays a more important role in submerged systems (shown in Fig. 12b) for anti-fouling. Distilled water, culture medium, filtrate, air, or fermentation gas are usually applied as the backflushing fluid [204].

Figure 16 depicts a filtration of LA fermentation broth using the immersed flat membrane as shown in Fig. 13c, in which a backflush is used. In order to maintain the constant dilution rate (0.4 h^{-1}), the TMP must keep rising while the membrane fouls during the filtration. Once the TMP reaches its maximum operating pressure ($TMP_{max} \approx 0.8$ bar), the backflushing with air is carried out for 2 s. The foulants are partly removed from the membrane surface and the permeability is therefore

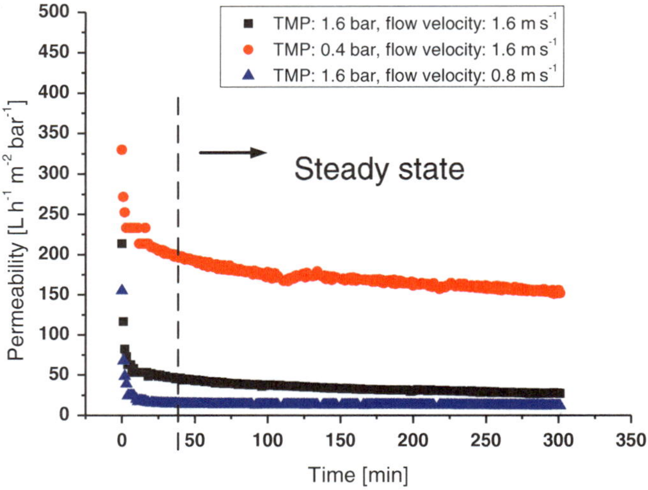

Fig. 15 Filtration of lactic acid fermentation broth using ceramic membrane at different transmembrane pressures and flow velocities. membrane shape: tubular, monochannel, $\Phi_{in} = 6$ mm, $\Phi_{out} = 10$ mm, membrane material: Al_2O_3, MWCO: 100 kDa, cell dry weight: ~2.5 g·L^{-1}, lactate concentration: ~50 g·L^{-1}, temperature: 53 °C

recovered. However, due to the continuous increase of filtration cake formation, the time interval between each backflushing becomes shorter and shorter. To avoid the consolidation of the filtration cake and keep the membrane permeability more stable, frequent backshocking could be applied, whereby flow reversal occurs every few seconds at high pressure but very briefly. The effectiveness of this technique has already been demonstrated by many studies for controlling membrane fouling and maintaining acceptable membrane flux in long-term filtration [206–208].

Promising results for LA production were achieved with integrated tangential flow filtration/cross-flow membrane filtration [176, 194, 196–198, 200]. The membrane devices were used in these processes for retention of the cells as well as for removal of the LA. Because the LA is continuously removed from the fermentation broth, the growth inhibition in the reaction system could be then alleviated. In this case, the cell density could reach a higher level, which results in increased LA productivity. Contamination risks from unexpected microorganisms are reduced [209]. Moreover, the steps of downstream processing could be simplified through in situ recovery of LA using MRB, thus lowering production costs even further. In comparison to conventional batch fermentations, LA productivity is expected to increase 5–40-fold [176, 210] (Fig. 16).

Cell growth can be enhanced by increasing the dilution rate; thus, the volumetric LA productivity can be increased [194]. However, high dilution rates might reduce the LA concentration in the permeate so that the cost of downstream processing is driven up; thus, the advantages of this process might be counteracted.

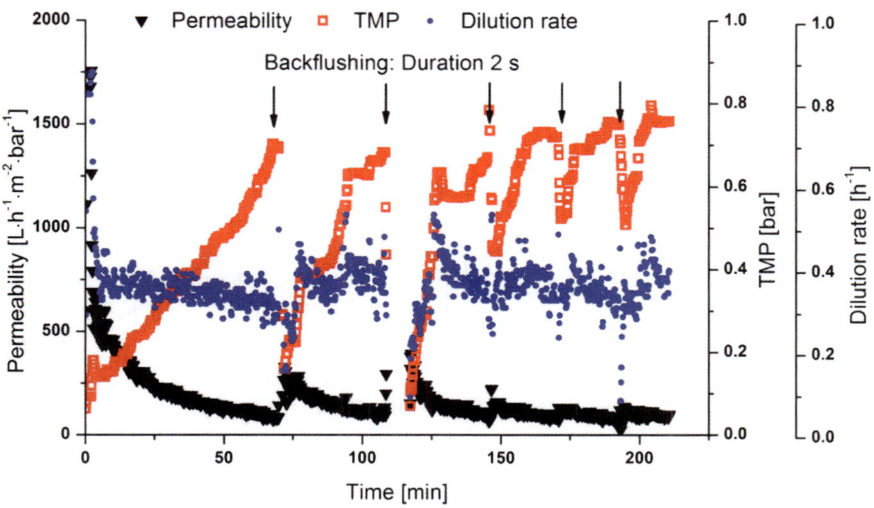

Fig. 16 Filtration of lactic acid fermentation broth using immersed ceramic membrane with backflushing. Membrane shape: flat, membrane material: Al_2O_3, pore size: 0.2 μm, cell dry weight: ~4.3 g·L^{-1}, lactate concentration: ~70 g·L^{-1}, temperature: 53 °C, $D_{set} = 0.4$ h^{-1}. TMP, transmembrane pressure

A possible solution is to feed the fermentation broth back into the fermenter after removal of LA; therefore, the substrate could be adequately utilized.

Another important issue is the membrane cleaning and regeneration. Due to the membrane fouling and the concentration polarization, the treatment capacity and system efficiency drop over time. An economical process for membrane cleansing needs to be worked out to ensure the maintenance of the membrane flux in the industrially acceptable range. This is not unlike the periodic maintenance well known in many industrial unit operations for which slurries are treated, such as fouling in heat exchange, extraction, and distillation.

6 Future Perspectives for Biotechnological Organic Acid Production

Research on the biotechnological production of organic acids continues. Environmental awareness, customers who desire natural products, and the trend towards sustainability point towards replacement of petrochemical products. Biotechnological alternatives already exist for some organic acids that are currently chemically synthesized. Further optimization of biotechnology-based approaches will likely enable economically viable biotechnological production of acid additives for food and feed. In regard to optimization of biotechnological routes to acids, much can be learned from fermentative lactic acid production (see Sect. 5). Research into this process continues despite decades of experience. One exceptionally important lesson from biotechnology-based lactic acid is that the production process as a whole must be considered: substrate(s), microorganisms, bioreactor systems and fermentation conditions, together with the recovery and separation processes, form one entity.

An example of the interconnectedness in biotechnological processing is that a change of the carbon source or the minerals in the medium may impact productivity and growth of the microorganisms as well as the downstream processing (e.g. fouling characteristics of the product stream). The development of raw material prices has to been taken into account because this often constitutes a large portion of the price of production. What is currently considered to be biological waste material could become costly in the future depending on market conditions.

The choice of microorganism is crucial. Substrate has to be used effectively with high yield. Byproducts should be minimized to simplify product recovery. Furthermore, the microorganism has to be tolerant to high acidity and product concentrations to achieve high titers, which are essential for an economic recovery process. If substrate or product inhibitions occur, an improvement could be achieved by using an optimized bioreactor system and a suitable process control strategy. This may require changes in product recovery. Current research results regarding inexpensive separation methods have to been taken into account. A well-thought-out recycling strategy for waste streams could raise efficiency. For example, mineral salts separated in the downstream process could be reused in the fermentation as medium components. Alternatively, waste streams could be monetized as fertilizer or used for energy production (e.g. biomass to biogas).

Biocatalysts (whole cells or purified enzymes) can be used as in the example of ascorbic acid production. As with microorganisms, reactor systems and biocatalysts have to be optimized together with downstream processing and the feedstock. Enzyme stability and ease of recovery are crucial in biocatalysis. Immobilization and enzyme recovery and/or recycling via membrane processes address these issues. Enzymatic catalysis at elevated temperature or in nonaqueous media discourages microbial contamination, but it is enabled only if proper enzymatic activity is available.

The main drivers of customer demand for natural food ingredients, sustainable processing, and regulatory activity motivate ongoing research towards acids for food and feed based on biotechnology. New products are being targeted, acids that are inexpensively produced by chemistry are being considered for conversion to biotechnological pathways (sorbic acid), and scale-up of products such as biotechnology-based lactobionic acid is being attempted. The market opportunities for biotechnology-based production of acid additives for food and feed are extremely encouraging. An integrated system view of biotechnological processes will likely lead to new products with attractive economic parameters.

References

1. Association of American Feed Control Officials (2008) Pet food and specialty pet food labeling guide
2. Kanehisa M, Goto S, Sato Y, Furumichi M, Tanabe M (2012) KEGG for integration and interpretation of large-scale molecular data sets. Nucleic Acids Res 40:D109–D114. doi: 10.1093/nar/gkr988
3. He H, Wang H, Fang C, Wu H, Guo X, Liu C, Lin Z, Lin W (2012) Barnyard grass stress up regulates the biosynthesis of phenolic compounds in allelopathic rice. J Plant Physiol 169:1747–1753. doi:10.1016/j.jplph.2012.06.018
4. Thakur BR, Singh RK, Handa AK, Rao MA (1997) Chemistry and uses of pectin—a review. Crit Rev Food Sci Nutr 37:47–73
5. Ames WM (1952) The conversion of collagen to gelatin and their molecular structures. J Sci Food Agric 3:454–463. doi:10.1002/jsfa.2740031004
6. Wienen WJ, Shallenberger RS (1988) Influence of acid and temperature on the rate of inversion of sucrose. Food Chem 29:51–55. doi:10.1016/0308-8146(88)90075-1
7. Soccol CR, Vandenberghe LPS, Rodrigues C, Pandey A (2006) New perspectives for citric acid production and application. Food Technol Biotechnol 44:141
8. Buck DF (1991) Antioxidants. In: Smith J (ed) Food additive user's handbook. Springer, New York, p 1–46
9. Pokorny J, Yanishlieva N, Gordon M (2001) Antioxidants in food. CRC press, Florida
10. Aruoma OI, Murcia A, Butler J, Halliwell B (1993) Evaluation of the antioxidant and prooxidant actions of gallic acid and its derivatives. J Agric Food Chem 41:1880–1885. doi: 10.1021/jf00035a014
11. Yen GC, Duh PD, Tsai HL (2002) Antioxidant and pro-oxidant properties of ascorbic acid and gallic acid. Food Chem 79:307–313. doi:10.1016/S0308-8146(02)00145-0
12. Graf E (1992) Antioxidant potential of ferulic acid. Free Radical Biol Med 13:435–448. doi: 10.1016/0891-5849(92)90184-I
13. Biebaut D (1991) Flour improvers and raising agents. In: Smith J (ed) Food additive user's handbook. Springer, New York, p 242–256
14. Grosch W, Wieser H (1999) Redox reactions in wheat dough as affected by ascorbic acid. J Cereal Sci 29:1–16. doi:10.1006/jcrs.1998.0218
15. Wehrle K, Grau H, Arendt EK (1997) Effects of lactic acid, acetic acid, and table salt on fundamental rheological properties of wheat dough. Cereal Chem J 74:739–744. doi: 10.1094/cchem.1997.74.6.739
16. Nakamura M, Kurata T (1997) Effect of l-ascorbic acid on the rheological properties of wheat flour-water dough. Cereal Chem J 74:647–650. doi:10.1094/cchem.1997.74.5.647
17. McGraw-Hill PSP (2002) McGraw-Hill dictionary of scientific and technical terms. The McGraw-Hill Companies Inc, New York
18. Codex Alimentarius Commission (1989) Class names and the international numbering system for food additives. CAC/GL (36-1989)
19. Smith J, Hong-Shum L (2011) Food additives data book. Wiley, Oxford
20. Gordon RJ (1991) Flavourings. In: Smith J (ed) Food additive user's handbook. Springer, New York, p 75–88
21. Arrigoni O, De Tullio MC (2002) Ascorbic acid: much more than just an antioxidant. Biochim Biophys Acta 1569:1–9. doi:10.1016/S0304-4165(01)00235-5
22. Roig MG, Rivera ZS, Kennedy JF (1993) L-ascorbic acid: an overview. Int J Food Sci Nutr 44:59–72. doi:10.3109/09637489309017424
23. Alonso S, Rendueles M, Díaz M (2013) Bio-production of lactobionic acid: current status, applications and future prospects. Biotechnol Adv. doi:10.1016/j.biotechadv.2013.04.010
24. Gutiérrez LF, Hamoudi S, Belkacemi K (2012) Lactobionic acid: a high value-added lactose derivative for food and pharmaceutical applications. Int Dairy J 26:103–111. doi: 10.1016/j.idairyj.2012.05.003

25. Schaafsma G (2008) Lactose and lactose derivatives as bioactive ingredients in human nutrition. Int Dairy J 18:458–465. doi:10.1016/j.idairyj.2007.11.013
26. Ou S, Kwok KC (2004) Ferulic acid: pharmaceutical functions, preparation and applications in foods. J Sci Food Agric 84:1261–1269. doi:10.1002/jsfa.1873
27. Wee YJ, Kim JN, Ryu HW (2006) Biotechnological production of lactic acid and its recent applications. Food Technol Biotechnol 44:163–172
28. Sloan AE, Labuza TP (1976) Prediction of water activity lowering ability of food humectants at high aw. J Food Sci 41:532–535. doi:10.1111/j.1365-2621.1976.tb00664.x
29. Lück E, Jager M (1997) Antimicrobial action of preservatives. In: Antimicrobial food additives. Springer, Berlin, p 36–57
30. Leavening agent (2013) Encyclopædia britannica online. Accessed 24 Jun 2013 http://www.britannica.com/EBchecked/topic/334116/leavening-agent
31. Nitin K (2008) Longman Science Chemistry 10. Pearson Education. Delhi, India
32. Bender DA (2009) A dictionary of food and nutrition. Oxford University Press, Oxford
33. Giménez B, Turnay J, Lizarbe MA, Montero P, Gómez-Guillén MC (2005) Use of lactic acid for extraction of fish skin gelatin. Food Hydrocolloids 19:941–950. doi: 10.1016/j.foodhyd.2004.09.011
34. Nauta T (1991) Chelating agents. In: Smith J (ed) Food additive user's handbook. Springer, Oxford, p 273–279
35. Phadungath (2005) The mechanism and properties of acid-coagulated milk gels. Songklanakarin. J Sci Technol 27:433–448
36. Revis C, Payne GA (1907) The acid coagulation of milk. J Hyg 7:216–231
37. Guyomarc'h F, Renan M, Chatriot M, Gamerre V, Famelart MH (2007) Acid gelation properties of heated skim milk as a result of enzymatically induced changes in the micelle/serum distribution of the whey protein/κ-casein aggregates. J Agric Food Chem 55:10986–10993. doi:10.1021/jf0722304
38. Donato L, Alexander M, Dalgleish DG (2007) Acid gelation in heated and unheated milks: interactions between serum protein complexes and the surfaces of casein micelles. J Agric Food Chem 55:4160–4168. doi:10.1021/jf063242c
39. Sahul JK, Das H (2010) A continuous heat-acid coagulation unit for continuous production of chhana. Assam Univ J Sci Technol 4:40–45
40. Liu Z, Chang SK (2004) Effect of soy milk characteristics and cooking conditions on coagulant requirements for making filled tofu. J Agric Food Chem 52:3405–3411. doi: 10.1021/jf035139i
41. Puppo MC, Añón MC (1998) Effect of pH and protein concentration on rheological behavior of acidic soybean protein gels. J Agric Food Chem 46:3039–3046. doi: 10.1021/jf971092n
42. Theron MM, Lues JFR (2010) Application of organic acids in food preservation. In: Organic acids and food preservation, CRC PressI Llc, p 51–95, Bosa Roca, USA
43. Lückstädt C (2008) The use of acidifiers in fish nutrition. CAB Rev 3:1–8
44. Sauli I, Danuser J, Geeraerd AH, Van Impe JF, Rüfenacht J, Bissig-Choisat B, Wenk C, Stärk KDC (2005) Estimating the probability and level of contamination with Salmonella of feed for finishing pigs produced in Switzerland—the impact of the production pathway. Int J Food Microbiol 100:289–310. doi:10.1016/j.ijfoodmicro.2004.10.026
45. Freitag M (2007) Organic acids and salts promote performance and health in animal husbandry. In: Lückstädt C (ed) Acidifiers in animal nutrition: a guide for feed preservation and acidification to promote animal performance. p 1–11, Packington, GB
46. Papatsiros VG, Cristodoulopoulos C, Filippopoulos LC (2012) The use of organic acids in monogastric animals (swine and rabbits). J Cell Anim Biol 6:154–159
47. Suryanarayana MVAN, Suresh J, Rajasekhar MV (2012) Organic acids in swine feeding—a review. Agric Sci Res J 2:523–533
48. Tung CM, Pettigrew JE (2008) Critical review of acidifiers. Report NPB:05-169
49. Kim YY, Kil DY, Oh HK, Han IK (2005) Acidifier as an alternative material to antibiotics in animal feed. Asian-Australas J Anim Sci 18:1048

50. Desai DN, Patwardhan DS, Ranade AS, Lückstädt C (2007) Acidifiers in poultry diets and poultry production. In: Lückstädt C (ed) Acidifiers in animal nutrition: a guide for feed preservation and acidification to promote animal performance. Nottingham University Press, Nottingham, pp 63–69
51. Haque MN, Chowdhury R, Islam KMS, Akbar MA (2009) Propionic acid is an alternative to antibiotics in poultry diet. Bangladesh J Anim Sci 38:115–122
52. Menconi A, Reginatto AR, Londero A, Pumford NR, Morgan M, Hargis BM, Tellez G (2013) Effect of organic acids on salmonella typhimurium infection in broiler chickens. Int J Poult Sci 12:72–75
53. Rosyidah MR, Loh TC, Foo HL, Cheng XF, Bejo MH (2011) Effect of feeding metabolites and acidifier on growth performance, faecal characteristics and microflora in broiler chickens. J Anim Vet Adv 10:2758–2764. doi:10.3923/javaa.2011 2758.2764
54. Lückstädt C, Kühlmann KJ (2011) The use of diaformates in tilapia - ways to improve performance sustainably: a short review. In: Lückstädt C (ed) Standards for acidifiers. Nottingham, GB
55. Ganguly S, Dora KC, Sarkar S, Chowdhury S (2013) Supplementation of prebiotics in fish feed: a review. Rev in Fish Biol and Fisheries 23:195–199. doi:10.1007/s11160-012-9291-5
56. Mine S, Boopathy R (2011) Effect of organic acids on shrimp pathogen, vibrio harveyi. Curr Microbiol 63:1–7. doi:10.1007/s00284-011-9932-2
57. Ajiboye OO, Yakubu AF, Adams TE (2012) A perspective on the ingestion and nutritional effects of feed additives in farmed fish species. World J Fish Mar Sci 4:87–101. doi: 10.5829/idosi.wjfms.2012.04.01.56264
58. Papatsiros VG, Christodoulopoulos G (2011) The use of organic acids in rabbit farming. Online J Anim Feed Res 1:434–438
59. Falcão-e-Cunha L, Castro-Solla L, Maertens L, Marounek M, Pinheiro V, Freire J (2010) Alternatives to antibiotic growth promoters in rabbit feeding: a review. World Rabbit Sci 15:127–140
60. Ribeiro MD, Pereira JC, Queiroz AC, Cecon PR, Detmann E, Azevêdo JAG (2009) Performance of dairy calves fed milk, milk replacer or post-weaning concentrate with acidifiers. Rev Bras Zootec 38:956–963
61. Alibaba.com (2013) http://www.alibaba.com. Accessed 18 Sept 2013
62. Sauer M, Porro D, Mattanovich D, Branduardi P (2008) Microbial production of organic acids: expanding the markets. Trends Biotechnol 26:100–108. doi: 10.1016/j.tibtech.2007.11.006
63. Burdock GA (2004) In: Fenaroli's handbook of flavor ingredients. CRC press, Boca Raton
64. IHS chemical (2013) Formic acid
65. Xu Z, Shi Z, Jiang L (2011) 3.18 - Acetic and Propionic Acids. In: Editor-in-Chief: Murray M-Y (ed) Comprehensive biotechnology, 2nd edn. Academic Press, Burlington, p 189–199
66. Lück E, Jager M, Raczek N (2000) Sorbic acid. In: Ullmann's encyclopedia of industrial chemistry. Wiley-VCH Verlag GmbH & Co. KGaA, Weinheim
67. Sengun IY, Karabiyikli S (2011) Importance of acetic acid bacteria in food industry. Food Control 22:647–656. doi:http://dx.doi.org/10.1016/j.foodcont.2010.11.008
68. Kocher GS, Dhillon HK (2013) Fermentative production of sugarcane vinegar by immobilized cells of acetobacter aceti under packed bed conditions. Sugar Tech 15:71–76. doi:10.1007/s12355-012-0179-4
69. Jiménez-Hornero JE, Santos-Dueñas IM, García-García I (2009) Optimization of biotechnological processes. The acetic acid fermentation. Part I: the proposed model. Biochem Eng J 45:1–6. doi:10.1016/j.bej.2009.01.009
70. Awad HM, Malek RA, Othman NZ, Aziz RA, El Enshasy HA (2012) Efficient production process for food grade acetic acid by acetobacter aceti in shake flask and in bioreactor cultures. E-J Chem 9:2275–2286. doi:10.1155/2012/965432
71. Wang Z, Yan M, Chen X, Li D, Qin L, Li Z, Yao J, Liang X (2013) Mixed culture of saccharomyces cerevisiae and acetobacter pasteurianus for acetic acid production. Biochem Eng J 79:41–45. doi:10.1016/j.bej.2013.06.019

72. Matsutani M, Nishikura M, Saichana N, Hatano T, Masud-Tippayasak U, Theergool G, Yakushi T, Matsushita K (2013) Adaptive mutation of acetobacter pasteurianus SKU1108 enhances acetic acid fermentation ability at high temperature. J Biotechnol 165:109–119. doi:10.1016/j.jbiotec.2013.03.006
73. Reutemann W, Kieczka H (2000) Formic acid. In: Ullmann's encyclopedia of industrial chemistry. Wiley-VCH Verlag GmbH & Co. KGaA, Weinheim
74. Eguchi SY, Nishio N, Nagai S (1985) Formic acid production from H2 and bicarbonate by a formateutilizing methanogen. Appl Microbiol Biotechnol 22:148–151. doi: 10.1007/bf00250036
75. Shams Yazdani S, Gonzalez R (2008) Engineering Escherichia coli for the efficient conversion of glycerol to ethanol and co-products. Metab Eng 10:340–351. doi: 10.1016/j.ymben.2008.08.005
76. Murarka A, Dharmadi Y, Yazdani SS, Gonzalez R (2008) Fermentative utilization of glycerol by Escherichia coli and its implications for the production of fuels and chemicals. Appl Environ Microbiol 74:1124–1135. doi:10.1128/AEM.02192-07
77. Liu L, Zhu Y, Li J, Wang M, Lee P, Du G, Chen J (2012) Microbial production of propionic acid from propionibacteria: current state, challenges and perspectives. Crit Rev Biotechnol 32:374–381. doi:10.3109/07388551.2011.651428
78. Zhu Y, Li J, Tan M, Liu L, Jiang L, Sun J, Lee P, Du G, Chen J (2010) Optimization and scale-up of propionic acid production by propionic acid-tolerant Propionibacterium acidipropionici with glycerol as the carbon source. Bioresour Technol 101:8902–8906. doi: 10.1016/j.biortech.2010.06.070
79. Jin Z, Yang S (1998) Extractive fermentation for enhanced propionic acid production from lactose by Propionibacterium acidipropionici. Biotechnol Prog 14:457–465. doi: 10.1021/bp980026i
80. Kagliwal LD, Survase SA, Singhal RS, Granström T (2013) Wheat flour based propionic acid fermentation: an economic approach. Bioresour Technol 129:694–699. doi: 10.1016/j.biortech.2012.12.154
81. Wang Z, Yang ST (2013) Propionic acid production in glycerol/glucose co-fermentation by Propionibacterium freudenreichii subsp. shermanii. Bioresour Technol 137:116–123. doi: 10.1016/j.biortech.2013.03.012
82. Zhang A, Yang S (2009) Propionic acid production from glycerol by metabolically engineered Propionibacterium acidipropionici. Process Biochem 44:1346–1351. doi: 10.1016/j.procbio.2009.07.013
83. Suwannakham S, Huang Y, Yang S (2006) Construction and characterization of ack knock-out mutants of Propionibacterium acidipropionici for enhanced propionic acid fermentation. Biotechnol Bioeng 94:383–395. doi:10.1002/bit.20866
84. Zhang A, Yang S (2009) Engineering Propionibacterium acidipropionici for enhanced propionic acid tolerance and fermentation. Biotechnol Bioeng 104:766–773. doi: 10.1002/bit.22437
85. Claypool JT, Raman DR (2012) A coarse techno-economic model of a combined fermentation-catalysis route to sorbic acid. In: Agricultural and biosystems engineering presentations, posters and proceedings, Dallas
86. Xie D, Shao Z, Achkar J, Zha W, Frost JW, Zhao H (2006) Microbial synthesis of triacetic acid lactone. Biotechnol Bioeng 93:727–736. doi:10.1002/bit.20759
87. Tang S, Qian S, Akinterinwa O, Frei CS, Gredell JA, Cirino PC (2013) Screening for enhanced triacetic acid lactone production by recombinant Escherichia coli expressing a designed triacetic acid lactone reporter. J Am Chem Soc 135:10099–10103. doi: 10.1021/ja402654z
88. Polen T, Spelberg M, Bott M (2013) Toward biotechnological production of adipic acid and precursors from biorenewables. J Biotechnol. doi:10.1016/j.jbiotec.2012.07.008
89. Merchant Research and Consulting LTD (2013) Adipic Acid (ADPA): 2013 World market outlook and forecast up to 2017

90. Yang ST, Zhang K, Zhang B, Huang H (2011) 3.16 - Fumaric acid. In: Editor-in-Chief: Murray M-Y (ed) Comprehensive Biotechnology, 2nd edn. Academic Press, Burlington, p 163–177
91. Roa Engel CA, van Gulik WM, Marang L, van der Wielen LAM, Straathof AJJ (2011) Development of a low pH fermentation strategy for fumaric acid production by Rhizopus oryzae. Enzyme Microb Technol 48:39–47. doi:10.1016/j.enzmictec.2010.09.001
92. Cukalovic A, Stevens CV (2008) Feasibility of production methods for succinic acid derivatives: a marriage of renewable resources and chemical technology. Biofuels, Bioprod Biorefin 2:505–529. doi:10.1002/bbb.105
93. Musser MT (2000) Adipic acid. In: Ullmann's encyclopedia of industrial chemistry. Wiley-VCH Verlag GmbH & Co. KGaA, Weinheim
94. van Duuren JBJH, Wijte D, Karge B, Martins dos Santos VAP, Yang Y, Mars AE, Eggink G (2012) pH-stat fed-batch process to enhance the production of cis, cis-muconate from benzoate by Pseudomonas putida KT2440-JD1. Biotechnol Prog 28:85–92. doi:10.1002/btpr.709
95. Niu W, Draths KM, Frost JW (2002) Benzene-free synthesis of adipic acid. Biotechnol Prog 18:201–211. doi:10.1021/bp010179x
96. Moon TS, Yoon SH, Lanza AM, Roy-Mayhew JD, Prather KL (2009) Production of glucaric acid from a synthetic pathway in recombinant Escherichia coli. Appl Environ Microbiol 75:589–595. doi:10.1128/AEM.00973-08
97. Moon TS, Dueber JE, Shiue E, Prather KLJ (2010) Use of modular, synthetic scaffolds for improved production of glucaric acid in engineered E. coli. Metab Eng 12:298–305. doi:10.1016/j.ymben.2010.01.003
98. Lippow SM, Moon TS, Basu S, Yoon S, Li X, Chapman BA, Robison K, Lipovšek D, Prather KLJ (2010) Engineering enzyme specificity using computational design of a defined-sequence library. Chem Biol 17:1306–1315. doi:10.1016/j.chembiol.2010.10.012
99. Thomas JM, Raja R, Johnson BFG, O'Connell TJ, Sankar G, Khimyak T (2003) Bimetallic nanocatalysts for the conversion of muconic acid to adipic acid. Chem Commun 10:1126–1127. doi:10.1039/b300203a
100. Picataggio S, Rohrer T, Deanda K, Lanning D, Reynolds R, Mielenz J, Eirich LD (1992) Metabolic Engineering of Candida Tropicalis for the Production of Long-Chain Dicarboxylic Acids. Nat Biotechnol 10:894–898
101. Hara A, Ueda M, Matsui T, Arie M, Saeki H, Matsuda H, Furuhashi K, Kanai T, Tanaka A (2001) Repression of fatty-acyl-CoA oxidase-encoding gene expression is not necessarily a determinant of high-level production of dicarboxylic acids in industrial dicarboxylic-acid-producing Candida tropicalis. Appl Microbiol Biotechnol 56:478–485. doi:10.1007/s002530000543
102. Xu Q, Li S, Huang H, Wen J (2012) Key technologies for the industrial production of fumaric acid by fermentation. Biotechnol Adv 30:1685–1696. doi:10.1016/j.biotechadv.2012.08.007
103. Song CW, Kim DI, Choi S, Jang JW, Lee SY (2013) Metabolic engineering of Escherichia coli for the production of fumaric acid. Biotechnol Bioeng 110:2025–2034. doi:10.1002/bit.24868
104. Nakajima-Kambe T, Nozue T, Mukouyama M, Nakahara T (1997) Bioconversion of maleic acid to fumaric acid by Pseudomonas alcaligenes strain XD-1. J Ferment Bioeng 84:165–168. doi:10.1016/S0922-338X(97)82549-4
105. Ichikawa S, Iino T, Sato S, Nakahara T, Mukataka S (2003) Improvement of production rate and yield of fumaric acid from maleic acid by heat treatment of Pseudomonas alcaligenes strain XD-1. Biochem Eng J 13:7–13. doi:10.1016/S1369-703X(02)00080-3
106. Lee JW, Han MS, Choi S, Yi J, Lee TW, Lee SY (2011) 3.15 - Organic acids: succinic and malic acids. In: Editor-in-Chief: Murray M-Y (ed) Comprehensive biotechnology, 2nd edn. Academic Press, Burlington, p 149–161

107. Meynial-Salles I, Dorotyn S, Soucaille P (2008) A new process for the continuous production of succinic acid from glucose at high yield, titer, and productivity. Biotechnol Bioeng 99:129–135. doi:10.1002/bit.21521
108. Cheng K, Wang G, Zeng J, Zhang J (2013) Improved succinate production by metabolic engineering. BioMed Res Int 2013:12. doi:10.1155/2013/538790
109. Chen Y, Nielsen J (2013) Advances in metabolic pathway and strain engineering paving the way for sustainable production of chemical building blocks. Curr Opin Biotechnol. doi: 10.1016/j.copbio.2013.03.008
110. Okino S, Noburyu R, Suda M, Jojima T, Inui M, Yukawa H (2008) An efficient succinic acid production process in a metabolically engineered Corynebacterium glutamicum strain. Appl Microbiol Biotechnol 81:459–464. doi:10.1007/s00253-008-1668-y
111. Kirimura K, Honda Y, Hattori T (2011) 3.13 - Citric acid. In: Editor-in-Chief: Murray M-Y (ed) Comprehensive biotechnology, 2nd edn. Academic Press, Burlington, p 135–142
112. Singh Dhillon G, Kaur Brar S, Verma M, Tyagi RD (2011) Recent advances in citric acid bio-production and recovery. Food Bioprocess Technol 4:505–529. doi: 10.1007/s11947-010-0399-0
113. Global Industry Analysts (2012) Lactic acid: a global strategic business report. MCP-2089
114. Mujtaba IM, Edreder EA, Emtir M (2012) Significant thermal energy reduction in lactic acid production process. Appl Energy 89:74–80. doi:10.1016/j.apenergy.2010.11.031
115. Verhoff FH (2000) Citric acid. In: Ullmann's encyclopedia of industrial chemistry. Wiley-VCH Verlag GmbH & Co. KGaA, Weinheim
116. Ikram-ul H, Ali S, Qadeer MA, Iqbal J (2004) Citric acid production by selected mutants of Aspergillus niger from cane molasses. Bioresour technol 93:125–130. doi: http://dx.doi.org/10.1016/j.biortech.2003.10.018
117. Buque-Taboada EM, Straathof AJJ, Heijnen JJ, Wielen LAM (2006) In situ product recovery (ISPR) by crystallization: basic principles, design, and potential applications in whole-cell biocatalysis. Appl Microbiol Biotechnol 71:1–12. doi:10.1007/s00253-006-0378-6
118. Chibata I, Tosa T, Takata I (1983) Continuous production of L-malic acid by immobilized cells. Trends Biotechnol 1:9–11. doi:10.1016/0167-7799(83)90019-7
119. Neufeld RJ, Peleg Y, Rokem JS, Pines O, Goldberg I (1991) l-Malic acid formation by immobilized Saccharomyces cerevisiae amplified for fumarase. Enzyme Microb Technol 13:991–996. doi:10.1016/0141-0229(91)90122-Q
120. Zelle RM, de Hulster E, van Winden WA, de Waard P, Dijkema C, Winkler AA, Geertman JM, van Dijken JP, Pronk JT, van Maris AJ (2008) Malic acid production by Saccharomyces cerevisiae: engineering of pyruvate carboxylation, oxaloacetate reduction, and malate export. Appl Environ Microbiol 74:2766–2777. doi:10.1128/AEM.02591-07
121. Singh OV, Kumar R (2007) Biotechnological production of gluconic acid: future implications. Appl Microbiol Biotechnol 75:713–722. doi:10.1007/s00253-007-0851-x
122. Kirimura K, Honda Y, Hattori T (2011) 3.14 - Gluconic and itaconic acids. In: Editor-in-Chief: Murray M-Y (ed) Comprehensive biotechnology, 2nd edn. Academic Press, Burlington, p 143–147
123. Global Industry Analysts (2008) Tartaric acid - a global strategic business report. MCP-2151
124. Oster B, Fechtel U (2000) Vitamins, 7. Vitamin C (l-ascorbic acid). In: Ullmann's encyclopedia of industrial chemistry. Wiley-VCH Verlag GmbH & Co. KGaA, Weinheim
125. Ramachandran S, Fontanille P, Pandey A, Larroche C (2006) Gluconic acid: properties, applications and microbial production. Food Technol Biotechnol 44:185–195
126. Anastassiadis S, Rehm H (2006) Continuous gluconic acid production by the yeast-like Aureobasidium pullulans in a cascading operation of two bioreactors. Appl Microbiol Biotechnol 73:541–548. doi:10.1007/s00253-006-0499-y
127. Anastassiadis S, Rehm H (2006) Continuous gluconic acid production by Aureobasidium pullulans with and without biomass retention. Electron J Biotechnol 9. doi: 10.2225/vol9-issue5-fulltext-18

128. Sankpal NV, Joshi AP, Sutar II, Kulkarni BD (1999) Continuous production of gluconic acid by Aspergillus niger immobilized on a cellulosic support: study of low pH fermentative behaviour of Aspergillus niger. Process Biochem 35:317–325. doi:10.1016/S0032-9592(99)00074-6
129. van Hecke W, Bhagwat A, Ludwig R, Dewulf J, Haltrich D, van Langenhove H (2009) Kinetic modeling of a bi-enzymatic system for efficient conversion of lactose to lactobionic acid. Biotechnol Bioeng 102:1475–1482. doi:10.1002/bit.22165
130. Murakami H, Seko A, Azumi M, Ueshima N, Yoshizumi H, Nakano H, Kitahata S (2003) Fermentative production of lactobionic acid by Burkholderia cepacia. J Appl Glycosc 50:117–120
131. Alonso S, Rendueles M, Díaz M (2013) Feeding strategies for enhanced lactobionic acid production from whey by Pseudomonas taetrolens. Bioresour Technol 134:134–142. doi: http://dx.doi.org/10.1016/j.biortech.2013.01.145
132. Malvessi E, Carra S, Pasquali FC, Kern DB, Silveira MM, Ayub MAZ (2013) Production of organic acids by periplasmic enzymes present in free and immobilized cells of Zymomonas mobilis. J Ind Microbiol Biotechnol 40:1–10. doi:10.1007/s10295-012-1198-6
133. Kassaian JM (2000) Tartaric acid. In: Ullmann's encyclopedia of industrial chemistry. Wiley-VCH Verlag GmbH & Co. KGaA, Weinheim
134. Klasen R, Bringer-Meyer S, Sahm H (1992) Incapability of Gluconobacter oxydans to produce tartaric acid. Biotechnol Bioeng 40:183–186. doi:10.1002/bit.260400126
135. Chandrashekar K, Felse P, Panda T (1999) Optimization of temperature and initial pH and kinetic analysis of tartaric acid production by Gluconobacter suboxydans. Bioprocess Eng 20:203–207. doi:10.1007/pl00009044
136. Mantha D, Aslam Basha Z, Panda T (1998) Optimization of medium composition by response surface methodology for the production of tartaric acid by Gluconobacter suboxydans. Bioprocess Eng 19:285–288. doi:10.1007/pl00009020
137. Matzerath I, Kläui W, Klasen R, Sahm H (1995) Vanadate catalysed oxidation of 5-keto-d-gluconic acid to tartaric acid: the unexpected effect of phosphate and carbonate on rate and selectivity. Inorg Chim Acta 237:203–205. doi:http://dx.doi.org/10.1016/0020-1693(95)04665-V
138. Natella F, Nardini M, Di Felice M, Scaccini C (1999) Benzoic and cinnamic acid derivatives as antioxidants: structure-activity relation. J Agric Food Chem 47:1453–1459. doi:10.1021/jf980737w
139. World Health Organization (2005) Benzoic acid and sodium benzoate. Organization WH, Geneva
140. Maki T, Takeda K (2000) Benzoic acid and derivatives. In: Ullmann's encyclopedia of industrial chemistry. Wiley-VCH Verlag GmbH & Co. KGaA, Weinheim
141. Kikuzaki H, Hisamoto M, Hirose K, Akiyama K, Taniguchi H (2002) Antioxidant properties of ferulic acid and its related compounds. J Agric Food Chem 50:2161–2168. doi: 10.1021/jf011348w
142. Figueroa-Espinoza MC, Morel MH, Surget A, Asther M, Moukha S, Sigoillot JC, Rouau X (1999) Attempt to cross-link feruloylated arabinoxylans and proteins with a fungal laccase. Food Hydrocolloids 13:65–71. doi:http://dx.doi.org/10.1016/S0268-005X(98)00072-1
143. Priefert H, Rabenhorst J, Steinbüchel A (2001) Biotechnological production of vanillin. Appl Microbiol Biotechnol 56:296–314. doi:10.1007/s002530100687
144. Bajpai B, Patil S (2008) A new approach to microbial production of gallic acid. Braz J Microbiol 39:708–711
145. Noda S, Kitazono E, Tanaka T, Ogino C, Kondo A (2012) Benzoic acid fermentation from starch and cellulose via a plant-like beta-oxidation pathway in Streptomyces maritimus. Microb Cell Fact 11:49
146. Taniguchi H, Hosoda A, Tsuno T, Maruta Y, Nomura E (1999) Preparation of ferulic acid and its application for the synthesis of cancer chemopreventive agents. Anticancer Res 19:3757–3761

147. Couteau D, Mathaly P (1997) Purification of ferulic acid by adsorption after enzymic release from a sugar-beet pulp extract. Ind Crops Prod 6:237–252. doi:10.1016/S0926-6690(97)00014-9
148. Couteau D, Mathaly P (1998) Fixed-bed purification of ferulic acid from sugar-beet pulp using activated carbon: optimization studies. Bioresour Technol 64:17–25. doi: http://dx.doi.org/10.1016/S0960-8524(97)00152-1
149. Madhavi DL, Smith MAL, Linas AC, Mitiku G (1997) Accumulation of ferulic acid in cell cultures of Ajuga pyramidalis metallica crispa. J Agric Food Chem 45:1506–1508. doi: 10.1021/jf9607831
150. Kar B, Banerjee R, Bhattacharyya BC (1999) Microbial production of gallic acid by modified solid state fermentation. J Ind Microbiol Biotechnol 23:173–177. doi: 10.1038/sj.jim.2900713
151. Banerjee D, Mahapatra S, Pati BR (2007) Gallic acid production by submerged fermentation of Aspergillus aculeatus DBF9. Res J Microbiol 2:462–468
152. Chávez-González M, Rodríguez-Durán L, Balagurusamy N, Prado-Barragán A, Rodríguez R, Contreras J, Aguilar C (2012) Biotechnological Advances and Challenges of Tannase: An Overview. Food Bioprocess Technol 5:445–459. doi:10.1007/s11947-011-0608-5
153. Gonçalves HB, Jorge JA, Pessela BC, Lorente GF, Guisán JM, Guimarães LHS (2013) Characterization of a tannase from Emericela nidulans immobilized on ionic and covalent supports for propyl gallate synthesis. Biotechnol Lett 35:591–598. doi: 10.1007/s10529-012-1111-4
154. Gao T, Wong Y, Ng C, Ho K (2012) L-lactic acid production by Bacillus subtilis MUR1. Bioresour Technol 121:105–110. doi:10.1016/j.biortech.2012.06.108
155. Castillo Martinez FA, Balciunas EM, Salgado JM, Domínguez González JM, Converti A, Oliveira RP (2013) Lactic acid properties, applications and production: a review. Trends Food Sci Technol 30:70–83. doi:10.1016/j.tifs.2012.11.007
156. Miller C, Fosmer A, Rush B, McMullin T, Beacom D, Suominen P (2011) 3.17—Industrial production of lactic acid. In: Editor-in-Chief: Murray M-Y (ed) Comprehensive biotechnology, 2nd edn. Academic Press, Burlington, p 179–188
157. Holten CH (1971) Lactic acid: properties and chemistry of lactic acid and derivatives. Verlag Chemie GmbH, Weinheim
158. U.S. Food and Drug Administration (FAO) (2012) Code of federal regulations—lactic acid. 21CFR184.1061
159. Hazards EPoB, EFSA Panel on Food Contact Materials E, Flavourings and Processing Aids (2011) Scientific opinion on the evaluation of the safety and efficacy of lactic acid for the removal of microbial surface contamination of beef carcasses, cuts and trimmings. In: Opinion of the scientific committee/scientific panel. European Food Safety Authority (EFSA), Parma, p 33
160. Scientific Committee on Food (1991) First series of food additives of various technological functions. Communities CotE, Luxembourg. CD-NA13416-EN-C
161. FAO (1974) Toxicological evaluation of some food additives including anticaking agents, antimicrobials, antioxidants, emulsifiers and thickening agents. FAO nutrition meetings report series, p 1–520
162. Gupta B, Revagade N, Hilborn J (2007) Poly(lactic acid) fiber: an overview. Prog Polym Sci 32:455–482. doi:10.1016/j.progpolymsci.2007.01.005
163. Hofvendahl K, Hahn-Hägerdal B (2000) Factors affecting the fermentative lactic acid production from renewable resources. Enzyme Microb Technol 26:87–107. doi: 10.1016/S0141-0229(99)00155-6
164. Axelsson L (2004) Lactic acid bacteria: classification and physiology. In: Salminen S, von Wright A, Ouwehand A (eds) Lactic acid bacteria: microbiological and functional aspects, 3rd edn. Marcel Dekker, New York, p 1–66
165. Moon SK, Wee YJ, Choi GW (2012) A novel lactic acid bacterium for the production of high purity l-lactic acid, Lactobacillus paracasei subsp. paracasei CHB2121. J Biosci Bioeng 114:155–159. doi:10.1016/j.jbiosc.2012.03.016

166. Meng Y, Xue Y, Yu B, Gao C, Ma Y (2012) Efficient production of l-lactic acid with high optical purity by alkaliphilic Bacillus sp. WL-S20. Bioresour Technol 116:334–339. doi: 10.1016/j.biortech.2012.03.103
167. Ye L, Zhou X, Hudari MSB, Li Z, Wu JC (2013) Highly efficient production of l-lactic acid from xylose by newly isolated Bacillus coagulans C106. Bioresour Technol 132:38–44. doi: 10.1016/j.biortech.2013.01.011
168. Osawa F, Fujii T, Nishida T, Tada N, Ohnishi T, Kobayashi O, Komeda T, Yoshida S (2009) Efficient production of L-lactic acid by Crabtree-negative yeast Candida boidinii. Yeast 26:485–496. doi:10.1002/yea.1702
169. Datta R, Henry M (2006) Lactic acid: recent advances in products, processes and technologies — a review. J Chem Technol Biotechnol 81:1119–1129. doi:10.1002/jctb.1486
170. Litchfield JH (1996) Microbiological production of lactic acid. Adv Appl Microbiol 42:45–95. doi:10.1016/S0065-2164(08)70372-1
171. John RP, Nampoothiri KM, Pandey A (2007) Fermentative production of lactic acid from biomass: an overview on process developments and future perspectives. Appl Microbiol Biotechnol 74:524–534. doi:10.1007/s00253-006-0779-6
172. Tay A, Yang ST (2002) Production of L(+)-lactic acid from glucose and starch by immobilized cells of Rhizopus oryzae in a rotating fibrous bed bioreactor. Biotechnol Bioeng 80:1–12. doi:10.1002/bit.10340
173. Xu G, Chu J, Wang Y, Zhuang Y, Zhang S, Peng H (2006) Development of a continuous cell-recycle fermentation system for production of lactic acid by Lactobacillus paracasei. Process Biochem 41:2458–2463. doi:10.1016/j.procbio.2006.05.022
174. Wee YJ, Ryu HW (2009) Lactic acid production by Lactobacillus sp. RKY2 in a cell-recycle continuous fermentation using lignocellulosic hydrolyzates as inexpensive raw materials. Bioresour Technol 100:4262–4270. doi:10.1016/j.biortech.2009.03.074
175. Joglekar HG, Rahman I, Babu S, Kulkarni BD, Joshi A (2006) Comparative assessment of downstream processing options for lactic acid. Sep Purif Technol 52:1–17. doi: 10.1016/j.seppur.2006.03.015
176. Pal P, Sikder J, Roy S, Giorno L (2009) Process intensification in lactic acid production: a review of membrane based processes. Chem Eng Process 48:1549–1559. doi: 10.1016/j.cep.2009.09.003
177. Jiang S, Zheng Z, Zhu Y, Wu X, Pan L, Luo S, Du W (2008) Repeated intermittent L-lactic acid fermentation technology by self-immobilized Rhizopus oryzae. Chinese J Biotechnol 24:1729–1733
178. Ning SY, Li SZ (2006) Primary study on fermentation of L(+)-lactic acid in fungal immobilized-bed bioreactor. Food Ferment Ind 32:22–25
179. Park EY, Kosakai Y, Okabe M (1998) Efficient production of l-(+) -lactic acid using mycelial cotton-like flocs of Rhizopusoryzae in an air-lift bioreactor. Biotechnol Prog 14:699–704. doi:10.1021/bp9800642
180. Hujanen M, Linko YY (1996) Effect of temperature and various nitrogen sources on L(+)-lactic acid production by Lactobacillus casei. Appl Microbiol Biotechnol 45:307–313. doi: 10.1007/s002530050688
181. Audet P, Paquin C, Lacroix C (1989) Sugar utilization and acid production by free and entrapped cells of streptococcus salivarius subsp. thermophilus, lactobacillus delbrueckii subsp. bulgaricus, and lactococcus lactis subsp. lactis in a whey permeate medium. Appl Environ Microbiol 55:185–189
182. Tiwari KP, Pandey A, Mishra N (1979) Lactic acid production from molasses by mixed population of lactobacilli. Zentralblatt fur Bakteriologie, Parasitenkunde, Infektionskrankheiten und Hygiene. Zweite naturwissenschaftliche Abteilung: Mikrobiologie der Landwirtschaft der Technologie und des Umweltschutzes 134:544–546
183. Porro D, Bianchi MM, Brambilla L, Menghini R, Bolzani D, Carrera V, Lievense J, Liu CL, Ranzi BM, Frontali L, Alberghina L (1999) Replacement of a metabolic pathway for large-scale production of lactic acid from engineered yeasts. Appl Environ Microbiol 65:4211–4215

184. Valli M, Sauer M, Branduardi P, Borth N, Porro D, Mattanovich D (2006) Improvement of lactic acid production in Saccharomyces cerevisiae by cell sorting for high intracellular pH. Appl Environ Microbiol 72:5492–5499. doi:10.1128/AEM.00683-06
185. Ilmen M, Koivuranta K, Ruohonen L, Suominen P, Penttila M (2007) Efficient production of L-lactic acid from xylose by Pichia stipitis. Appl Environ Microbiol 73:117–123. doi: 10.1128/AEM.01311-06
186. Michelson T, Kask K, Jõgi E, Talpsep E, Suitso I, Nurk A (2006) l(+)-Lactic acid producer bacillus coagulans SIM-7 DSM 14043 and its comparison with Lactobacillus delbrueckii ssp. lactis DSM 20073. Enzyme Microb Technol 39:861–867. doi:10.1016/j.enzmictec.2006.01.015
187. Stiles ME, Holzapfel WH (1997) Lactic acid bacteria of foods and their current taxonomy. Int J Food Microbiol 36:1–29. doi:10.1016/S0168-1605(96)01233-0
188. Chopin A (1993) Organization and regulation of genes for amino acid biosynthesis in lactic acid bacteria. FEMS Microbiol Rev 12:21–37. doi:10.1016/0168-6445(93)90056-F
189. Kwon S, Yoo IK, Lee WG, Chang HN, Chang YK (2001) High-rate continuous production of lactic acid by Lactobacillus rhamnosus in a two-stage membrane cell-recycle bioreactor. Biotechnol Bioeng 73:25–34
190. M G, Michiteru K, Rie G, Hirokazu T, Makoto H, Tadashi H (2005) Development of a continuous electrodialysis fermentation system for production of lactic acid by Lactobacillus rhamnosus. Process Biochem 40:1033–1036. doi:10.1016/j.procbio.2004.02.028
191. Cotton J, Pometto A, Gvozdenovic-Jeremic J (2001) Continuous lactic acid fermentation using a plastic composite support biofilm reactor. Appl Microbiol Biotechnol 57:626–630. doi:10.1007/s002530100820
192. Iyer PV, Lee YY (1999) Simultaneous saccharification and extractive fermentation of lignocellulosic materials into lactic acid in a two-zone fermentor-extractor system. In: Davison B, Finkelstein M (eds) Twentieth symposium on biotechnology for fuels and chemicals, Humana Press, p 409–419
193. Monteagudo JM, Aldavero M (1999) Production of L-lactic acid by Lactobacillus delbrueckii in chemostat culture using an ion exchange resins system. J Chem Technol Biotechnol 74:627–634. doi:10.1002/(sici)1097-4660(199907)74:7<627:aid-jctb84>3.0.co;2-k
194. Dey P, Pal P (2012) Direct production of l (+) lactic acid in a continuous and fully membrane-integrated hybrid reactor system under non-neutralizing conditions. J Membr Sci 389:355–362. doi:10.1016/j.memsci.2011.10.051
195. Vaidya AN, Pandey RA, Mudliar S, Kumar MS, Chakrabarti T, Devotta S (2005) Production and recovery of lactic acid for polylactide—an overview. Crit Rev Environ Sci Technol 35:429–467. doi:10.1080/10643380590966181
196. Wasewar KL (2005) Separation of lactic acid: recent advances. Chem Biochem Eng Q 19:159–172
197. Gao M, Shimamura T, Ishida N, Nagamori E, Takahashi H, Umemoto S, Omasa T, Ohtake H (2009) Extractive lactic acid fermentation with tri-n-decylamine as the extractant. Enzyme Microb Technol 44:350–354. doi:10.1016/j.enzmictec.2008.12.001
198. Wang Y, Huang C, Xu T (2011) Which is more competitive for production of organic acids, ion-exchange or electrodialysis with bipolar membranes? J Membr Sci 374:150–156. doi: 10.1016/j.memsci.2011.03.026
199. Strathmann H (2010) Electrodialysis, a mature technology with a multitude of new applications. Desalination 264:268–288. doi:10.1016/j.desal.2010.04.069
200. Huang C, Xu T, Zhang Y, Xue Y, Chen G (2007) Application of electrodialysis to the production of organic acids: state-of-the-art and recent developments. J Membr Sci 288:1–12. doi:10.1016/j.memsci.2006.11.026
201. Eurodia production of organic or amino acids by bipolar membrane electrodialysis. http://www.eurodia.com/html/ebc.html. Accessed 19 Jun 2013

202. Fernandes P, Prazeres DF, Cabral JS (2003) Membrane-Assisted Extractive Bioconversions. In: Stockar U, Wielen LAM (eds) Process integration in biochemical engineering. Springer, Berlin, pp 115–148
203. Sanchez Marcano JG, Tsotsis TT (2000) Membrane reactors. In: Ullmann's encyclopedia of industrial chemistry. Wiley-VCH Verlag GmbH & Co. KGaA, Weinheim
204. Melin T, Rautenbach R (2007) Membranverfahren grundlagen der modul- und anlagenauslegung. Springer, Berlin
205. Carstensen F, Apel A, Wessling M (2012) In situ product recovery: submerged membranes versus external loop membranes. J Membr Sci 394–395:1–36. doi:10.1016/j.memsci.2011.11.029
206. Sondhi R, Bhave R (2001) Role of backpulsing in fouling minimization in crossflow filtration with ceramic membranes. J Membr Sci 186:41–52. doi:10.1016/S0376-7388(00)00663-3
207. Heran M, Eimaleh S (2000) Cross-flow microfiltration with high frequency reverse flow. In: International conference on membrane technology in environmental management. International Water Association, Tokyo, p 337–343
208. Silalahi SH, Leiknes T (2011) High frequency back-pulsing for fouling development control in ceramic microfiltration for treatment of produced water. Desalin Water Treat 28:137–152. doi:10.5004/dwt.2011.2482
209. Lu Z, Wei M, Yu L (2012) Enhancement of pilot scale production of l(+)-lactic acid by fermentation coupled with separation using membrane bioreactor. Process Biochem 47:410–415. doi:10.1016/j.procbio.2011.11.022
210. Singh N, Cheryan M (1998) Membrane technology in corn refining and bioproduct-processing. Starch/Staerke 50:16–23. doi:10.1002/(sici)1521-379x(199801)50:1<16:aid-star16>3.0.co;2-d

Adv Biochem Eng Biotechnol (2014) 143: 143–188
DOI: 10.1007/10_2013_243

Published Online: 21 November 2013

Industrial Production of L-Ascorbic Acid (Vitamin C) and D-Isoascorbic Acid

Günter Pappenberger and Hans-Peter Hohmann

Abstract L-ascorbic acid (vitamin C) was first isolated in 1928 and subsequently identified as the long-sought antiscorbutic factor. Industrially produced L-ascorbic acid is widely used in the feed, food, and pharmaceutical sector as nutritional supplement and preservative, making use of its antioxidative properties. Until recently, the Reichstein–Grüssner process, designed in 1933, was the main industrial route. Here, D-sorbitol is converted to L-ascorbic acid via 2-keto-L-gulonic acid (2KGA) as key intermediate, using a bio-oxidation with *Gluconobacter oxydans* and several chemical steps. Today, industrial production processes use additional bio-oxidation steps with *Ketogulonicigenium vulgare* as biocatalyst to convert D-sorbitol to the intermediate 2KGA without chemical steps. The enzymes involved are characterized by a broad substrate range, but remarkable regiospecificity. This puzzling specificity pattern can be understood from the preferences of these enyzmes for certain of the many isomeric structures which the carbohydrate substrates adopt in aqueous solution. Recently, novel enzymes were identified that generate L-ascorbic acid directly via oxidation of L-sorbosone, an intermediate of the bio-oxidation of D-sorbitol to 2KGA. This opens the possibility for a direct route from D-sorbitol to L-ascorbic acid, obviating the need for chemical rearrangement of 2KGA. Similar concepts for industrial processes apply for the production of D-isoascorbic acid, the C5 epimer of L-ascorbic acid. D-isoascorbic acid has the same conformation at C5 as D-glucose and can be derived more directly than L-ascorbic acid from this common carbohydrate feed stock.

Keywords Vitamin C · Ascorbic acid · Isoascorbic acid · L-Sorbosone · Microbial oxidation · *Gluconobacter oxydans* · *Ketogulonicigenium vulgare*

Contents

G. Pappenberger · H.-P. Hohmann (✉)
Biotechnology R&D, DSM Nutritional Products, P.O. Box 32554002,
Basel, Switzerland
e-mail: hans-peter.hohmann@dsm.com

1 Introduction and Scope

The year 2012 saw the celebration of the 100th anniversary of the vitamin concept. The Polish biochemist Casimir Funk coined the term "vitamin" in 1912 to refer to micronutrients which are indispensable components of the human diet [1]. Funk isolated the first of these components, a nitrogen-containing small organic molecule later designated thiamine or vitamin B1. Thiamine was identified as the missing factor in polished white rice causing the nutritional deficiency syndrome beriberi. Based on the assumption that all essential microcomponents of the human diet such as thiamine would contain amino groups, Funk coupled the chemical term amine, that is, a nitrogen-containing organic compound, with the Latin word vita to indicate that these compounds are amines essential for life. However, the general relevance of amino groups for vitamins was soon put into question and indeed when the micronutrient responsible for the prevention of scurvy was identified in 1927–1932, it turned out to be a sugar acid lacking any amine group. This compound was originally designated as hexuronic acid and is today known as L-ascorbic acid or vitamin C. Soon after its discovery a demand for pure vitamin C began to be seen, which triggered the development of industrial production processes in the early 1930s. It is the aim of the current review to trace the evolution of these processes from their early beginnings to the current status delivering more than 100,000 tons pure vitamin C per year. Right from the start, the key step in the industrial production of the vitamin was a microbial oxidation converting D-sorbitol to L-sorbose. This is followed by two further oxidation reactions, chemical or biotechnological. Until today, the immediate product of these oxidations is 2-keto-L-gulonic acid, which is isolated, purified, and chemically rearranged to vitamin C. Our review will point out the increasing importance that biotechnology attained during the development of industrial vitamin C production and highlights the prospect of converting D-sorbitol directly to vitamin C purely by biotechnological means, avoiding 2-keto-L-gulonic acid as an intermediate.

Some of the major developments of processes for industrial vitamin C production have been achieved in China, where currently most of this vitamin is produced. The present review gives an overview of the key developments and the major research trends being pursued in China, but the selection of publications from China presented here is far from being exhaustive. A broader coverage of the Chinese literature on vitamin C production can be found in a recent review by Zhou et al. [2].

This review aims to provide a comprehensive overview of the relevant dehydrogenases involved in the oxidation reactions from D-sorbitol to 2-keto-L-gulonate or vitamin C. Most of these dehydrogenases are characterized by a broad specificity for their substrates, for which they have a relatively low affinity. The regiospecificity of these enzymes, however, is quite remarkable. It is our aim not just to delineate the reaction sequences catalyzed by these enzymes, but to emphasize how the observed preferences for certain substrates (or certain isomeric structures of the substrates) are in line with and can explain the selectivity of these enzymes toward oxidation of specific hydroxyl groups of their carbohydrate substrates. In doing so, the coherent basic principles of the sometimes puzzling substrate and product spectrum of these enzymes should become visible.

Many routes can be conceived that convert a carbohydrate feed-stock to vitamin C, but only those involving L-sorbose as intermediate achieved commercial application, this is to say demonstrated their superior technical and economic efficiency. Another route to 2-keto-L-gulonic acid, starting from D-glucose with 2,5-diketo-D-gluconate as central intermediate, attracted much attention in the 1980s, but processes based on this route were not used on industrial scale. Microbial routes directly affording vitamin C, based on the natural biosynthetic pathways, were proposed or attempted numerous times. These routes build on the conversion of a common carbohydrate feed-stock, e.g., D-glucose, to L-gulose or L-galactose, which then serve as substrates for suitable oxidases for the oxidation to vitamin C. None of these concepts, however, advanced beyond a conceptual or feasibility phase. These conceivable but not implemented vitamin C production routes via 2,5-diketo-D-gluconate, L-gulose, or L-galactose are not covered in the present review. The interested reader is referred to previous reviews from Bremus et al. [3] and Hancock [4].

This review also features a chapter on D-isoascorbic acid, the C5 epimer of L-ascorbic acid. This molecule has antioxidative characteristics similar to L-ascorbic acid, but lacks most or all of the anti-scorbutic vitamin C activity. Having the same conformation at the C5 stereocenter as D-glucose, it can be more directly derived than L-ascorbic acid from this common carbohydrate feed stock. It is, therefore, interesting to compare the similarities and differences of biosynthesis and industrial production processes between L-ascorbic acid and D-isoascorbic acid.

2 Relevance of Vitamin C for Humans and Guinea Pigs

The great naval explorers of the sixteenth to eighteenth century pursued their epic voyages facing many perils, such as severe weather, uncharted waters, aggressive indigenous peoples, hostile ships from belligerent European neighbors, and wavering support by the ship's crew. They would have been acutely aware of all these based on centuries of naval experience. The most deadly of all perils, however, took the first naval explorers unprepared. It only emerged after many weeks of continuous voyaging that was required to span the Indian and Pacific oceans. Scurvy claimed the lives of more than half of the crew of Vasco da Gama's first exploration of the Indian Ocean in 1499 and three-quarters of Ferdinand Magellan's crew during the first circumnavigation of the planet in 1519–1522. Cures for scurvy by certain fresh fruits were reported early on and from today's perspective, the link to the restricted diet aboard these ships would seem obvious. Still, even more than two centuries after Magellan, no reliable prevention or treatment was established, either because of wrong beliefs or unavailability. In 1741, Vitus Bering, the Danish explorer in the service of Russia, died of scurvy after his mission had been stranded on an island near the Kamchatkan peninsula. He spent his last days half buried in the soil, possibly due to the widespread superstition that contact with solid ground would be the surest cure against this sea-borne disease. His crew member, the naturalist Georg Wilhelm Steller, was well aware of the curative power of certain fresh herbs, but could not make sufficient amounts available on this barren island. In 1747, the Scottish naval surgeon James Lind subjected groups of scurvy-stricken sailors on a ship with different substances (e.g., cider, seawater, vinegar, and citrus fruits) that were considered to be curative. This effort to create irrefutable evidence is now perceived as the first-ever clinical trial [5]. The results clearly showed the unique benefit of citrus fruits in treating scurvy, but still it took until the end of that century until the British navy began to distribute regular rations of lime juice during long sea voyages.

With the growing understanding of the biochemical basics of physiology and the shaping of the vitamin concept by Christiaan Eijkman, Frederick Gowland Hopkins, and Casimir Funk around 1900 [6, 7], the search began for the molecular component called "Vitamin C," responsible for the scurvy-preventing property of fresh fruits and vegetables. In 1927, Albert Szent-Györgyi, a Hungarian scientist with a focus on energy metabolism and respiration, isolated a compound (originally called "hexuronic acid") from adrenal glands, orange juice, and later red pepper, which he found to prevent the oxidative browning of plant tissues. An initial presumption that hexuronic acid could be related to vitamin C was refuted by tests in a different laboratory. In 1931, Joseph Svirbley joined Szent-Györgyi, bringing with him the expertise in guinea pig trials (which, curiously, also develop scurvy when put on a restricted diet). He previously worked with Charles Glen King on childhood scurvy and with this interest he retested hexuronic acid, unequivocally demonstrating its anti-scurvy effect. At the same time, King independently purified and tested hexuronic acid from lemon juice, reaching the same conclusion. Both findings were published at the same time [8, 9], but dispute on the credits for the discovery remained [10]. The chemical structure of L-ascorbic acid (as hexuronic acid was renamed after discovery of its physiological role) was determined by Walter Norman Harworth and Edmund Hirst in 1933 [11]. They also achieved the chemical synthesis via L-xylosone, starting from D-galacturonic acid derived from citrus pectic acid [12]. At the same time, Tadeus Reichstein and Anton Grüssner had developed a much simpler procedure starting from readily available D-glucose [13], which was subsequently implemented for commercial production.

From a chemical perspective (Fig. 1), L-ascorbic acid (hereinafter designated as Asc) is the 1,4-lactone of 2-keto-L-gulonic acid (hereinafter designated as 2KGA). The enol isomer is highly preferred over the keto-isomer due to the possibility to form hydrogen bonds in the five-membered ring of the former.

The chemical structure of Asc acid still attests to its derivation from hexose carbohydrates, but its characteristic conjugated enediol-carbonyl results in very specific chemistry. The stabilization provided by this conjugated system results in deprotonation of the enolic hydroxy group at C3 at a pK_A comparable to carboxylic acids (pK_A = 4.25). The conjugated system also allows Asc to readily donate one or two electrons and thus act as reducing agent. This reductive property is also key to its biological role [14]. It is implied as the major antioxidant in physiological redox homeostasis, particularly at the mitochondria, where cellular respiration leads to the formation of reactive oxygen species. In addition, it has a specific role as redox cofactor in few but critical enzymes, primarily mono- and

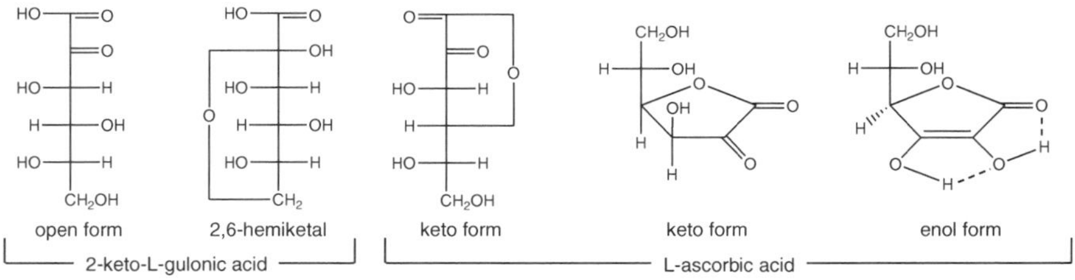

Fig. 1 Chemical structures of L-ascorbic acid (Asc) and 2-keto-L-gulonic acid (2KGA), the immediate precursor in industrial Asc synthesis. Structures are shown as Fischer projections, for Asc also as the Haworth projection. Asc and 2KGA have the same oxidation state, but differ in ring connectivity; 2KGA in aqueous solution occurs predominantly as the 2,6 hemiketal and Asc can be considered as the 1,4 lactone of 2KGA. Asc preferentially occurs as the enol, stabilized by intramolecular hydrogen bonds

di-oxygenase involved in hydroxylations to generate, e.g., carnitine, norepinephrine, or 4-hydroxy-proline. Hydroxylation of proline and lysine side chains in certain transcription factors may have an impact on the regulation of cellular responses. 4-Hydroxy-proline is an essential constituent of collagen, responsible for stabilizing interactions in connective tissues. Some of the outstanding symptoms of scurvy such as subcutaneous bleedings and loss of teeth can be directly attributed to weakening of connective tissues caused by lack of proline hydroxylation.

3 Biosynthesis of Asc in Plants and Mammals: Two Ways to Achieve the L-Configuration at C5

Monosaccharides are classified as "D-" or "L-" stereoisomers, depending on the configuration of the asymmetric carbon furthest away from the carbonyl group (C5 in glucose). Most hexoses commonly found in nature have the D-configuration at C5, while Asc has the eponymous L-configuration. Starting from D-glucose the L-configuration at C5 is accomplished in nature by two rather different routes [15].

In animals, the carbon skeleton of D-glucose is inverted, equivalent to turning the molecule in the Fischer projection upside down. C2 of D-glucose with its D-configuration (hydroxy group facing to the right) is thereby reassigned as C5 with L-configuration (hydroxy group facing to the left). This inversion became evident early on due to the incorporation pattern of isotopically labeled glucose [16]. Biochemically, it is achieved by first oxidizing the hydroxy function at C6 of the UDP activated form of D-glucose to give UDP-D-glucuronic acid (Fig. 2). After UDP removal by hydrolysis, the aldehyde function at C1 is reduced and with this the sugar molecular is reclassified as L-gulonic acid, the former C1 becoming C6 and the carboxylic acid functionality becoming the new C1. The 1,4 lactone of L-gulonic acid already possessing Asc-like ring-connectivity is afforded by the catalytic activity of a specific lactonase encoded by the SMP30 gene [17].

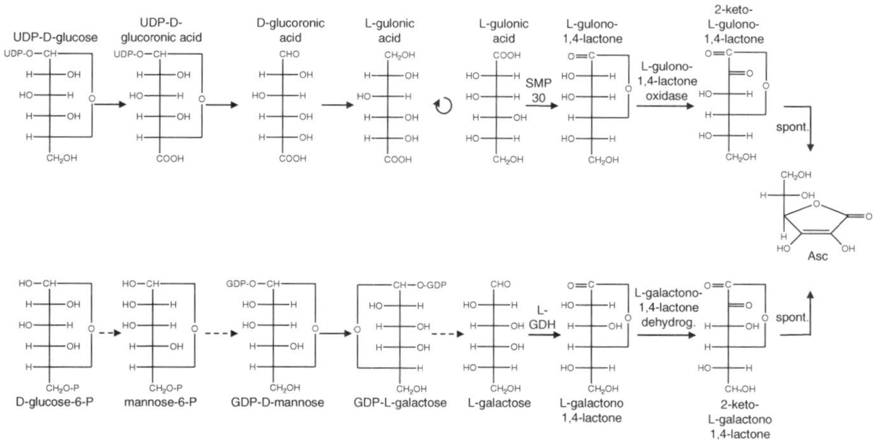

Fig. 2 Asc biosynthesis pathways in animals (*top*) and plants (*bottom*). The chemical structures of key intermediates are shown as Fischer projections. Both biosynthetic routes are highly different: In animals, the L-configuration of Asc is the result of the inversion of the carbon skeleton, while in plants it results from epimerization at C5. Only the ultimate step of oxidation of the 1,4 lactone precursor to Asc is similar in both plant and animal pathways

Oxidation of the lactone at C2 by L-gulono-1,4-lactone oxidase yields 2-keto-L-gulono-1,4-lactone, which enolizes to Asc [18].

In plants, labeling experiments had shown that the carbon skeleton is not inverted and C1 of glucose will stay as C1 of Asc [19]. It was further known that L-galactono-1,4-lactone, the C3-epimer of the mammalian Asc-precursor, is readily converted to Asc in plant tissue [20]. It was, however, not until the late 1990s when the pathway leading from D-glucose to L-galactono-1,4-lactone could be identified [21]. It consists of two isomerization reactions converting D-glucose to D-mannose (linked to GDP) followed by two epimerization reactions at C3 and C5 to yield L-galactose which is eventually oxidized to L-galactonic acid (Fig. 2). Only the final reaction step from the 1,4 lactone to Asc is similar in the animal and plant biosynthesis routes. Further biosynthetic routes toward Asc exist in plants, but probably have minor relevance only [22].

Ascorbic acid biosynthesis is a property of animals and plants, but not found to be naturally occurring in prokaryotes. Fungi also do not make Asc, but some have a biosynthetic route toward the pentose analog D-erythroascorbic acid [23] or the C5-epimer of Asc, D-isoascorbic acid. The biosynthesis of Asc or Asc-like molecules has apparently been invented four times by nature, underlying its importance in eukaryotes. On the other hand, biosynthetic capabilities have frequently been lost independently, not only in primates and guinea pigs, but also in bats, certain birds, reptiles, fishes, and insects, prompting speculations if losing this biosynthesis capability could offer benefits, provided there is sufficient supply in the diet [14].

4 The Vitamin C Market

Among the 13 vitamins, Asc has the largest industrial production volume. Roughly 110 kilotons vitamin C per year are industrially produced, which is primarily applied in the pharmaceutical industry (50 %), followed by the use as antioxidant in the food (25 %) and beverages sector (15 %). Only about 10 % of the vitamin is used for animal feed applications. This is in contrast to all other vitamins, where the feed sector is the major application area of vitamins. More than 80 % of today's world demand for vitamin C is satisfied by Chinese producers. In the late 1950s, Chinese production capacity was only 30 tons per annum. Supply was controlled by big European and Japanese manufacturers such as Roche, BASF, Merck, and Takeda Pharmaceutical. By the early 1990s, 26 Chinese manufacturers had already gained one-third of the world's vitamin C market. Recognizing the threat, the established producers tried to prevent Chinese companies from further entering the market by several rounds of severe price cuts. By 2002, only four Chinese manufacturers had survived and the price had hit record lows that year. After a short recovery in 2003 and the first half of 2004, vitamin C prices kept at a very low level of under 4 USD/kg for years. This resulted in most of the western companies abandoning vitamin C production. In 2008, the price of vitamin C rose sharply above 10 USD/kg partly as a result of increases in basic food prices but also in anticipation of a stoppage of two Chinese plants, situated at Shijiazhuang near Beijing, as part of a general shutdown of polluting industry in China over the period of the Olympic Games. The high-price period lasted until the middle of 2010 when Shandong Tianli Pharmaceutical Company, a major sorbitol producer in China, announced plans to enter the vitamin C market.

Since 2002, the production and export activities of Chinese vitamin C manufacturers had been under the supervision of the trade association CCCMHPIE (China Chamber of Commerce for Import and Export of Medicines and Health Products), which is considered to be close to the Ministry of Commerce of the central Chinese Government. The close interaction of the Chinese vitamin C producers coordinated by CCMHPIE sparked the suspicion of price-fixing. In 2005, two American companies in the feed and food sectors filed a complaint accusing the Chinese companies of conspiring to inflate prices for bulk vitamin C. The vitamin companies did not deny the accusation, but argued that the Chinese Government forced them to fix prices. This contention was rejected in a recent ruling of an American court. The case is still pending (status end 2012).

Today's leading Chinese producers are: (i) Northeast Pharmaceutical Group Co., Ltd (NEPG) of Shenyang, Liaoning Province, (ii) Weisheng Pharmaceutical Company (CSPC) of Shijiazhuang, Hebei Province, (iii) Welcome Pharmaceutical Company (NCPC), Shijiazhuang, Hubei Province, and (iv) Jiangsu Jiangshan Pharmaceutical (new name is Aland Nutraceutical Co., Ltd.) of Jingjiang, Jiangsu Province.

After the fierce price battles during the last 20 years, DSM Nutritional Products of Switzerland (formerly Roche Vitamins) operating a production plant in Dalry,

UK, remained as the sole Western vitamin C producer. With the Quali®-C brand for its vitamin C products, the company positioned itself in the premium segment of the market, which is less affected by the high price-volatility of bulk vitamin C.

5 The Reichstein Process: The Major Industrial Asc Process Until the Late 1990s

In the early 1930s, researchers began to consider commercial production processes to satisfy an emerging market demand for Asc. The German company Merck was first on the market with an Asc product called Cebion obtained from plant material. The product is still available today. From the various efforts to obtain synthetic Asc from a readily available feed stock, the process conceived by Reichstein and Grüssner prevailed as the industrial standard for Asc production until the late 1990s.

5.1 From D-Sorbitol to L-Sorbose: Inversion of the Carbon Skeleton

Realizing the relationship between 2KGA and Asc—the enolized 1,4 lactone of 2-KGA—Reichstein and Grüssner presumed that 2-keto sugar acids subjected to reaction conditions that are conductive for lactonization should convert to their 1,4 lactones and enolize [13]. In a preliminary trial, upon heating the readily available 2-keto-D-galactonic acid under acidic conditions they obtained a reductive product that was tentatively denoted as D-ascorbic acid. Taking this as a preliminary proof of their hypothesis, Reichstein and Grüssner then proceeded to synthesize 2KGA that should rearrange to Asc. As starting material for 2KGA synthesis (Fig. 3) they selected L-sorbose, already containing the critical L-configuration at C5. Thirty years earlier Bertrand had published a series of reports [24–26], where he demonstrated the conversion of D-sorbitol to L-sorbose making use of a microorganism designated as "la bacterie du sorbose," now designated as *Acetobacter aceti* subsp. *xylinum* [27]. For their initial experiments, Reichstein and Grüssner did not use the isolate of Bertrand, but mother of vinegar, a biofilm of *Acetobacteraceae* formed during microbial oxidation of ethanol to acetic acid. A yield of 60 % L-sorbose yield was obtained "mit einem wild eingefangenen Stamm" (with an undomesticated strain). D-sorbitol was readily available by catalytic hydrogenation of D-glucose.

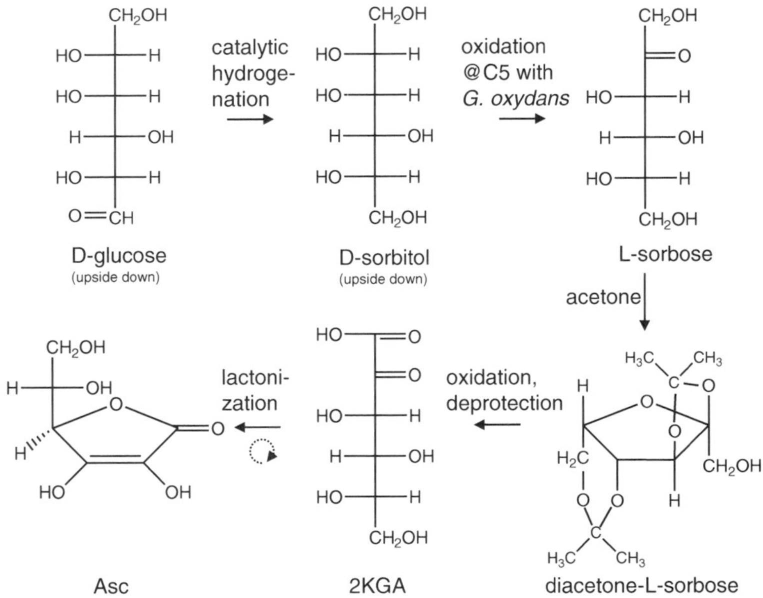

Fig. 3 Conceptual outline of the chemical Asc synthesis according to Reichstein and Grüssner. The chemical structures of key intermediates are shown as Fischer projections. The L-configuration of Asc is derived from D-glucose by inversion of the carbon skeleton. For this, D-sorbitol is regioselectively oxidized at C5 using *Gluconobacter oxydans*. Further chemical oxidation of the adjacent primary alcohol to the acid is done after protection of the remaining hydroxyl groups with acetone. The resulting 2KGA is then rearranged to yield Asc

5.2 Oxidation of L-Sorbose to 2KGA and Rearrangement to Asc

Oxidation of L-sorbose to 2KGA was achieved by potassium permanganate treatment after protection of the hydroxy groups at C2 and C3 and C4 and C6 as acetone-ketals. After this oxidation of the primary hydroxy group at C1 to the carboxy group, the resulting diacetone-2KGA was hydrolyzed to the free acid, which was isolated and crystallized as the 2,6-hemiketal isomer. Upon heating under acidic conditions, the compound rearranged into the 1,4-lactone, which then enolized to Asc. The rearrangement proceeds with higher yields, if 2KGA is esterified with methanol to afford methyl-2KGA, which, after treatment with sodium methoxide in methanol followed by acidification, produces Asc. The overall yield of the early process variants reached 15–18 %. In succeeding years, the synthesis was optimized so that each individual step could be carried out in greater than 90 % yield providing Asc in greater than 50 % overall yield from glucose. As a result of these process improvements, the Reichstein/Grüssner synthesis became the industrial standard for the production of Asc until the 1990s [28]. The replacement of permanganate or bleach (NaOCl) oxidation by palladium-catalyzed air oxidation is possible and should result in a more environmental friendly process variant [29, 30]. The palladium-catalyzed air oxidation even showed some preference for the C1 hydroxy group of L-sorbose [31], but the selectivity was not sufficient to allow the oxidation to proceed without the need to protect the other hydroxy groups in the molecule.

5.3 *Polyol Oxidation by* Gluconobacter

For a cost-efficient synthesis of Asc, which is basically a hexose-derived acid with L-configuration at C5, it is most critical to be backward-integrated to the D-sugars commonly found in nature. The Reichstein/Grüssner synthesis accomplishes this task by the "inversion of the carbon skeleton" of D-glucose such that its C2 carbon becomes C5 of L-sorbose by an oxidation step exclusively at C5 of the intermediate D-sorbitol. The exquisite regioselectivity required for this reaction was achieved by acetic acid bacteria serving as biocatalyst for this reaction. Acetic acid bacteria are obligate aerobes, which are well known as vinegar producers (ethanol oxidation) and also for their ability to only partially oxidize various sugars and sugar alcohols such as D-glucose, glycerol, and D-sorbitol [32]. Of the various genera of acetic acid bacteria *Gluconobacter* (*Acetobacter suboxydans* is an older designation) is particularly effective in partially oxidizing sugars and sugar alcohols. Following the empirical Bertrand-Hudson rule "polyols with a cis-arrangement of two secondary hydroxy groups in D-configurations to the adjacent primary alcohol group are oxidized to the corresponding ketones" [33] the secondary alcohol at C5, but not at C2, of D-sorbitol is oxidized to the carbonyl group affording L-sorbose with an almost 100 % yield.

5.4 *Sorbitol Dehydrogenase: The Key Enzyme of* Gluconobacter *for L-Sorbose Production*

From *Gluconobacter oxydans* IFO3255, an 80-kDa dehydrogenase depending on the redox cofactor pyrroloquinoline quinone (PQQ) was purified and designated as sorbitol dehydrogenase (Sldh) [34]. Sldh oxidizes D-sorbitol in accordance with the Bertrand-Hudson rule exclusively at the C5 hydroxy group, affording L-sorbose (Fig. 4). D-mannitol with twofold rotational symmetry is oxidized by Sldh at the C2 and C5 hydroxy groups, both reactions leading to D-fructose formation. 5-Keto-fructose could be expected, but has not been reported. Sldh-catalyzed glycerol and D-gluconic acid oxidation afford dihydroxyacetone and 5-keto-D-gluconic acid, respectively [34–36]. On the basis of peptide sequences obtained from purified Sldh, the corresponding gene, designated as *sldA*, was cloned by reverse genetics and sequenced [37]. *sldA* encodes a 740-residues polypeptide comprising a 24-residues signal sequence. The gene shares significant sequence homology to the membrane-

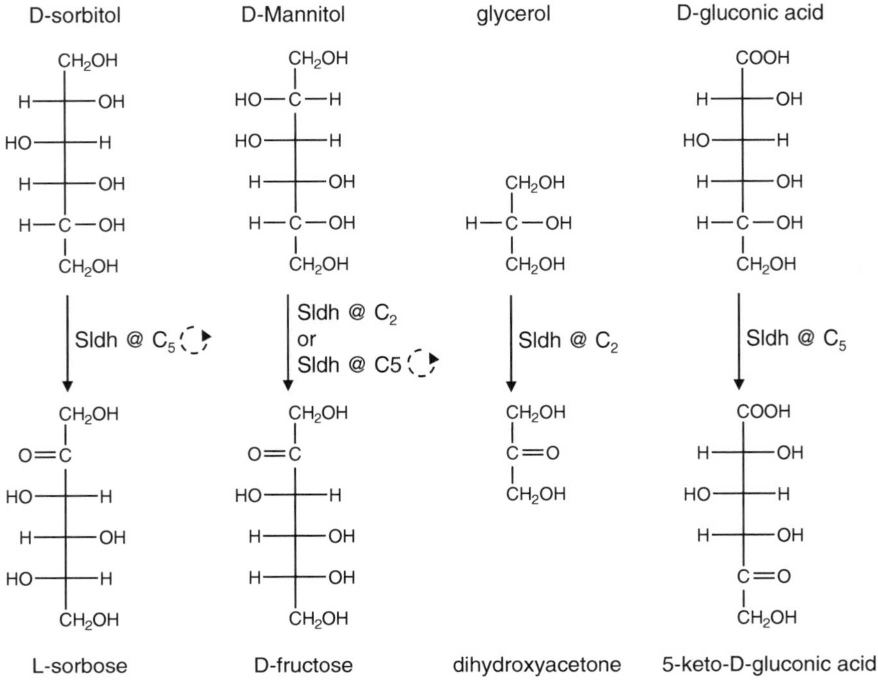

Fig. 4 Regioselectivity of polyol/sugar oxidation by PQQ-linked sorbitol dehydrogenase from *Gluconobacter oxydans* (Sldh). The chemical structures of substrates and products are shown as Fischer projections. The regioselectivity in Sldh-catalyzed oxidation is described by the Bertrand-Hudson rule (see text)

bound quinoprotein glucose dehydrogenases from *E. coli*, *G. oxydans*, and *Acinetobacter calcoaceticus*. Immediately upstream of *sldA,* the *sldB* gene was localized encoding a polypeptide of 126 hydrophobic amino acids predicted to form four transmembrane helices. *sldA* could be functionally expressed in *E. coli* provided that *sldB* was co-expressed and the cultivation medium was supplemented with PQQ (*E. coli* is lacking the intrinsic capability to synthesize PQQ). A *sldA* deletion mutant in *G. oxydans* IFO3255 could neither convert D-sorbitol to L-sorbose, nor could the mutant oxidize the other polyols to the corresponding ketones [38, 39]. SldB is required for Sldh activity in its natural host probably acting as a chaperone and membrane anchor [38, 39].

Prior to Sldh, other D-sorbitol oxidizing enzymes from other *Gluconobacter strains* were purified and characterized, among them an enzyme consisting of a subunit covalently bound to FAD and a cytochrome c subunit [40]. Furthermore, a PQQ-dependent D-sorbitol dehydrogenase was isolated from *G. oxydans* ATCC 621 consisting of 75, 50, and 14-kDa subunits. The 50-kDa subunit turned out to be a cytochrome [41]. However, Sldh is now generally believed to be the major polyol dehydrogenase of *G. oxydans* [35]. It is the key enzyme of the Reichstein process affording in a highly effective manner the L-configuration at C5 of the Asc molecule.

6 Early Attempts to Produce 2KGA by Microbial Oxidation of D-Sorbitol

In 1968 Isono et al. from Takeda Chemical Industries Ltd., Osaka, Japan published an extended study on microbial oxidation of D-sorbitol and L-sorbose involving 12 genera of bacteria with about 500 strains [42]. Microbial oxidative conversion of the two substrates was found to be quite abundant. Among the tested strains *Gluconobacter melanogenus*, a subspecies of *G. oxydans* was the most proficient 2KGA producer.

6.1 Elucidation of the Sorbosone Pathway

The conversion of L-sorbose to 2KGA comprises two oxidation steps at C1 and a sequential progression of such oxidation steps with L-sorbosone as intermediate would seem most plausible. L-sorbosone, however, was not identified by the Takeda scientists among the sorbitol oxidation products of *G. melanogenus*. Instead, L-idonic acid was detected [43, 44]. Based on the identified oxidation products, the authors suggested a metabolic pathway from D-sorbitol to 2KGA via L-sorbose, L-idose, and L-idonic acid (Fig. 5). In this reaction scheme, L-sorbose is converted to L-idose by an isomerization reaction followed by two oxidation reactions at C1 and C2 leading to L-idonic acid and 2KGA. For L-sorbose oxidation to 2KGA by *Pseudomonas*, a similar reaction scheme was proposed [45]. In contrast to *G. melanogenus*, *P. aeruginosa* could not start the oxidative path from D-sorbitol.

In a series of publications from 1972, Perlman and colleagues showed that (1) growing and resting cultures and cell-free preparations of *G. melanogenus* IFO 3293 converted L-sorbose to 2KGA, but also to D-sorbitol, D-fructose, and 5-keto-D-fructose [46]; (2) the 2KGA-forming activity, which was localized in the particulate fraction, was inducible, as only extracts of *G. melanogenus* cells cultivated with L-sorbose but not with glycerol could produce 2KGA [47]; and (3) 2KGA formation did not proceed via L-idose and L-idonic acid as pathway intermediates, since among other evidences no isotopic dilution of the formed ^{14}C 2KGA was observed if the reaction mix containing ^{14}C-labeled L-sorbose as substrate was spiked with unlabeled L-idose [48]. The authors postulated a direct oxidation of L-sorbose to 2KGA in *G. melanogenus* without accumulation of the L-sorbosone intermediate (Fig. 5).

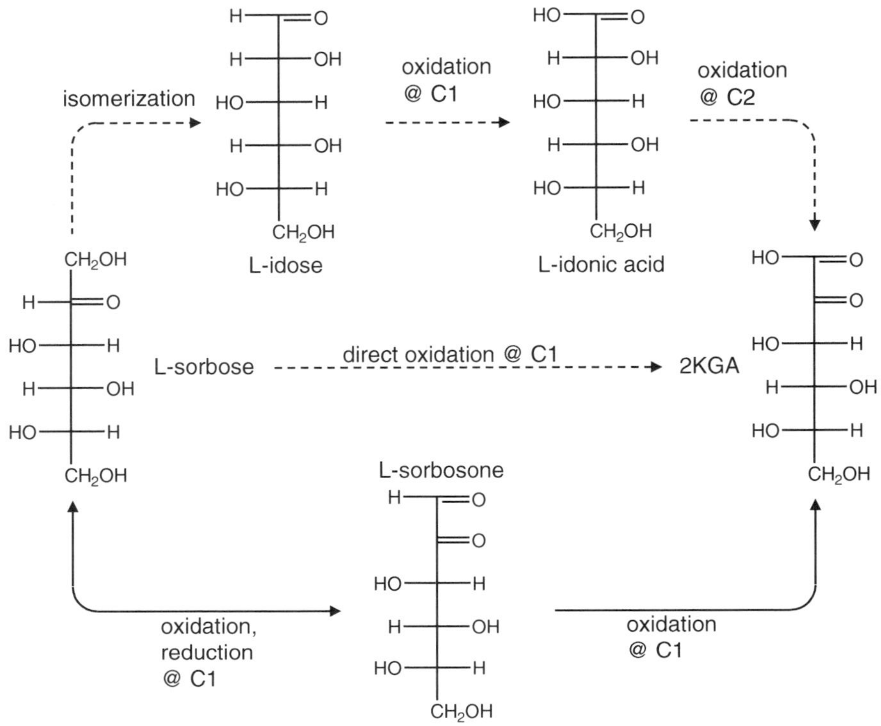

Fig. 5 Proposed routes for the oxidation of L-sorbose to 2KGA in *Gluconobacter oxydans*. The chemical structures are shown as Fischer projections. 2KGA formation via L-sorbosone as intermediate (bottom route) was found to be in best agreement with experimental evidence

The intuitively most obvious pathway to 2KGA, a two-step oxidation of the primary alcohol via L-sorbosone as intermediate (Fig. 5) to the carboxyl group escaped attention until 1975, probably due to the lack of L-sorbosone required as reference standard. Once this key substance became available, a research team at Roche Inc., Nutley N.J., identified L-sorbosone as one of the oxidation products generated from L-sorbose, which was treated with a *Pseudomonas putida* ATCC 21812 cell-free extracts or a soluble fraction thereof [49]. Strong isotopic dilution of the formed ^{14}C-labeled 2KGA was observed if the reaction mix with ^{14}C-labeled L-sorbose as substrate was spiked with unlabeled L-sorbosone providing evidence for L-sorbosone being a pathway intermediate. Based on experimental results on related 2-keto aldoses [50], the primary carbonyl function of L-sorbosone should be hydrated to a geminal diol, thus providing an alcohol functionality for further oxidation to 2KGA. The rate-limiting reaction in the sorbosone pathway was the oxidation of L-sorbose to L-sorbosone. *P. putida* and *G. melanogenus* contained an enzyme involved in the further metabolism of 2KGA to L-idonic acid. This enzyme, referred to as 2KGA reductase, was found in the soluble fraction of cell-free extracts and was dependent on NADH or NADPH. Thus, L-idonic acid is not a

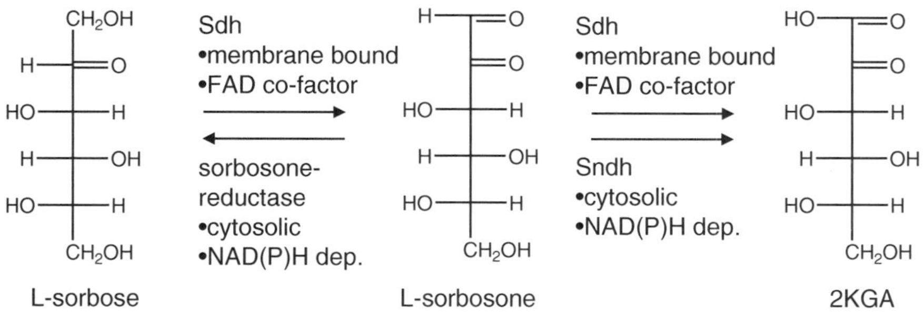

Fig. 6 Biochemistry of the oxidation of L-sorbose to 2KGA in *Gluconobacter oxydans*. The chemical structures are shown as Fischer projections. Involved enzymes and cofactors are given at the reaction arrows. FAD-linked sorbose dehydrogenase (Sdh) is responsible for L-sorbose to L-sorbosone oxidation. NAD(P)-linked sorbosone dehydrogenase (Sndh) is the key enzyme for L-sorbosone to 2KGA oxidation, but Sdh was also found to have activity for this reaction

precursor of 2KGA generated by L-idose oxidation as suggested by the Takeda researchers, but the product of 2KGA reduction catalyzed by 2KGA-reductase.

The sorbosone pathway and the reversible reduction of 2KGA to L-idonic acid was confirmed by Perlman's group [51] and by researchers from the Nippon Roche Research Center in Kamakura Japan [52]. It turned out that in *G. melanogenus* the L-sorbose oxidizing activity, i.e., L-sorbose dehydrogenase Sdh, was mainly recovered in the particulate fraction and the L-sorbosone to 2KGA converting activity, i.e., sorbosone dehydrogenase Sndh, requiring reduced nucleotides as co-substrates, was in the cytosolic fraction. An L-sorbosone to L-sorbose reducing activity was also identified in this fraction.

6.2 Purification and Characterization of Sorbose Dehydrogenase Sdh and Sorbosone Dehydrogenase Sndh

Purification of *G. melanogenus* membrane-bound L-sorbose dehydrogenase (Sdh) to near homogeneity was accomplished by the Roche researchers in Japan [53] after Triton X-100 solubilization of the enzyme from a membrane fraction of UV10 strain, a *G. melanogenus* IFO3293 mutant selected for increased 2KGA productivity [54]. L-idonic acid dehydrogenase (2KGA reductase) also present in the UV10 membrane fraction had to be removed first by pre-extraction with Tween 80, since this enzyme showed a very similar chromatographic behavior to L-sorbose dehydrogenase.

Sdh with an apparent molecular mass of about 58 kDa accepted both 2,6-dichlorophenolindophenol or phenazine methosulfate as artificial electron acceptors. The enzyme showed high substrate specificity for L-sorbose. L-sorbosone as product of the L-sorbose dehydrogenase-catalyzed oxidation of L-sorbose was confirmed (Fig. 6). In addition, L-sorbosone also serves as a substrate for Sdh, affording 2KGA as the oxidation product (our unpublished results). The presence of FAD as a prosthetic group in Sdh was indicated by fluorescence of the Sdh band in SDS PAGE gels for samples where the cofactor was cross-linked to the enzyme by UV exposure prior to SDS treatment. The enzyme activity in a membrane fraction of the UV10 mutant was 20-fold elevated over the IFO3293 parent strain. In accordance with earlier results from Perlman's group [47], the activity was found to be inducible, as cultivation of the UV10 mutant with L-sorbose and D-sorbitol, which is rapidly oxidized to L-sorbose, afforded tenfold higher specific Sdh activities than cultivation in glycerol, mannitol, or fructose.

Simultaneously with Sdh purification, the purification of the second enzyme of the sorbosone pathway of *Gluconobacter melanogenes*, L-sorbosone dehydrogenase Sndh, was also reported by the Roche researchers in Japan [55]. The NAD(P)-dependent enzyme was purified to near homogeneity from the cytosolic fraction of *G. melanogenus* UV10. The 50-kDa enzyme was most active with L-sorbosone as substrate, but other alpha-keto or alpha-hydroxy aldehydes (glyoxal, methylglyoxal, and glycolaldehyde) were accepted as substrates, albeit with lower preference.

6.3 Cloning of the Genes Encoding Sorbose Dehydrogenase Sdh and Sorbosone Dehydrogenase Sndh

In a Roche patent application filed in 1987 in the USA [56], the cloning of DNA fragments comprising the *sdh* gene was described. The inventors isolated mutants from a *G. melanogenus* UV10 Tn5 transposon library, which were unable to produce 2KGA from L-sorbose. Since the mutants were able to produce 2KGA from L-sorbosone, it was presumed that the isolated mutants contained a functional Sndh encoding gene, but lacked an intact Sdh gene. This was confirmed by enzymatic assays for Sdh. Using DNA around the Tn5 insertion points in the mutant genomes as probes, broad-host-range plasmids (RSF1010 based) comprising the wild type *sdh* locus of UV10 were isolated. Upon transformation of these plasmids into *G. melanogenus,* strains were obtained that were comparable to the 2KGA overproducing UV10 mutant with regard to 2KGA productivity and Sdh activity.

Cloning of the *sdh* locus from *G. oxydans* was also accomplished by scientists from Fujisawa Pharmaceutical Co. Ltd., Japan based on peptide sequences obtained from *G. oxydans* T-100 Sdh [57]. One of the DNA fragments identified by reverse genetics comprised two overlapping open reading frames, which encoded Sdh peptides (downstream 1,599 kb ORF) and Sndh peptides (upstream 1,497 kb ORF). Hence, *sndh* and *sdh* form an operon in *G. oxydans* T-100 as well as in IFO3293 (our own unpublished results).

6.4 Gluconobacter *Production Strains for 2KGA*

Subsequent to Reichstein's seminal design of an industrially feasible Asc synthesis and in parallel to the elucidation of the sorbosone pathway, attempts were repeatedly made to replace the chemical oxidation of L-sorbose at C1 by a regioselective microbial oxidation step. The costly protection group chemistry mandatory to limit the oxidation to the C1 hydroxy group in the chemical process would then become obsolete. A US patent from Pfizer & Co Inc. [58] claimed various species of *Pseudomonas*, which produced 2KGA from L-sorbose. Figures on product yield and productivity were not provided. A 1961 publication from the Pfizer research laboratories reported the conversion of L-sorbose to 2KGA with a UV-irradiated mutant strain of an unidentified species of *Pseudomonas*. At 10-liter scale little more than 3 g/l 2KGA was obtained from 20 g/l L-sorbose after 120 h [59].

In 1962, Takeda filed patent claims on *Pseudomonas* and *Acetobacter* strains able to convert D-sorbitol to 2KGA [60]. Best results were obtained with *G. melanogenus* ATCC15163 (now *G. oxydans* NBRC3292, formerly also IFO3292). From 50 g/l sorbitol 6.5 g/l 2KGA were obtained after a 150-h cultivation time.

In a study at the Roche Research Center in Japan comparing various *G. melanogenus* IFO isolates, IFO3293 turned out to be superior to all other strains including IFO3292 (ATCC15163). A culture of IFO3293 produced 2.77 g/l 2KGA from 25 g/l L-sorbose in 7 days [54]. From a mass culture of IFO3293 cultivated on agar plates, a single colony designated SPO1 was isolated that produced 13 g/l of 2KGA from 50 g/l L-sorbose. Based on SPO1, an extended classical strain improvement program ensued comprising treatment with the mutagens UV irradiation, NTG, ICR170, ICR194, and acridine orange. Spheroplast fusion was employed to combine beneficial traits of different strains into one strain. The screening of mutant strains with improved 2KGA productivity was performed in 50-ml shake-flask cultures. Acidification of the culture broth caused by 2KGA excretion was prevented by solid calcium carbonate in the broth which acted as a buffer substance. A number of highly improved strains were isolated producing 50–60 g/l 2KGA in 3 l jar fermentations from 100 g/l D-sorbitol or 100 g/l L-sorbose during 80–100 h. The transient accumulation of L-idonic acid was observed and interpreted as the result of 2KGA reduction at C2 followed by reoxidation in a later phase of the fermentation. Transient accumulation of L-sorbosone was not reported. Biomass accumulated after 30 h to an optical density of 20 corresponding to roughly 8 g/l dry cell mass.

An intermediate of this improvement program was strain UV10, from which Sdh and Sndh were isolated (see previous chapter) and N44-1, which was used to study the peculiar growth characteristics and the central metabolism of *Gluconobacter* [61].

The next step in the development of *Gluconobacter*-based 2KGA production strains followed a recombinant approach at Fujisawa Pharmaceutical Co. Ltd. enabled by the cloning of the Sndh and Sdh encoding genes (see previous chapter).

The Fujisawa researchers constructed an *E. coli*/*Gluconobacter* shuttle vector from a conventional cloning plasmid and the 4.6 kb plasmid pF4 isolated from *G. oxydans* [57]. Into this vector, a 4.6-kb DNA fragment from *G. oxydans* T-100 was introduced carrying the genes for Sndh and Sdh including the endogenous promoter and the terminator up- and downstream of the genes, respectively. In the further developed plasmid *psdh-tufB1*, the endogenous promoter driving *sndh* and *sdh* transcription was replaced by the *E. coli tufB* promoter. *G. oxydans* G624, an L-sorbose-accumulating strain isolated from a Japanese peach, was transformed with *psdh-tufB1*. Before transformation, the host strain had been selected for suppressed L-idonic acid production after treatment with a chemical mutagen. The recombinant *Gluconobacter* strain produced 130 g/l 2KGA from 150 g/l D-sorbitol during 72 h cultivation in a 30 l jar fermenter. Beside D-sorbitol, the cultivation medium contained 2.0 % corn steep liquor, 0.5 % calcium carbonate, and 6 % glycerol.

7 Superior 2KGA Production Processes Based on *Ketogulonicigenium* Strains

Despite the impressive performance increase that was achieved through classical strain improvement and genetic engineering, 2KGA production processes based on *Gluconobacter* strains were never implemented. This might be partly due to the still insufficient efficiency of these processes compared to the established Reichstein procedure. But the main reason was that a novel, very efficient 2KGA production organism had entered the stage.

7.1 Ketogulonicigenium vulgare

In 1980, a group of Chinese scientists around Yin Guang-Lin of the Institute of Microbiology, CAS (Chinese Academy of Sciences), Beijing, published the results of an extended screening program during which 1,615 microbial strains able to grow on L-sorbose as sole carbon source were isolated out of 670 biological samples [62]. The isolate N1197A which produced 2KGA from L-sorbose turned out to consist of a mixture of *Pseudomonas striata* which did not produce 2KGA and a species tentatively assigned as *Gluconobacter oxydans* that formed much smaller colonies compared to *P. striata*. The small colonies forming strain was the actual 2KGA producer. Its productivity in pure culture was very low, but greatly stimulated by co-cultivation with *P. striata*. In mixed culture fermentation comprising both strains, L-sorbose was converted to 2KGA with a yield of 40 % reaching 30 g/l product [63]. In a patent filed in 1987 [64], the CAS researchers described a 2KGA mixed culture fermentation process starting from L-sorbose using the 2KGA producer strain DSM 4025 according to the designation of the "Deutsche Sammlung für Mikroorganismen und Zellkulturen" in Braunschweig, Germany. As helper strain *Bacillus megaterium* was employed. The initial ratio between the two strains in the fermenter were reported not to be critical and could range between 1:10 and 1:300 (helper strain:2KGA-forming strain). After 46 h, the mixed culture produced in a batch fermentation 60 g/l 2KGA from 70 g/l L-sorbose. In the mid 1980s, recognizing its enormous potential, Hoffmann-La Roche, then the leading Asc producer worldwide, licensed-in the microbial 2KGA-forming strain and in parallel to Chinese researchers further developed the production technology based on that strain.

The classification of DSM4025 as *G. oxydans* was erroneous, a misconception that was maintained particularly in the Chinese literature until recently. It became clear soon after its discovery that the 2KGA-producing strain did not belong to the genus *Gluconobacter* (T. Sugisawa, personal communication). Phylogenetic and phenotypic data contradict the original identification and supported reclassification of the strain within the newly proposed genus *Ketogulonicigenium* [65]. The name *Ketogulonigenium* in Urbance et al. was due to a misspelling and later corrected. *Ketogulonicigenium* belongs to the α-subclass of the Proteobacteria phylogenetically closest related to *Roseobacter* (92 % 16S rDNA identity). There is only a distant relationship to *Gluconobacter* or *Acetobacter* (81 % to 83 % 16S rDNA identity). DSM4025 belongs to the approved species *K. vulgare* [66].

The genomic sequence of DSM4025 has not yet been published, but a 1,371 bp partial sequence of the small subunit ribosomal RNA gene of DSM4025 shows a 99.9 % (1370 bp/1371 bp) match to the corresponding sequences of *K. vulgare* Y25 and WSH-001, whose genome sequences have recently been published [67, 68].

7.2 Ketogulonicigenium robustum

An independent broad screening by ADM researchers for microorganisms converting L-sorbose to 2KGA resulted in the isolation of a 2KGA-producing strain designated as B-21627 [69]. They used environmental specimens sampled from different US habitats including moist soil, sand, sediment, fruit, berries, and humus. In single culture fermentation, the novel isolate performed significantly better than DSM4025 as revealed by side-by-side comparison. Nevertheless, productivity (not yield) of B-21627 was greatly enhanced by co-cultivation with a helper strain, for instance various species of the genera *Corynebacterium*, *Brevibacterium*, or *Aureobacterium*.

B-21627 shares nearly identical 16S rDNA sequences with DSM4025, but both strains have distinct phenotypic differences [65]. This suggests that B-21627 belongs to a different *Ketogulonicigenium* species, which, because of the more robust growth characteristics, was designated as *K. robustum* [66].

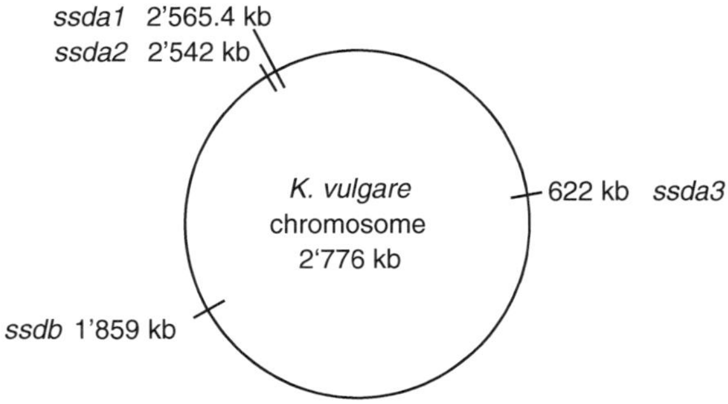

Fig. 7 Mapping of the localization of the four *ssdh* genes (*ssda1*, *ssda2*, *ssda3*, *ssdb*) on the genome of *Ketogulonicigenium vulgare*. The genes are highly homologous, but are dispersed over the genome and not linked

7.3 Sorbose/Sorbosone Dehydrogenase Ssdh of K. vulgare *Consists of Subunits Encoded by the ssda and ssdb Genes*

The unique 2KGA-forming capability of *K. vulgare* is due to the enzyme sorbose/sorbosone dehydrogenase Ssdh, which was purified by a team of Roche scientists to homogeneity from a soluble fraction of *K. vulgare* as a 135-kDa dimer consisting from 62.5- and 64.5-kDa subunits [70]. The purified enzyme employing PQQ as prosthetic group oxidized L-sorbose to 2KGA with some L-sorbosone accumulation. Phenazine methosulfate was required as artificial electron acceptor. Obviously, Ssdh cannot directly transfer the electrons gathered during L-sorbose dehydrogenation to oxygen. The reaction was also facilitated by cytochrome c from *K. vulgare*, while cytochrome c from other species was not effective. In vivo, the cytochrome might link the dehydrogenase to the *K. vulgare* respiratory chain.

The Ssdh-encoding genes were isolated by expression cloning as described in a patent publication filed by Roche in 1996 [71]. For this purpose, cosmid libraries comprising *K. vulgare* DSM4025 DNA fragments were transferred from the original *E. coli* host into *P. putida* by conjugal mating. The resulting *P. putida* colonies were screened with a polyclonal antibody raised against purified Ssdh. Several clones encoding 62- to 64-kDa polypeptides, which were reactive toward the Ssdh antibody, were isolated. Restriction analysis indicated that the identified polypeptides originated from genes that were dispersed over 4 loci of the *K. vulgare* DSM4025 genome. The genes were designated as *ssda1*, *ssda2*, *ssda3*, and *ssdb* (Fig. 7). Based on the recently published genome sequences of two *K. vulgare* strains [67, 68] *ssda1*, *ssda2*, *ssda3*, and *ssdb* can be localized at map position 2,565.4, 2,542.0, 622.4, and 1,859.0 kb, respectively, of the 2,776 kb *K. vulgare* chromosome.

The Ssda polypeptides comprising signatures typical for PQQ-dependent dehydrogenases share more than 85 % sequence identity. The identity between this group and Ssdb, also containing a PQQ signature, is 80 %. The 135-kDa Ssdh isolated from its natural host consists from homo- or heterodimers of *ssda1*, *ssda2*, *ssda3*, or *ssdb*, but the exact subunit composition of Ssdh has not been reported.

The *ssda1*, *ssda2*, *ssda3*, or *ssdb* DNA sequences encode signal peptides of 23 amino acids, typical signal peptide cleavage sites for periplasmic proteins of

Gram-negative bacteria, and no obvious membrane spanning regions indicating that the various 135-kDa dimers of Ssda1, Ssda2, Sada3, and Ssdb collectively designated as Ssdh exist as soluble proteins in the periplasm of *K. vulgare*.

In a publication from 2008, a research team from the Chinese Asc producer Welcome/NCPC reported the purification of Ssda and Ssdb polypeptides from *K. vulgare* WBO104 [72]. According to the sequence information disclosed in the Chinese patent CN101085987 [73] submitted in 2006, these enzymes are identical to the Ssda polypeptides and Ssdb from *K. vulgare* DSM4025. Before the Welcome/NCPC, but years after the Roche patent application an application from the Chinese Academy of Military Sciences was published disclosing the isolation of Ssdh and the cloning of the *ssda1* gene encoding one of the constituent polypeptides of Ssdh [74].

7.4 The ssda-Encoded Enzymes Are Rather Unspecific Alcohol/ Aldehyde Dehydrogenases

The four *K. vulgare ssd* genes were separately expressed in the *P. putida* host and the encoded polypeptides were further characterized [71]. The Ssda polypeptides are rather unspecific alcohol/aldehyde dehydrogenases converting D-sorbitol to D-glucose and L-gulose [71] (Fig. 8). Obviously, the enzymes do not discriminate between the primary alcohol groups at C1 and C6 of the substrate. Further oxidation to the corresponding sugar acids, D-gluconic acid and L-gulonic acid, was not observed, probably because of the 1,5 hemiacetal formation of D-glucose and L-gulose had taken place immediately after the oxidation reaction. D-glucuronic acid and L-guluronic acid possibly resulting from an attack of Ssda on the exocyclic primary alcohol at C6 of the two 1,5 hemiacetals are not reported.

L-sorbose and L-sorbosone were converted to 2KGA by oxidation of the primary alcohol and one of the geminal hydroxy groups of the presumably hydrated carbonyl group at C1, respectively. L-sorbose and L-sorbosone mainly adopt the 2,6-hemiketal configuration in aqueous solution with the primary alcohol group at C6, also a potential substrate, involved in the six-membered ring formation and therefore protected against oxidation.

7.5 The ssdb Gene Encodes a Dehydrogenase with Preference for Secondary Alcohols

Despite the high sequence identity to the Ssda polypeptides, Ssdb has a rather different substrate specificity. Ssdb-catalyzed oxidation of D-sorbitol follows the empirical Bertrand-Hudson rule (like Sldh of *G. oxydans*) leading to L-sorbose, but

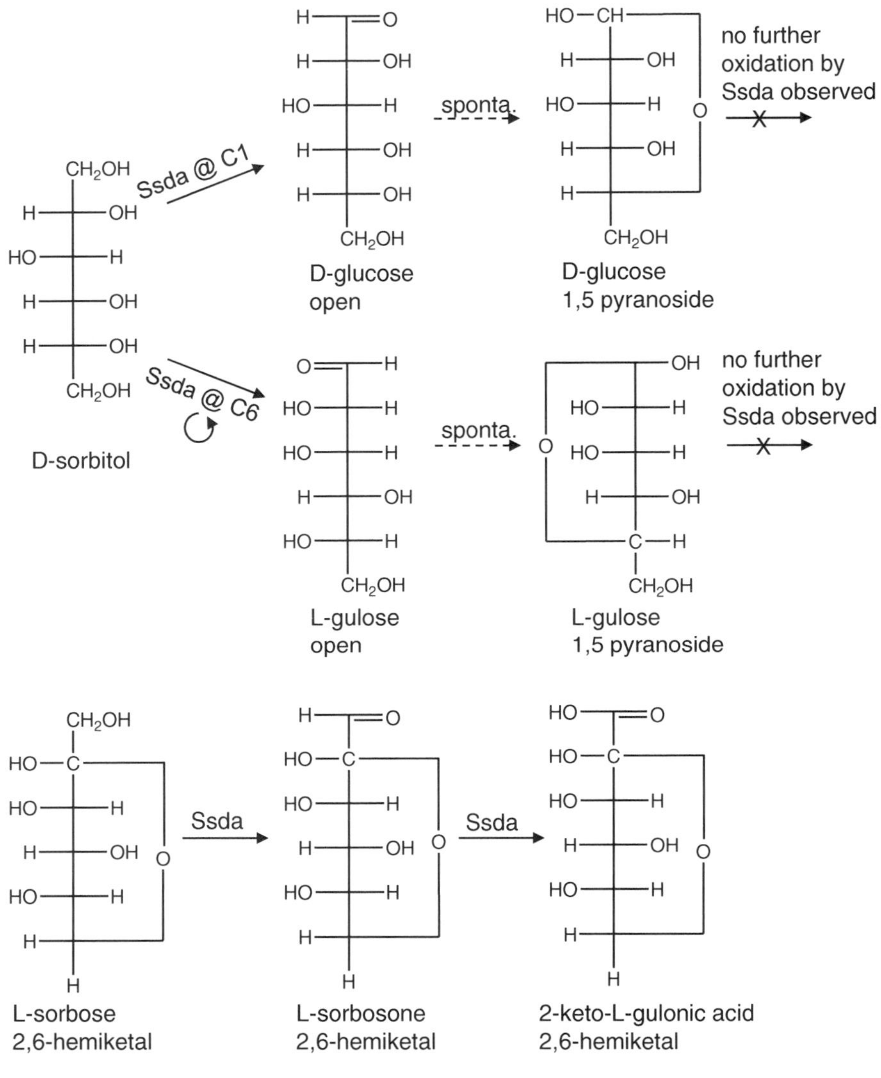

Fig. 8 Substrate specificity of the Ssda enzyme family. The chemical structures are shown as Fischer projections. Ssda enzymes oxidize primary hydroxy groups of sugars or sugar alcohols, such as in D-sorbitol, L-sorbose, and D-fructose. In the case of D-sorbitol, the primary hydroxy groups at either end can be oxidized, yielding either D-glucose or L-gulose. In L-sorbose and D-fructose, the primary alcohol at C6 is unavailable due to hemiketal formation, leaving only C1 for reaction. Usually, the reaction proceeds to the corresponding acid (for L-sorbose and D-fructose as substrates). In the case of D-sorbitol as substrate, the reaction stops at D-glucose/L-gulose, presumably because of rapid sequestering of the new aldehyde functionality as the hemiacetal

not D-fructose (Fig. 9). D-Gluconic acid and D-mannoic acid are oxidized to the corresponding 5-keto compounds. These reactions reveal Ssdb as a dehydrogenase with preference for secondary alcohols. In apparent contradiction to this preference

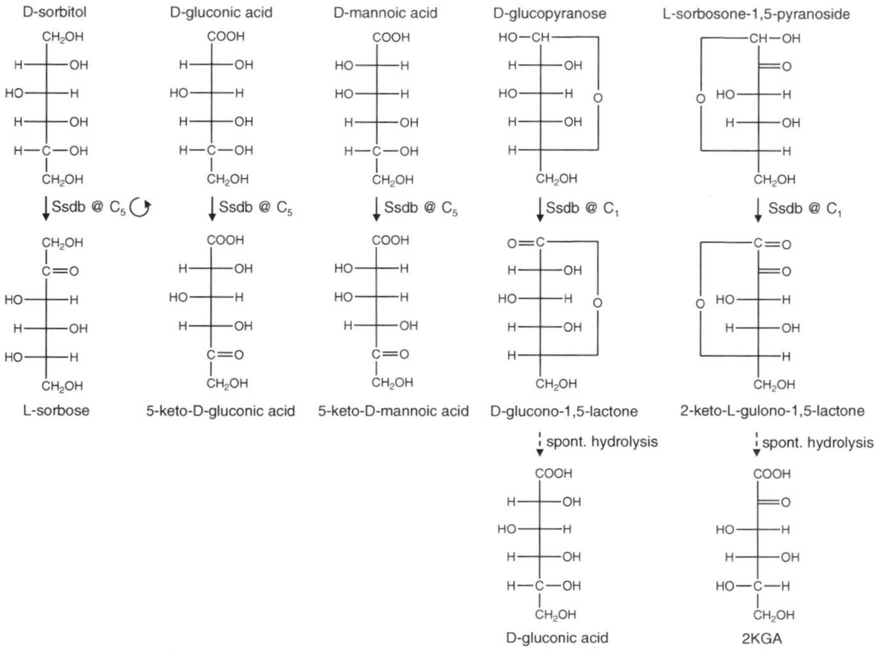

Fig. 9 Substrate specificity of Ssdb. The chemical structures are shown as Fischer projections. Ssdb enzyme oxidizes secondary hydroxy groups in D-sorbitol, D-gluconic acid, and D-mannonic acid to the corresponding ketones according to the Bertrand-Hudson rule. The aldehyde functionalities at C1 of D-glucose and L-sorbosone are oxidized to the corresponding acids, presumably because in the prevalent hemiacetal configurations of these substrates this functionality resembles a secondary hydroxy group with the appropriate steric requirements

is the Ssdb-catalyzed oxidation of the aldehyde functions of D-glucose to D-gluconic acid and of L-sorbosone to 2KGA. This is reconciled, however, when the 1,5 pyranoside of D-glucose, which is by far the dominant form in aqueous solution, is considered as the real substrate of the reaction and the immediate product is the 1,5-lactone of D-gluconic acid. Instead of a secondary alcohol group, Ssdb obviously also accepts the similar hemiacetal group as substrate. In analogy, the real substrate of the Ssdb-catalyzed oxidation of L-sorbosone to 2KGA should be the 1,5 pyranoside isomer. Ab initio calculations (our own unpublished results) indicated that in fact in an equilibrated pH neutral aqueous solution of L-sorbosone the 1,5-pyranoside isomer is present to a significant extent, whereas the majority of the molecules adopt the 2,6-pyranoside structure. The immediate product of the reaction should then be the 2-keto-L-gulono-1,5-lactone, which will finally hydrolyze at neutral pH to 2KGA (Fig. 9).

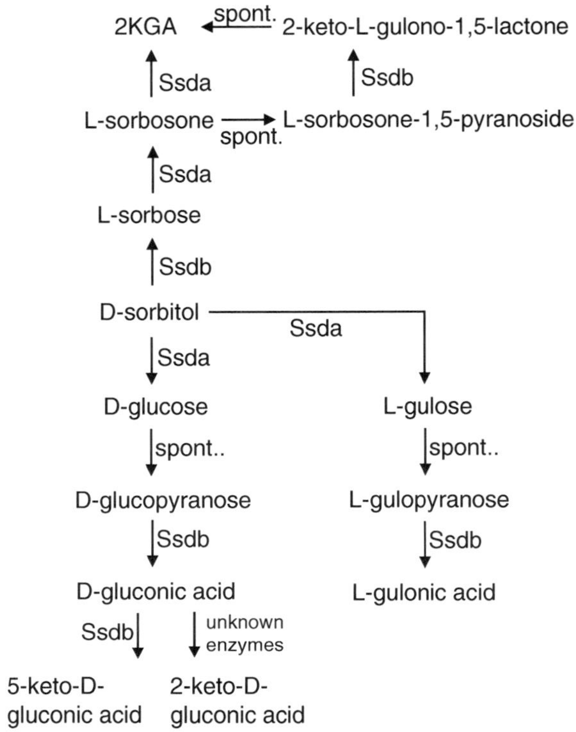

Fig. 10 Conversions occurring during fermentation of D-sorbitol with *Ketogulonicigenium vulgare*, based on the individual substrate specificities of Ssda and Ssdb enzymes. Ssdb is the critical enzyme for D-sorbitol conversion to L-sorbose and also contributes to L-sorbosone conversion to 2KGA. Ssda is the critical enzyme for L-sorbose oxidation to L-sorbosone and 2KGA, but it leads to L-gulose and D-glucose dead-end products from D-sorbitol (which are further oxidized by Ssdb)

7.6 The ssdb and ssdal Genes Encode the Key Enzymes of K. vulgare *for D-Sorbitol to L-Sorbose and L-Sorbose to 2KGA Oxidation, Respectively*

The relative importance of the various Ssdh enzymes for the oxidative capabilities of *K. vulgare* was revealed by deletion analysis (Masako Shinjoh, unpublished). A *ssdA1/ssdA3* double knock out *K. vulgare* mutant is deficient in conversion of L-sorbose to 2KGA. Furthermore, the mutant is unable to oxidize D-sorbitol to D-glucose or L-gulose. 2KGA-formation with a *ssdA1* single mutant is significantly impaired. Only traces of the aldoses are formed from D-sorbitol. The single *ssdA2* and *ssdA3* single knockouts had no phenotype. The *ssdB* single knockout could not produce L-sorbose from D-sorbitol. From the substrate specificities and the knockout impacts, one could conclude that Ssdb is the key enzyme for D-sorbitol oxidation to L-sorbose, while Ssda1 is the key enzyme for L-sorbose oxidation to 2KGA.

7.7 The Two-Step 2KGA Fermentation Process

Figure 10 illustrates the reaction cascade ensuing during cultivation of *K. vulgare* and a suited helper strain, e.g., *B. megaterium* with D-sorbitol. Conversion of the sugar alcohol to L-sorbose by Ssdb is followed by further oxidation to L-sorbosone and 2KGA by the Ssda dehydrogenases, mainly Ssda1. L-sorbosone oxidation is also catalyzed by Ssdb via the cyclic 1,5 pyranoside isomer and the 2-keto-L-gulono-1,5-lactone as immediate oxidation products. In addition to this desired reaction scheme, the Ssda dehydrogenases oxidize L-sorbitol to D-glucose and L-gulose. After cyclization to the corresponding pyranosides, the aldoses are further oxidized by Ssdb and other enzymes.

To prevent D-sorbitol oxidation to the aldoses by the Ssda dehydrogenases, the current industrial microbial 2KGA processes follow a two-step regime (Fig. 11). As in the Reichstein process, D-sorbitol is oxidized in a first step to L-sorbose by *G. oxydans*. After complete consumption of the substrate which takes not more than 24 h resulting in product titers above 200 g/l, the pasteurized and appropriately water-diluted broth is inoculated with a *K. vulgare* culture together with the helper strain, frequently *B. megaterium,* to initiate the second step. L-sorbose is converted within 40 h to 2KGA reaching final 2KGA titers of 100 g/l [75]. Sodium hydroxide is used as a titrant to neutralize the acid produced during the fermentation run. The overall yield of 2KGA production from D-sorbitol is around 90 %. Because of the higher volumetric productivity of the first compared to the second step (ca. twofold higher product concentration reached in half the time), the fermenter volume dedicated to the former is much smaller than that required for the latter. The oxygen transfer rate to be realized during both fermentation runs stays well below 100 mmol/l $\times$ h facilitating the usage of air-lift fermenters with beneficial effects on investment and running costs.

The workup of the fermentation broth consists of biomass removal frequently applying nanofiltration techniques, cation exchange to obtain the product in the protonated form (acidic 2KGA), and crystallization of the product. The chemical conversion to Asc is achieved as in the Reichstein process, mainly via the 2KGA methyl ester.

According to two recent publications, the second conversion step can also be carried out in continuous mode. In single culture, but stimulated with a very high content of complex components (3 % corn steep liquor, 7 % baker's yeast) in the culture broth *K. vulgare* DSM4025 produced 2KGA at a steady-state concentration of 112.2 g/l 2KGA for 140 h [76]. Thereafter, the productivity declined. The dilution rate was kept between 0.035 and 0.043 per hour resulting in a volumetric 2KGA productivity of 3.90 to 4.80 g/l/h. The average molar conversion yield of 2KGA from L-sorbose was 91.3 %.

In continuous mixed culture fermentations with *Xanthomonas maltophilia* IFO 12692 as a helper strain 2KGA production from L-sorbose with DSM4025 could be kept in a stable, continuous mode for more than 1,300 h [77]. Instead of a larger stirred-tank reactor, two smaller, similar sized fermenters were connected in series.

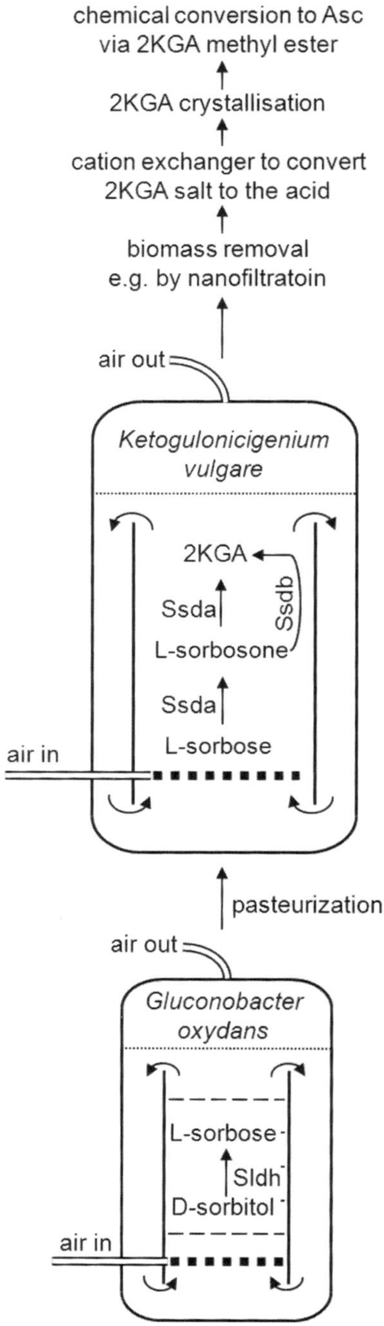

Fig. 11 Conceptual outline of the industrial Asc synthesis based on 2-KGA fermentation with *Ketogulonicigenium vulgare*. To avoid dead-end products from D-sorbitol by *K. vulgare*, L-sorbose is used a starting material. L-sorbose is generated in a separate preceding step from D-sorbitol with help of *Gluconobacter oxydans* as in the original Reichstein process. Also the final conversion of KGA to Asc follows the original Reichstein process

At a dilution rate of 0.0380 per hour and a steady concentration of 113 g/l 2KGA, the volumetric productivity (considering both fermenter volumes) was 2.15 g/l/h. The molar conversion yield was 90.1 %. About 70 % of the conversion was reached in the first of the two serially connected fermenters.

7.8 The One-Step 2KGA Fermentation Process

The specific activity of *Gluconobacter* to convert D-sorbitol to L-sorbose is much higher than the specific activity of *Ketogulonicigenium* to convert the sugar alcohol to unwanted aldoses and follow-up oxidation products. Therefore the simultaneous inoculation of a D-sorbitol containing culture broth with both strains is possible, provided that the right ratio between the two strains in the inoculum is met. It must be ensured that the substrate is already oxidized to L-sorbose by *Gluconobacter* before *Ketogulonicigenium* have the chance to convert D-sorbitol to D-glucose and L-gulose in considerable amounts. The advantages of this procedure are twofold. First, fewer fermenters have to be installed and fewer unit operations are required compared to two-step processes, resulting in obvious economic benefits. Furthermore, an additional helper strain to stimulate the oxidizing capabilities of *Ketogulonicigenium* is not required, since *Gluconobacter* can take over this function.

In a patent application filed in 1991 [78], a team of Roche scientists reported for the first time on the one-step 2KGA process. In one example, simultaneously inoculating *K. vulgare* DSM4025 together with *G. oxydans* IFO3291 in a broth containing D-sorbitol 140 g/l 2KGA was obtained at a molar conversion yield of 89 % during a process time of only 48 h.

According to a patent application filed in 1996 by a research team of the ADM Company, the one-step 2KGA process can also be carried out with a combination of *K. robustum* B-21630, a derivative of *K. robustum* B-21627 isolated after NTG mutagenesis, and *G. oxydans* IFO3293 or ATTC621 [69]. According to the patent description, 110 g/l 2KGA from D-sorbitol was produced during 70 h process time at a conversion yield of close to 100 %. With 31.5 g/l corn steep liquor (as dry solids), the cultivation medium was very rich in complex components.

An alternative approach achieves rapid D-sorbitol to L-sorbose conversion with a *K. vulgare* DSM4025 derivative strain that has the *ssdb* gene amplified on a vector. This overexpression of the Ssdb enzyme reduces the loss of D-sorbitol as D-glucose/L-gulose caused by the undesired activity of the Ssda enzymes on D-sorbitol. Using a complex growth medium containing autoclaved yeast, efficient D-sorbitol to 2KGA conversion could be achieved with this *K. vulgare* strain alone, without co-cultivation with a helper strain. 99 g/l 2KGA were reported from 150 g/l L-sorbitol in a 51 h fermentation run [71], and further improvements of the conversion yield by using stronger promoters for Ssdb expression are indicated.

7.9 *Significance of the Helper Strain for 2KGA Production by* K. vulgare

The impressive 2KGA productivity of *Ketogulonicigenium* processes at high intensity, industrial-scale level has only been achieved upon cultivation of *K. vulgare* with complex media components, for cost reasons preferably corn steep liquor [79, 80] and upon co-cultivation with a helper strain, today frequently *B. megaterium*. Numerous publications mainly from Chinese authors are dedicated to the elucidation of the role of the helper strain. Examination of the available *K. vulgare* genome sequences [67, 68] indicates that the genes for a number of biosynthetic pathways toward essential cellular metabolites are missing. For instance, no *fol* homologs encoding the genes of the de novo folate biosynthetic pathway are annotated in the KEGG database. Fortification with folate compounds, particularly dihydrofolate, of an already rather complex cultivation medium (yeast extract, pancreatic meat digest, corn steep liquor) boosted both growth and 2KGA production of monocultured *K. vulgare* [81]. In line with this earlier finding is the observation that expression of the *Lactococcus lactis fol* genes, including *folB, folKE, folP, folQ,* and *folC* increased growth and productivity of *K. vulgare* monocultured in a complex medium (yeast extract, peptone) [82].

A *cysH* homolog encoding phosphoadenosine phosphosulfate (PAPS) reductase, which is involved in sulfide recruitment from sulfate, is also missing in the *K. vulgare* genome [83]. Upon complementation with glutathion or DTT, monocultured *K. vulgare* reached 60–80 % of the growth and 2KGA production efficiency compared to a *K. vulgare* and *B. megaterium* mixed culture. Cystein complementation was less effective. As complex components, the culture medium contained corn steep liquor [75, 84].

Furthermore, the *K. vulgare* Y25 genome lacks homologs to genes involved in the biosynthesis of various amino acids. Affected are the biosynthetic pathways toward his, gly, lys, pro, thr, met, leu, and ile [85]. The possibly resulting auxotrophies for the corresponding amino acids were not demonstrated, but growth and 2KGA production-stimulating effects in a mixed culture by a blend of the amino acids gly, prol, thr, and isoleucine or by gelatine rich in gly and pro were reported.

K. vulgare Y25 but not WSH001 is afflicted with an out-of-frame deletion in *purH* encoding the ultimate and penultimate steps toward inosine monophosphate biosynthesis, which should result in purine auxotrophy of the former strain. The auxotrophy has not been experimentally demonstrated, but purine supplementation of the Y25 strain stimulated its growth and 2KGA productivity [75]. A similar experiment with WSH001 has not been reported.

The available literature suggests that in contrast to other microorganisms used as production strains in large-scale industrial processes, *K. vulgare* is particularly dependent on the external supply of growth auxiliaries. The limited genetic potential revealed by the recently published genome sequences of two industrially used *K. vulgare* strains seems a plausible explanation for this dependency, but an experimental confirmation of the auxotrophies implied by the inspection of the genome sequences has not yet been reported. Complementing the presumed auxotrophies with complex media components like yeast extract, peptone, beef extracts, and others should be possible from a technical view point and might lead to high-performing processes. However, for economic reasons only comparably cheap corn steep liquor is available at industrial scale, only partly satisfying the growth requirements of *K. vulgare*. The function of the helper strain, which can be exerted by various very different microorganisms, e.g., *P. striata*, various *Bacillus* strains and *G. oxydans*, might be to bridge the nutritional gap corn steep liquor cannot provide, rather than to supply a specific growth factor. Therefore, the relationship between the production and the helper strain might not be regarded as symbiotic in the sense of an evolved way of living together of two different species to their mutual advantage.

7.10 2KGA Strain Improvement

Early reports on 2KGA production with *K. vulgare* showed product yields of 40 % reaching 30 g/l 2KGA [63] with *P. striata* as helper strain. 60 g/l 2KGA during 40 h was reached at 86 % conversion yield with *B. megaterium* as helper strain [64]. Since then several reports have been published on strain improvement efforts by random mutagenesis and selection. For the generation of mutants conventional procedures such as treatment of a *K. vulgare* culture with UV irradiation or nitrosoguanidine were applied, but also technologies which are rarely used in Western laboratories such as ion beam implementation or space mutation during flights with the spaceships Shenzhou IV and Shenzhou VII. For reviews in the Mandarin language see [86, 87]. An article in English on ion beam mutagenesis of *K. vulgare* and the helper strain *B. megaterium* was published in 2004 [88]. In 1997 the isolation of the strain-pair *G. oxydans* SCB329 (presumably a *Ketogulonicigenium* species) and *Bacillus thuringiensis* SCB933 was published. During 40 to 50 h fermentation runs, 2KGA at titers between 115 and 130 g/l was produced at a conversion rate of 88 mol % [89].

8 Direct Microbial Production of Asc

After more than three decades of strain and process optimization, the 2KGA fermentation by *K. vulgare* has reached a performance level that makes it increasingly difficult to achieve further cost-relevant improvements. Instead, opportunities can be seen in the succeeding step of 2KGA rearrangement to ascorbic acid, which still follows the same concept as laid out in the 1930s by Reichstein and Grüssner. This chemical step contributes significantly to the overall process costs. A process concept that could convert sorbitol directly to ascorbic acid would therefore be most attractive. In theory, this could build on the established 2KGA fermentation with an enzyme-catalyzed 2KGA to Asc rearrangement (2,6-hemiacetal to 1,4-lactone) as extension. Ab initio energy calculations as well as experimental results (own unpublished results) indicate that in aqueous environment, Asc is thermodynamically far more stable than 2KGA and (nearly) quantitative conversion should be possible. However, no enzyme efficiently catalyzing this reaction has so far been identified. The few publications of enzyme catalysis for this reaction so far shows only trace activity [90, 91] and no significant improvements have been reported. 2KGA may represent a kinetic trap in an aqueous environment and biotechnological reaction pathways all the way to Asc may need to avoid 2KGA. Accordingly, 2KGA is also not part of natural biosynthetic routes, where Asc formation directly results from the oxidation of precursor molecules (see Fig. 2) with appropriately preformed 1,4-lactone linkage (L-gulono-1,4-lactone in animals, L-galactono-1,4-lactone in plants). Enzymes converting L-gulono-1,4-lactone to Asc are also known from bacteria, even from *Ketogulonicigenium* [92]. The biochemical description of the *Ketogulonicigenium* enzyme indicates that it belongs to the family of heterotrimeric periplasmic flavohemoproteins, of which several can be found in the published *Ketogulonicigenium* genomes. Besides sharing the same FAD cofactor, these enzymes bear no similarity to the mammalian gulono-1,4-lactone dehydrogenase. The use of these natural or nature-like Asc-forming enzymatic steps in biotechnological production processes is so far precluded by the rare nature of these L-sugar-derived lactone precursor molecules and the lack of efficient production methods for these compounds. It was, therefore, a tantalizing discovery when Asc formation directly from L-sorbosone, the intermediate of the efficient 2KGA formation route, was identified in those two species already in the focus for 2KGA production for decades: *K. vulgare* [93] and *G. oxydans* [94]. Besides an earlier report of L-sorbosone to Asc activity derived from plant tissue [95], which did not see consolidating follow-ups, the above observations are the first evidence of biological Asc formation from a molecule other than a 1,4-lactone.

8.1 L-Sorbosone Oxidation to Asc by Sndhak from K. vulgare

The enzyme from *K. vulgare* responsible for L-sorbosone to Asc conversion was identified by purification of this activity, N-terminal sequencing of the purified protein and subsequent cloning of the gene [96]. This sorbosone dehydrogenase yielding ascorbic acid and originating from *Ketogulonicigenium* (in the original publication designated as Sndh, now abbreviated as Sndhak) is a periplasmic soluble dimeric PQQ-enzyme with six-bladed *β*-propeller architecture (Fig. 12). It is related to the soluble D-glucose dehydrogenase from *A. calcoaceticus*, which has been well studied for the application in glucose sensors for diabetes care [97]. Indeed, the specific activity of Sndhak against D-glucose is even ten times higher than with L-sorbosone as substrate [96]. The physiological role of Sndhak is,

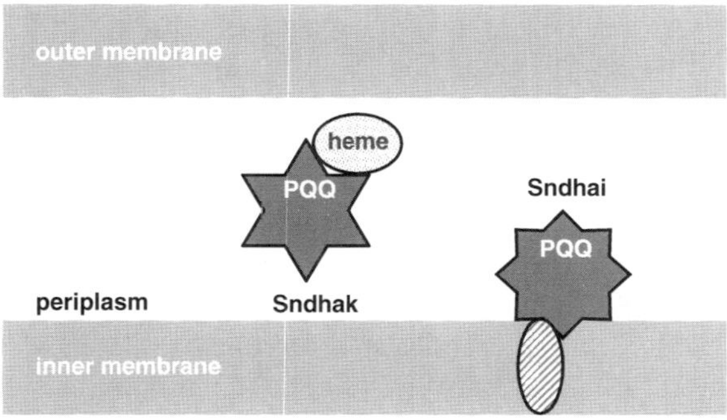

Fig. 12 Schematic representation of the two known Asc-forming sorbosone dehydrogenases. Both are PQQ enzymes localized in the periplasm. Sndhak from *Ketogulonicigenium vulgare* is a soluble enzyme with six-bladed *β*-propeller architecture and an additional heme-containing cytochrome domain. Sndhai from *Gluconobacter oxydans* is a membrane-bound enzyme with eight-bladed *β*-propeller architecture and an additional membrane anchor domain

therefore, presumably D-glucose oxidation and it is a fortuitous cross-reactivity with the non-physiological substrate L-sorbosone that allows biotechnological application for Asc formation. A fraction of Sndhak purified from *Ketogulonicigenium* was found to contain an additional C-terminal heme-containing cytochrome domain, which is encoded in the 3′ section of the gene. This cytochrome domain may play a role in electron transfer from the PQQ cofactor into the respiratory chain. The related *Acinetobacter* enzyme does not contain such a cytochrome domain and it is noteworthy that no per se activity of the soluble *Acinetobacter* D-glucose dehydrogenase can be detected [98, 99], in contrast to the *Ketogulonicigenium* enzyme. As a consequence of the dimerization of Sndhak, three different isoenzymes can be distinguished after purification, consisting of homodimers with or without a cytochrome domain and the mixed heterodimer (in the original publication designated as Sndh, Sndh2, and Sndh3). The cause and relevance of this heterogeneity are unknown.

8.2 L-Sorbosone Oxidation to Asc by Sndhai from G. oxydans

The second novel Asc-forming enzyme was discovered after small quantities of Asc were detected in reactions of L-sorbosone with *G. oxydans* IFO3293 [94]. The relevant gene was singled out by screening a transposone library of IFO3293 for loss of the capability to produce Asc. The sorbosone dehydrogenase yielding ascorbic acid from *G. oxydans* IFO3293 (abbreviated Sndhai) thereby identified is again a PQQ-enzyme, but shares no sequence similarity to Sndhak, the Asc-forming enzyme from *Ketogulonicigenium*. Sndhai is a membrane-bound periplasmic protein and consists of two domains, an N-terminal membrane anchor with five predicted transmembrane helices, followed by a catalytic domain with eight-bladed β-propellor architecture (Fig. 12). It shares sequence-homology with membrane-bound PQQ-linked D-glucose dehydrogenases from, e.g., *G. oxydans*, *Escherichia coli*, and *A. calcoaceticus*, as well as with the PQQ-linked sorbitol dehydrogenase from *G. oxydans*, the enzyme responsible for the rapid and selective oxidation of D-sorbitol to L-sorbose in the Reichstein–Grüssner process and all following D-sorbitol-based ascorbic acid production technologies. Sndhai shows no activity toward D-glucose or D-sorbitol, but is highly active (much more than for L-sorbosone) toward the cyclic polyol myoinositol as substrate (own unpublished results). Similar to Sndhak, L-sorbosone oxidation by Sndhai seems to be a non-physiological cross-reactivity.

Sndhai is virtually identical (one amino acid difference) to the gene annotated as "PQQ-containing dehydrogenase 1" in the published genome of *G. oxydans* ATCC621H [100], which was subsequently characterized as myo-inositol dehydrogenase [101]. Curiously, this enzyme had a past life as the subject of intense biochemical research in the 1940s and 1950s. Soon after the first report in 1938 of myo-inositol oxidation by *G. oxydans* ATCC621 (then called "*Acetobacter suboxydans*", [102]), it was shown that of the six hydroxy groups of myo-inositol only the hydroxy group at position 2 is oxidized by the enzyme [103–105] leading to myo-2-inosose formation. This exquisite selectivity is reminiscent of the D-sorbitol to L-sorbose oxidation by the related PQQ enzyme sorbitol dehydrogenase, but cannot be described by the Betrand-Hudson rules. This triggered thorough investigations by Erwin Chargaff and co-workers who studied the conversions and identified the exact products for a wide variety of different inositols and related compounds [106–108]. This culminated in the Chargaff rules of inositol oxidation [109, 110]. They provided a comprehensive description and predictive power for this enzyme's reactions, but never reached the same prominence as the Chargaff rules of DNA composition derived at the same time. Further biochemical work on inositol dehydrogenase/Sndhai was hampered by its resistance to purification [111, 112], which was overcome only in the 1990s [113].

8.3 Product Spectrum of L-Sorbosone Oxidation: Determined by the Structure of L-Sorbosone Isomers or by Metabolic Bifurcation?

Since both novel Asc-forming enzymes Sndhak and Sndhai have activity toward sugars or polyols it is not per se unusual to find activity toward oxidation of L-sorbosone. It is peculiar, however, to obtain Asc as main product and 2KGA only as side product [96], despite using the same substrate and achieving the same oxidation at C1 as the 2KGA-forming enzymes Ssda and Ssdb of *Ketogulonicigenium* and Sndh of *Gluconobacter*. How do the enzymes influence the type of product in this reaction? One key differentiator may be the structure of the substrate L-sorbosone. As commonly observed for sugar molecules, L-sorbosone cyclizes forming hemiacetal bonds between carbonyl and hydroxy functional groups. With its two carbonyl functions at C1 and C2, L-sorbosone can adopt an even wider diversity of isomeric structures than standard sugars based on different ring closures (Fig. 13a). The 2,6 hemiketal isomer of L-sorbosone, which is the dominant form of L-sorbosone in aqueous solution has the same ring structure as 2KGA in aqueous solution. Enzymes such as Ssda1 selectively oxidizing this L-sorbosone isomer at the exocyclic geminal diol at C1 (assuming hydration of the C1 carbonyl group) will directly yield 2KGA as product. The 1,4-hemiacetal isomer of L-sorbosone, which might occur as the minor isomer in aqueous solution, has the ring structure of Asc preformed. Oxidation of the acetalic hydroxy group at C1 in this sorbosone isomer by enzymes of appropriate selectivity such as Sndhak or Sndhai [93, 96] would (after rapid enolization) directly yield Asc. Ssdb, which shares with Sndhak and Sndhai the selectivity toward secondary alcohols, would yield 2KGA rather than Asc due to its selectivity toward oxidization of the possible additional minor isomer 1,5 L-sorbosone at its acetalic hydroxyl group at C1. The resulting 1,5 lactone intermediate, which has not yet been observed, presumably rapidly hydrolyzes and rearranges to the 2,6 pyranose form of 2KGA. According to this model, the nature of the product (Asc or 2KGA) formed from L-sorbosone is determined by the specificity of the enzymes toward the 2,6-hemiketal, 1,5-hemiacetal, or the 1,4-hemiacetal isomers of L-sorbosone. Some 2KGA side-product formation is observed for Sndhak and Sndhai, which could result from cross-reactivity with 2,6-L-sorbosone or, more likely (given the selectivity of these enzymes toward secondary alcohols), with 1,5-L-sorbosone.

As an alternative explanatory model (Fig. 13b) a pathway can be formulated, in which oxidation of the same 1,5-pyranose ring form of L-sorbosone can lead to either Asc or 2KGA, depending on the fate of the putative transient 1,5-lactone intermediate. Based on this alternative model, Sndhak and Sndhai (as well as Ssdb) could interact with the 1,5-pyranose L-sorbosone isomer affording the 1,5-lactone of 2KGA. Subsequent rearrangement of this labile compound either yields the stable 2,6-hemiketal of 2KGA implying the hydrolysis of the lactone ring, or the 1,4-lactone via a transesterification reaction with subsequent enolization to stable Asc. This metabolic bifurcation toward Asc or 2KGA would be influenced by the respective dehydrogenase enzymes, some like Ssdb favoring 2KGA formation while other enzymes like Sndhak or Sndhai favoring Asc formation. In this model, enzymes converting L-sorbosone to Asc are characterized not only by appropriate substrate selectivity (for the L-sorbosone 1,5-hemiacetal rather than the 2,6-hemiketal), but also by guiding the follow-up reactions of the immediate, labile oxidation product preferably toward the transesterification. Asc would become the major and 2KGA, the minor product of these dehydrogenases. Further work is required to consolidate this interpretation and to understand the differences between Ssdb and Sndhak/Sndhai responsible for the different products formed.

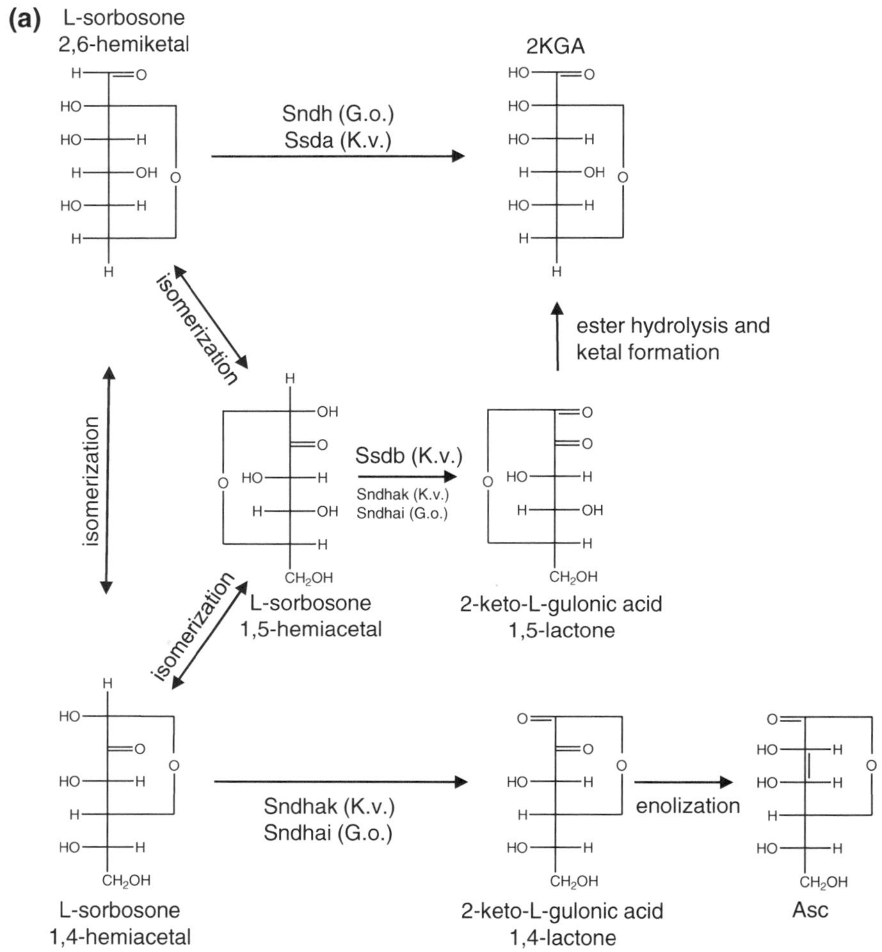

Fig. 13 Explanatory models for the conversion for L-sorbosone to either Asc or 2KGA by different enzymes. Both models build on the presence of several isomers of L-sorbosone with different ring connectivities, which may be recognized specifically by different enzymes and lead to different products. **a** 2,6-L-sorbosone is directly oxidized to 2KGA by Ssda or Sndh. 1,4-L-sorbosone is directly oxidized to Asc by Sndhak or Sndhai. 1,5-L-sorbosone is oxidized by Ssdb to the 1,5-lactone of 2KGA which hydrolyzes to yield 2KGA. 2KGA by-product formation by Sndhak and Sndhai could be due to trace activity toward 1,5-L-sorbosone. **b** 2,6 L-sorbosone is directly oxidized to 2KGA by Ssda or Sndh as in (**a**). 1,5-L-sorbosone is oxidized by Ssdb as well as Sndhak and Sndhai. The immediate 1,5-lactone product has two possible fates, either toward 2KGA by hydrolysis or toward Asc by transesterification. The enzymes can influence the partitioning between the two fates, Ssdb favoring 2KGA formation, Sndhak and Sndhai favoring Asc formation

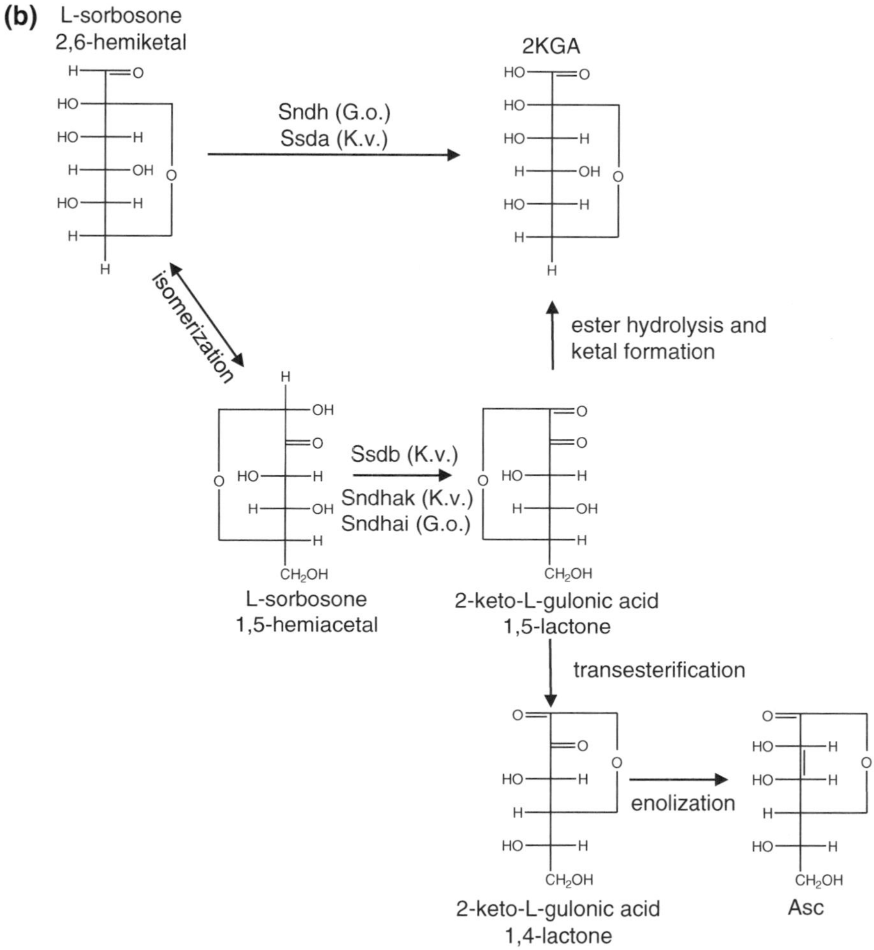

Fig. 13 continued

9 D-Isoascorbic Acid and D-Erythroascorbic Acid

Several compounds structurally similar to Asc were isolated from microbial cultures. Among them are D-isoascorbic acid (also referred as erythorbic acid) produced by various *Penicillium* species, e.g., *Penicillium notatum* [114] or *Penicillium cyaneo-fulvum* [115, 116] and the C5 sugar acid D-erythroascorbic acid from yeasts and other fungi.

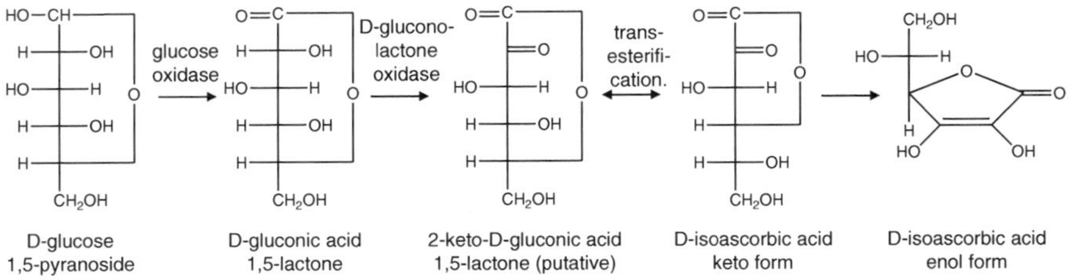

Fig. 14 Process for the conversion of D-glucose to D-isoascorbic acid by *P. notatum*. The chemical structures are shown as Fischer projections. The first oxidation reaction at C1, catalyzed by D-glucose oxidase, results in the D-glucono-1,5-lactone. The D-gluconolactone oxidase responsible for the second oxidation at C2 can use the 1,5-lactone as substrate, resulting in an uncharacterized intermediate, presumably the 2-keto-D-glucono-1,5-lactone. This intermediate spontaneously rearranges to form D-isoascorbic acid

9.1 Microbial Oxidation of D-Glucose to D-Isoascorbic Acid

The D-configuration of D-isoascorbic acid at C5 allows a short biosynthetic pathway (Fig. 14) from D-glucose, i.e., its 1,5-glucopyranoside, which is oxidized to D-glucono-1,5-lactone by glucose oxidase followed by oxidation at C2 by D-gluconolactone oxidase [117, 118]. The immediate oxidation product of D-glucono-1,5-lactone by gluconolactone oxidase already has reducing activity on, e.g., 2,6-dichlorphenolindophenol. It is rather stable at pH 4. Upon pH shift, this compound spontaneously converts to D-isoascorbic acid [119]. The unidentified immediate oxidation product could be 2-keto-D-glucono-1,5-lactone, which rearranges via a reversible transesterification reaction to the 1,4-lactone followed by an irreversible enolization to D-isoascorbic acid. The formation of 2-keto-D-gluconic acid as the result of 2-keto-D-glucono-1,5-lactone hydrolysis was not reported. The oxidation of the 1,4-lactone by D-gluconolactone oxidase might also occur to some extent, since D-glucono-1,5-lactone shows a tendency to slowly rearrange to the 1,4-lactone at pH > 4 [120] and the D-gluconolactone oxidase of *Penicillium cyaneofulvum* accepts both D-glucono-1,5-lactone and the corresponding 1,4-lactone [117]. This reaction would directly deliver the keto-isomer of D-isoascorbic acid. The sequence of the reactions from D-glucose to D-isoascorbic acid, first oxidation at C1, then oxidation at C2 (C1, C2), is similar to the naturally evolved Asc biosynthesis from L-galactose or L-gulose.

Oxidation of D-gluconolactone at C2 is also afforded by pyranose-2-oxidase from *Polyporus obtusus* [121]. In this reaction both D-isoascorbic acid and 2-keto-D-gluconic acid were obtained in a roughly 1:1 ratio. Obviously, following the natural C1, C2 oxidation sequence, transesterification and (iso)ascorbic acid formation are preferred over hydrolysis and 2-keto sugar acid formation or are at least possible to a significant extent.

If the sequence of oxidation reactions is reversed (C2, C1), i.e., D-glucopyranose is first oxidized by pyranose-2-oxidase to D-glucosone followed by glucose oxidase treatment, 2-keto-D-gluconate was reported as the only oxidation product [121].

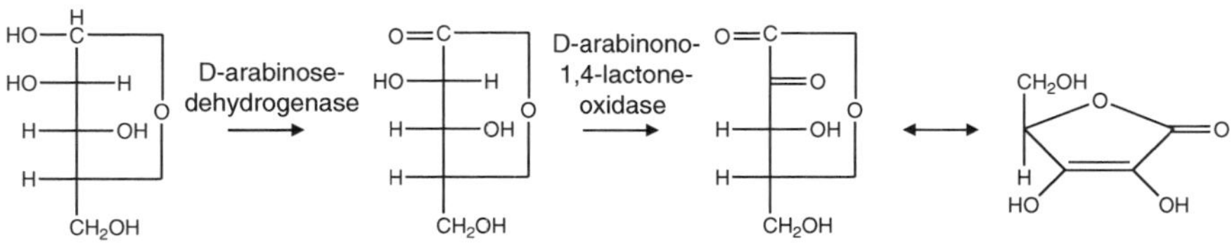

Fig. 15 Biosynthesis of D-erythroascorbic acid from D-arabinose in yeast. The chemical structures are shown as Fischer projections. The first oxidation reaction at C1, catalyzed by D-arabinose oxidase, results in the D-arabino-1,4-lactone. The second oxidation at C2, catalyzed by D-arabino-1,4-lactone oxidase, directly yields D-erythroascorbic acid

Though not explicitly reported, it is safe to assume that the later oxidation occurs with 2-keto-D-gluco-1,5-pyranose and delivers as the immediate reaction product 2-keto-D-glucono-1,5-lactone, which hydrolyzes affording 2-keto-D-gluconate. It is unclear why the spontaneous follow-up reaction of 2-keto-D-glucono-1,5-lactone delivers, at least to some extent, D-isoascorbic acid if obtained according to the C1, C2 reaction sequence, but only 2-keto-D-gluconate if obtained by the C2, C1 oxidation sequence.

9.2 *Biosynthesis of* D-*Erythroascorbic Acid*

Yeasts and other fungi synthesize the C5 sugar acid D-erythroascorbic acid which shares structural and physicochemical properties with Asc [122]. D-erythroascorbic acid serves similar protective functions in these microorganisms as Asc does in plants and animals, including the scavenging of reactive oxygen species. The biosynthesis of D-erythroascorbic acid (Fig. 15) starts from D-arabinose obtained by the microorganism from decaying plant material. D-arabinose, presumably in its 1,4-furanosidic isomeric form, is oxidized by $NAD(P)^+$ specific dehydrogenases [123, 124] to D-arabinono-1,4-lactone, which is further oxidized to D-erythroascorbic acid by D-arabinono-1,4-lactone oxidase [125]. Resting cells of *Saccharomyces cerevisiae* can synthesize Asc from L-galactose, L-galactono-1,4-lactone, or L-gulono-1,4-lactone via the pathway naturally used for D-erythroascorbic acid [126].

9.3 *Application, Market, and Industrial Production of* D-*Isoascorbic Acid*

From the two microbial Asc analogs only D-isoascorbic acid is of some economic interest. The compound, the C5 epimer of Asc, has no or only marginal anti-scurvy activity, but antioxidant properties similar to the latter. Hence, the compound has been used in the food industry for more than 30 years as an alternative to Asc to protect processed food against oxidation. A typical application is in meat products to support the formation and maintenance of the red color of cured meat. The addition of D-isoascorbate (like L-ascorbate) to sodium nitrite-cured meat increases the nitrosylation of the central iron atom of muscle myoglobin resulting in the formation of reddish-brown nitrosomyoglobin and the characteristic pink color of nitrosohemochrome or nitrosyl-heme upon cooking. Other food products that may contain D-isoascorbic acid are oils, beer, and fruit juices. D-isoascorbic acid and its salts are considered GRAS (generally recognized as safe) by the US Food and Drug Administration and are approved as food ingredients E315 (free acid) and E316 (sodium salt) in Europe.

With approximately 10,000 tons global demand per year supplied by more than 10 producers, all located in China, the D-isoascorbic acid world-market is much smaller than the market for Asc. PARCHN Sodium Isovitamin C Co. Ltd, Dexing City, Jiangxi Province is one of the bigger producers. D-isoascorbic acid market prices are less volatile than the prices for Asc, fluctuating around 5 USD/kg or below. In times of high Asc prices, D-isoascorbic acid is a cheap alternative in applications not relying on the vitamin activity of Asc.

In an effort to develop an industrial D-isoascorbic acid production process in the mid-1960s, Japanese Fujisawa Pharmaceutical Co. Ltd. performed an extended search for microorganisms with the ability to convert D-glucose into D-isoascorbic acid [114]. From all genera tested *Penicillium* was unique in being able to deliver the desired product. *P. notatum* FY 115 was a superior strain, which was further improved by mutagenesis and selection campaigns. In a lab-scale fermentation run for 136 h, for most of the time at $pH < 5$, 40 g/l D-isoascorbic acid was obtained from glucose with a yield of 45 % [127]. A transient accumulation of D-gluconic acid and to a lower extent D-gluconolactone was observed, in accordance with the biosynthetic pathway of D-isoascorbic acid in *Penicillium* (Fig. 14). D-Isoascorbic acid was recovered from the fermentation broth with high yield by anion exchange chromatography and crystallization.

Today the *Penicillium* process of D-glucose oxidation directly to D-isoascorbic acid is not industrially exploited, probably because of insufficient volumetric productivity and yield on glucose. Instead, D-isoascorbic acid is produced in a process designed in analogy to the Asc process. First, a 2-keto-sugar acid is obtained by microbial oxidation, which after isolation and purification is chemically rearranged to the desired product. The conversion of D-glucose to 2-keto-D-gluconic acid is achieved with *Pseudomonas fluorescens*, e.g., strain AR4 [128] and proceeds via D-gluconate. The enzymes catalyzing the two consecutive oxidation steps in *P. fluorescens* are glucose dehydrogenase [129] and gluconate dehydrogenase, respectively [130, 131]. *P. fluorescens* AR4 was selected for high glucose-tolerance (over 120 g/l) and low susceptibility to phage infection. With rice starch hydrolysate as source of D-glucose almost 170 g/l of 2-keto-D-gluconic acid was obtained in 24 h batch fermentation at a yield above 90 % [128].

10 Outlook

Nearly 80 years ago, Tadeus Reichstein first devised a practical Asc synthetic route for commercial application. With continuing optimizations, this remained the industry standard for more than 60 years, serving a vastly growing market and increasing production scale. Even today, the 2KGA fermentation route responsible for the majority of Asc produced still follows Reichstein's original concept: D-glucose as starting material, inversion of the carbon skeleton to L-sorbose, twofold oxidation to 2KGA, rearrangement to Asc. This continued success of Reichstein's synthesis route is based on its simplicity and low number of reaction steps, its basis in the common and renewable starting material D-glucose, and its amenability toward improvements. Building on these advantages, it remains the most promising route to also meet the challenges of the future, with production cost, product quality, and environmental sustainability as key drivers. Minimizing oil-based chemistry will be a critical hallmark of a sustainable future production process. The original Reichstein process had already made use of renewable starting materials and a biotechnological step to achieve the regioselectivity required to convert D-sorbitol to L-sorbose. The implementation of 2KGA fermentation further reduced the need for solvents and energy. Recent work outlined in this review opens the way toward a direct oxidation of L-sorbose to Asc, without the need for a chemical conversion of 2KGA to Asc, again lowering solvent and energy requirements. Additional opportunities may be found in ongoing developments for a more sustainable provision of the starting materials D-glucose [132] or D-sorbitol [133]. With D-glucose as starting material, the Asc production process will be tightly linked to the emerging portfolio of bio-based chemicals, for the good, through sharing technological advances, as well as for the bad, due to increasing competition for raw materials.

Acknowledgments For the preparation of the manuscript Grace Lee has been a great help in retrieving and translating numerous Chinese publications. The critical reading of the manuscript by Tom McClymont, Teruhide Sugisawa, Masako Shinjoh, and Dietmar Laudert is highly acknowledged.

References

1. Funk C (1912) The etiology of the deficiency diseases. Beri-beri, polyneuritis in birds, epidemic dropsy, scurvy, experimental scurvy in animals, infantile scurvy, ship beri-beri, pellagra. J State Med (London) 20:341–368
2. Zhou J, Du G, Chen J (2012) Metabolic Engineering of microorganisms for vitamin C production. In: Wang X, Chen J, Quinn P (eds) Reprogramming microbial metabolic pathways. Springer, Netherlands, pp 241–259
3. Bremus C, Herrmann U, Bringer-Meyer S, Sahm H (2006) The use of microorganisms in L-ascorbic acid production. J Biotechnol 124:196–205. doi:10.1016/j.jbiotec.2006.01.010

4. Hancock R (2009) Recent patents on vitamin C: opportunities for crop improvement and single-step biological manufacture. Recent Pat Food Nutr Agric 1:39–49
5. Bhatt A (2010) Evolution of clinical research: a history before and beyond James Lind. Perspect Clin Res 1:6–10
6. Zetterström R (2006) C. Eijkman (1858–1930) and Sir F.G. Hopkins (1861–1947): the dawn of vitamins and other essential nutritional growth factors. Acta Paediatr 95:1331–1333. doi: 10.1080/08035250600960036
7. Piro A, Tagarelli G, Lagonia P, Tagarelli A, Quattrone A (2010) Casimir Funk: his discovery of the vitamins and their deficiency disorders. Ann Nutr Metab 57:85–88
8. Svirbely J, Szent-Györgyi A (1932) The chemical nature of vitamin C. Biochem J 26:865–870
9. King C, Waugh W (1932) The chemical nature of vitamin C. Science 75:357–358
10. Jukes TH (1988) The identification of vitamin C, a historical summary. J Nutr 118:1290–1293
11. Herbert RW, Hirst EL, Percival EGV, Reynolds RJW, Smith F (1933) The constitution of ascorbic acid. J Chem Soc 299:1270–1290
12. Ault RG, Baird DK, Carrington HC, Haworth WN, Herbert R, Hirst EL, Percival EGV, Smith F, Stacey M (1933) Synthesis of D- and of L-ascorbic acid and of analogous substances. J Chem Soc (Resumed) 332:1419–1423
13. Reichstein T, Grüssner A (1934) Eine ergiebige Synthese der L-Ascorbinsäure (C-Vitamin). Helv Chim Acta 17:311–328. doi:10.1002/hlca.19340170136
14. Mandl J, Szarka A, Bánhegyi G (2009) Vitamin C: update on physiology and pharmacology. Br J Pharmacol 157:1097–1110. doi:10.1111/j.1476-5381.2009.00282.x
15. Valpuesta V, Botella MA (2004) Biosynthesis of L-ascorbic acid in plants: new pathways for an old antioxidant. Trends Plant Sci 9:573–577. doi:10.1016/j.tplants.2004.10.002
16. Horowitz HH, Doerschuk AP, King CG (1952) The origin of L-ascorbic acid in the albino rat. J Biol Chem 199:193–198
17. Kondo Y, Inai Y, Sato Y, Handa S, Kubo S, Shimokado K, Goto S, Nishikimi M, Maruyama N, Ishigami A (2006) Senescence marker protein 30 functions as gluconolactonase in L-ascorbic acid biosynthesis, and its knockout mice are prone to scurvy. Proc Natl Acad Sci 103:5723–5728. doi:10.1073/pnas.0511225103
18. Linster CL, Van Schaftingen E (2007) Vitamin C: biosynthesis, recycling and degradation in mammals. FEBS J 274:1–22. doi:10.1111/j.1742-4658.2006.05607.x
19. Loewus FA (1963) Tracer studies on ascorbic acid formation in plants. Phytochemistry 2:109–128. doi:10.1016/S0031-9422(00)82971-4
20. Isherwood FA, Chen YT, Mapson LW (1953) Synthesis of L-ascorbic acid in plants and animals. Nature 171:348–349
21. Smirnoff N, Wheeler G (2000) Ascorbic acid in plants: biosynthesis and function. Crit Rev Biochem Mol Biol 35:291–314
22. Endres S, Tenhaken R (2009) Myoinositol oxygenase controls the level of myoinositol in Arabidopsis, but does not increase ascorbic acid. Plant Physiol 149:1040–1042
23. Baroja-Mazo A, Valle Pd, Rúa J, de Cima S, Busto F, de Arriaga D, Smirnoff N (2005) Characterisation and biosynthesis of D-erythroascorbic acid in Phycomyces blakesleeanus. Fungal Genetics Biol 42:390–402. doi:10.1016/j.fgb.2005.01.005
24. Bertrand G (1896) Preparation biochimique du sorbose. C R Acad Sci 122:900–903
25. Bertrand G (1898) Recherches sur la production biochimique du sorbose. Ann Inst Pasteur (Paris) 12:385–399
26. Bertrand G (1904) Etude biochimique de la bacterie du sorbose. Ann Chim Phys 3:181–288
27. Gillis M, Kersters K, Gossele F, Swings J, De Ley J, MacKenzie A, Bousfield I (1983) Rediscovery of Bertrand's sorbose bacterium (*Acetobacter aceti* subsp. *xylinum*): proposal to designate NCIB 11664 in place of NCIB 4112 (ATCC 23767) as the type strain of *Acetobacter aceti* subsp. *xylinum*. Int J Syst Evol Microbiol 33:122–124
28. Crawford T, Crawford S (1980) Synthesis of L-ascorbic acid. Adv Carbohydr Chem Biochem 37:79–155

29. Dalmer O, Heyns K (1936) Process for the production of keto gulonic acid from sorbose. US 2190377
30. Tadamitsu K (1983) Process for the preparation of 2-keto-L-gulonic acid. US 4599446
31. Bronnimann C, Bodnar Z, Hug P, Mallat T, Baiker A (1994) Direct oxidation of L-sorbose to 2-keto-L-gulonic acid with molecular oxygen on platinum- and palladium-based catalysts. J Catal 150:199–211. doi:10.1006/jcat.1994.1336
32. Deppenmeier U, Ehrenreich A (2009) Physiology of acetic acid bacteria in light of the genome sequence of *Gluconobacter oxydans*. J Mol Microbiol Biotechnol 16:69–80
33. Hann RM, Tilden EB, Hudson CS (1938) The oxidation of sugar alcohols by *Acetobacter suboxydans*. J Am Chem Soc 60:1201–1203. doi:10.1021/ja01272a058
34. Sugisawa T, Hoshino T (2002) Purification and properties of membrane-bound D-sorbitol dehydrogenase from *Gluconobacter suboxydans* IFO 3255. Biosci Biotechnol Biochem 66:57–64
35. Matsushita K, Fujii Y, Ano Y, Toyama H, Shinjoh M, Tomiyama N, Miyazaki T, Sugisawa T, Hoshino T, Adachi O (2003) 5-keto-D-gluconate production is catalyzed by a quinoprotein glycerol dehydrogenase, major polyol dehydrogenase, in *Gluconobacter* species. Appl Environ Microbiol 69:1959–1966. doi:10.1128/aem.69.4.1959- 1966.2003
36. Salusjärvi T, Povelainen M, Hvorslev N, Eneyskaya EV, Kulminskaya AA, Shabalin KA, Neustroev KN, Kalkkinen N, Miasnikov AN (2004) Cloning of a gluconate/polyol dehydrogenase gene from *Gluconobacter suboxydans* IFO 12528, characterisation of the enzyme and its use for the production of 5-ketogluconate in a recombinant *Escherichia coli* strain. Appl Microbiol Biotechnol 65:306–314. doi:10.1007/s00253-004-1594-6
37. Miyazaki T, Tomiyama N, Shinjoh M, Hoshino T (2002) Molecular cloning and functional expression of D-sorbitol dehydrogenase from *Gluconobacter suboxydans* IFO3255, which requires pyrroloquinoline quinone and hydrophobic protein SldB for activity development in *E. coli*. Biosci Biotechnol Biochem 66:262–270
38. Shinjoh M, Tomiyama N, Miyazaki T, Hoshino T (2002) Main polyol dehydrogenase of *Gluconobacter suboxydans* IFO 3255, membrane-bound D-sorbitol dehydrogenase, that needs product of upstream gene, sldB, for activity. Biosci Biotechnol Biochem 66:2314–2322
39. Hoshino T, Sugisawa T, Shinjoh M, Tomiyama N, Miyazaki T (2003) Membrane-bound D-sorbitol dehydrogenase of *Gluconobacter suboxydans* IFO 3255 - enzymatic and genetic characterization. Biochimica et Biophysica Acta (BBA) Proteins and Proteomics 1647:278–288. doi:10.1016/S1570-9639(03)00071-2
40. Shinagawa E, Matsushita K, Adachi O, Ameyama M (1982) Purification and characterization of D-sorbitol dehydrogenase from membrane of *Gluconobacter suboxydans* var. α. Agric Biol Chem 46:135–141
41. Choi E-S, Lee E-H, Rhee S-K (1995) Purification of a membrane-bound sorbitol dehydrogenase from *Gluconobacter suboxydans*. FEMS Microbiol Lett 125:45–49. doi: 10.1111/j.1574-6968.1995.tb07333.x
42. Isono M, Nakanishi I, Sasajima K, Motizuki K, Kanzaki T, Okazaki H, Yoshino H (1968) 2-keto-L-gulonic acid fermentation. Part I. Paper chromatographic characterization of metabolic products from sorbitol and L-sorbose by various bacteria. Agric Biol Chem 35:424–431
43. Okazaki H, Kanzaki T, Doi M, Nara K, Motizuki K (1968) 2-keto-L-gulonic acid fermentation. Part II. Identification of metabolic products from sorbitol. Agric Biol Chem 32:1250–1255
44. Okazaki H, Kanzaki T, Sasajima K, Terada Y (1969) 2-keto-L-gulonic acid fermentation. Part III. Evaluation of the pathway of sorbitol metabolism in *Gluconobacter melanogenus*. Agric Biol Chem 33:207–211
45. Kanzaki T, Okazaki H (1970) 2-keto-L-gulonic acid fermentation. Part IV. L-sorbose metabolism in *Pseudomonas aeruginosa*. Agric Biol Chem 34:432–436

46. Tsukada Y, Perlman D (1972) The fermentation of L-sorbose by *Gluconobacter melanogenus*. I. General characteristics of the fermentation. Biotechnol Bioeng 14:799–810. doi:10.1002/bit.260140509
47. Tsukada Y, Perlman D (1972) The fermentation of L-sorbose by *Gluconobacter melanogenus*. II. Inducible formation of enzyme catalyzing conversion of L-sorbose to 2-keto-L-gulonic acid. Biotechnol Bioeng 14:811–818. doi:10.1002/bit.260140510
48. Tsukada Y, Perlman D (1972) The fermentation of L-sorbose by *Gluconobacter melanogenus*. III. Investigation of the metabolic pathway from sorbose to 2-keto-L-gulonic acid. Biotechnol Bioeng 14:1035–1038. doi:10.1002/bit.260140612
49. Makover S, Ramsey G, Vane F, Witt C, Wright R (1975) New mechanisms for the biosynthesis and metabolism of 2-keto-L-gulonic acid in bacteria. Biotechnol Bioeng 17:1485–1514
50. Köpper S, Freimund S (2003) The composition of keto aldoses in aqueous solution as determined by NMR spectroscopy. Helv Chim Acta 86:827–843. doi:10.1002/hlca.200390083
51. Kitamura I, Perlman D (1975) Metabolism of L-sorbose by enzymes from *Gluconobacter melanogenus* IFO 3293. Eur J Appl Microbiol 2:1–7. doi:10.1007/bf01385440
52. Hoshino T, Sugisawa T, Tazoe M, Shinjoh M, Fujiwara A (1990) Metabolic pathway for 2-keto-L-gulonic acid formation in *Gluconobacter melanogenus* IFO 3293. Agric Biol Chem 54:1211–1218
53. Sugisawa T, Hoshino T, Nomura S, Fujiwara A (1991) Isolation and characterization of membrane-bound L-sorbose dehydrogenase from *Gluconobacter melanogenus* UV10. Agric Biol Chem 55:363–370
54. Sugisawa T, Hoshino T, Masuda S, Nomura S, Setoguchi Y, Tazoe M, Shinjoh M, Someha S, Fujiwara A (1990) Microbial production of 2-keto-L-gulonic acid from L-sorbose and D-sorbitol by *Gluconobacter melanogenus*. Agric Biol Chem 54:1201–1209
55. Hoshino T, Sugisawa T, Fujiwara A (1991) Isolation and characterization of NAD(P)-dependent L-sorbosone dehydrogenase from *Gluconobacter melanogenus* UV10. Agric Biol Chem 55:665–670
56. Manning R, Kahn M (1987) Biosynthesis of 2 keto-L-gulonic acid. US 5082785
57. Saito Y, Ishii Y, Hayashi H, Imao Y, Akashi T, Yoshikawa K, Noguchi Y, Soeda S, Yoshida M, Niwa M, Hosoda J, Shimomura K (1997) Cloning of genes coding for L-sorbose and L-sorbosone dehydrogenases from *Gluconobacter oxydans* and microbial production of 2-keto-L-gulonate, a precursor of L-ascorbic acid, in a recombinant *G. oxydans* strain. Appl Environ Microbiol 63:454–460
58. Huang H (1960) Preparation of 2-keto-L-gulonic acid. US 3043749
59. Tengerdy R (1961) Redox potential changes in the 2-keto-L-gulonic acid fermentation—I. Correlation between redox potential and dissolved-oxygen concentration. J Biochem Microbiol Technol Eng 3:241–253
60. Motizuki K, Kanzaki T, Okazaki H, Yoshino H, Nara K, Isono M, Nakanishi I, Sasajima K (1962) Method for producing 2-keto-L-gulonic acid. US 3234105
61. Krajewski V, Simić P, Mouncey NJ, Bringer S, Sahm H, Bott M (2010) Metabolic engineering of *Gluconobacter oxydans* for improved growth rate and growth yield on glucose by elimination of gluconate formation. Appl Environ Microbiol 76:4369–4376. doi:10.1128/aem.03022-09
62. Yin G, Tao Z, Yu L, Wang D, Dan J, Yan Z, Ning W, Wang C, Wang S, Jiang H, Zhang D, Feng X, Zhao Q, Wei W (1980) Studies on the production of vitamin C precursor 2-keto-L-gulonic acid from L-sorbose by fermentation. I. isolation screening and identification of 2-keto-L-gulonic acid producing bacteria. Acta Microbiologica Sinica 20:246–251
63. Yan Z, Tao Z, Yu L, Yin G, Ning W, Wang C, Wang S, Jiang H, Yu J, Wang M, Yu X (1981) Studies on production of vitamin C precursor 2-keto-L-gulonic acid from L-sorbose by fermentation. II. Conditions for submerged fermentation of 2-keto-L-gulonic acid. Acta Mycrobiologica Sinica 21:185–191

64. Ning W, Tao Z, Wang C, Wang S, Yan Z, Yin G (1987) Fermentation process for producing 2-keto-L-gulonic acid. EP 0278447B1
65. Urbance J, Bratina B, Stoddard S, Schmidt T (2001) Taxonomic characterization of *Ketogulonigenium vulgare* gen. nov., sp. nov. and *Ketogulonigenium robustum* sp. nov., which oxidize L-sorbose to 2-keto-L-gulonic acid. Int J Syst Evol Microbiol 51:1059–1070
66. Anonymus (2001) Notification list. Int J Syst Evol Microbiol 51:1231–1233
67. Xiong X, Han S, Wang J, Jiang Z, Chen W, Jia N, Wei H, Cheng H, Yang Y, Zhu B, You S, He J, Hou W, Chen M, Yu C, Jiao Y, Zhang W (2011) Complete genome sequence of the bacterium *Ketogulonicigenium vulgare* Y25. J Bacteriol 193:315–316
68. Liu L, Li Y, Zhang J, Zhou Z, Liu J, Li X, Zhou J, Du G, Wang L, Chen J (2011) Complete genome sequence of the industrial strain *Ketogulonicigenium vulgare* WSH-001. J Bacteriol 193:6108–6109
69. Stoddard S, Liaw H, Eddington J, Yang Y (1996) Bacterial strains and use thereof in fermentation processes for 2-keto-L-gulonic acid protection. EP0939831 B1
70. Asakura A, Hoshino T (1999) Isolation and characterization of a new quinoprotein dehydrogenase, L-sorbose/L-sorbosone dehydrogenase. Biosci Biotechnol Biochem 63:46–53
71. Asakura A, Hoshino T, Ojima S, Shinjo M, Tomiyama N (1996) Alcohol/aldehyde dehydrogenase. US 6730503 B1
72. Hao A, Jia Q, Wu H, Zhou H, Geng W, Gao W, Zhao J, He J (2008) Isolation and characteristics research of L-sorbose dehydrogenase in *Ketogulonigenium* sp. WB0104. Ind Microbiol 38:10–14
73. Hao A, Jia Q, Wu H, Zhou H, Geng W, Gao W, Zhao J, He J (2006) L-sorbinose dehydrogenase and its coding gene and uses. CN 101085987 B
74. Zhan W, Jiao Y, Yuan H, Xie L (2003) A new L-sorbose dehydrogenase gene and the protein it encoded. CN 1521181
75. Ma Q, Zhou J, Zhang W, Meng X, Sun J, Yuan Y (2011) Integrated proteomic and metabolomic analysis of an artificial microbial community for two-step production of vitamin C. PLoS ONE 6:e26108. doi:10.1371/journal.pone.0026108
76. Takagi Y, Sugisawa T, Hoshino T (2009) Continuous 2-keto-L-gulonic acid fermentation from L-sorbose by *Ketogulonigenium vulgare* DSM 4025. Appl Microbiol Biotechnol 82:1049–1056. doi:10.1007/s00253-008-1822-6
77. Takagi Y, Sugisawa T, Hoshino T (2010) Continuous 2-keto-L-gulonic acid fermentation by mixed culture of *Ketogulonicigenium vulgare* DSM 4025 and *Bacillus megaterium* or *Xanthomonas maltophilia*. Appl Microbiol Biotechnol 86:469–480. doi: 10.1007/s00253-009-2312-1
78. Hoshino T, Ojima S, Sugisawa T (1991) Fermentation process for producing 2-keto-L-gulonic acid. EP0518136B1
79. Gao Y, Yuan Y-J (2011) Comprehensive quality evaluation of corn steep liquor in 2-keto-L-gulonic acid fermentation. J Agric Food Chem 59:9845–9853. doi:10.1021/jf201792u
80. Zhang J, Zhou J, Liu J, Chen K, Liu L, Chen J (2011) Development of chemically defined media supporting high cell density growth of *Ketogulonicigenium vulgare* and *Bacillus megaterium*. Biores Technol 102:4807–4814. doi:10.1016/j.biortech.2010.10.124
81. Leduc S, Troostembergh JC, Lebeault JM (2004) Folate requirements of the 2-keto-L-gulonic acid-producing strain *Ketogulonigenium vulgare* LMP P-20356 in L-sorbose/CSL medium. Appl Microbiol Biotechnol 65:163–167. doi:10.1007/s00253-004-1562-1
82. Cai L, Yuan M-Q, Li Z-J, Chen J-C, Chen G-Q (2012) Genetic engineering of *Ketogulonigenium vulgare* for enhanced production of 2-keto-L-gulonic acid. J Biotechnol 157:320–325. doi:10.1016/j.jbiotec.2011.12.004
83. Zou W, Liu L, Zhang J, Yang H, Zhou M, Hua Q, Chen J (2012) Reconstruction and analysis of a genome-scale metabolic model of the vitamin C producing industrial strain *Ketogulonicigenium vulgare* WSH-001. J Biotechnol 161:42–48. doi:10.1016/j.jbiotec.2012.05.015

84. Ma Q, Zhang W, Zhang L, Qiao B, Pan C, Yi H, Wang L, Yuan YJ (2012) Proteomic analysis of *Ketogulonicigenium vulgare* under glutathione reveals high demand for thiamin transport and antioxidant protection. PLoS ONE 7:e32156
85. Liu L, Chen K, Zhang J, Liu J, Chen J (2011) Gelatin enhances 2-keto-L-gulonic acid production based on *Ketogulonigenium vulgare* genome annotation. J Biotechnol 156:182–187. doi:10.1016/j.jbiotec.2011.08.007
86. Lv S, Zhao S, Yang Y, Zhang Z, Chen H (2011) Research progress on Vc precursor of 2-KGA production through mixed fermentation from L-sorbose. Biotechnol Bull 5:50–54
87. Yi H, Zhang H, Zhu W, Zeng Y (2003) Progresses of vitamin C productive technology. China Food Addit 6:76–81
88. Xu A, Yao J, Yu L, Lv S, Wang J, Yan B, Yu Z (2004) Mutation of *Gluconobacter oxydans* and *Bacillus megaterium* in a two-step process of L-ascorbic acid manufacture by ion beam. J Appl Microbiol 96:1317–1323
89. Song Q, He J, Ren S, Ye Q, Guo X, Cheng C, Yin G (1997) Production of vitamin C precursor-2-L-keto-gulonic acid from L-sorbose by a novel bacterial component system of SCB329 SCB933. III.The characteristics and control of 2-keto-L-gulonic fermentation. Ind Microbiol 27:6–10
90. Hoshino T, Kiyasu T, Shinjoh M (2001) Enzymatic process for the manufacture of L-ascorbic acid and D-erythorbic acid. US 2005019878
91. Asakura A, Hoshino T, Kiyasu T, Shinjoh M (1999) Manufacture of L-ascorbic acid and D-erythorbic acid. EP1026257 B1
92. Sugisawa T, Ojima S, Matzinger P, Hoshino T (1995) Isolation and characterization of a new vitamin C producing enzyme (L-gulono-γ-lactone dehydrogenase) of bacterial origin. Biosci Biotechnol Biochem 59:190–196
93. Sugisawa T, Miyazaki T, Hoshino T (2005) Microbial production of L-ascorbic acid from D-sorbitol, L-sorbose, L-gulose, and L-sorbosone by *Ketogulonicigenium vulgare* DSM 4025. Microbiol Ferment Technol Commun 69:659–662
94. Berry A, Lee C, Mayer A, Shinjoh M (2003) Microbial production of L-ascorbic acid. EP2348113
95. Loewus M, Bedgar D, Saito K, Loewus F (1990) Conversion of L-sorbosone to L-ascorbic acid by a NADP-dependent dehydrogenase in bean and spinach leaf. Plant Physiol 94:1492–1495
96. Miyazaki T, Sugisawa T, Hoshino T (2006) Pyrroloquinoline quinone-dependent dehydrogenases from *Ketogulonicigenium vulgare* catalyze the direct conversion of L-sorbosone to L-ascorbic acid. Appl Environ Microbiol 72:1487–1495. doi: 10.1128/aem.72.2.1487-1495.2006
97. Heller A, Feldman B (2008) Electrochemical glucose sensors and their applications in diabetes management. Chem Rev 108:2482–2505. doi:10.1021/cr068069y
98. Beardmore-Gray M, Anthony C (1986) The oxidation of glucose by *Acinetobacter calcoaceticus*: interaction of the quinoprotein glucose dehydrogenase with the electron transport chain. J Gen Microbiol 132:1257–1268. doi:10.1099/00221287-132-5-1257
99. Cleton-Jansen A, Goosen N, Vink K, van de Putte P (1989) Cloning of the genes encoding the two different glucose dehydrogenases from *Acinetobacter calcoaceticus*. Antonie Van Leeuwenhoek 56:73–79
100. Prust C, Hoffmeister M, Liesegang H, Wiezer A, Fricke WF, Ehrenreich A, Gottschalk G, Deppenmeier U (2005) Complete genome sequence of the acetic acid bacterium *Gluconobacter oxydans*. Nat Biotech 23:195–200. doi:http://www.nature.com/nbt/journal/v23/n2/suppinfo/nbt1062_S1.html
101. Hölscher T, Weinert-Sepalage D, Görisch H (2007) Identification of membrane-bound quinoprotein inositol dehydrogenase in *Gluconobacter oxydans* ATCC 621H. Microbiology 153:499–506. doi:10.1099/mic.0.2006/002196-0
102. Dunning JW, Fulmer EI, Guymon JF, Underkofler LA (1938) The growth and chemical action of *Acetobacter suboxydans* upon l-inositol. Science (New York, N.Y.) 87:72

103. Kluyver A, Boezaardt A (1939) Note on the biochemical preparation of inosose. Rec Trav Chim Pays-Bas 58:956–958
104. Posternak T (1941) Recherches dans la série des cyclites V. Sur un inosose préparé par voie biochimique. Helv Chim Acta 24:1045–1058. doi:10.1002/hlca.194102401127
105. Carter H, Belinskey C, Clark R, Flynn E, Lytle B, McCasland G, Robbins M (1948) Oxidation of inositol by *Acetobacter suboxydans*. J Biol Chem 174:415–426
106. Chargaff E, Magasanik B (1946) Oxidation of stereoisomers of the inositol group by *Acetobacter suboxydans*. J Biol Chem 165:379–380
107. Magasanik B, Chargaff E (1948) The stereochemistry of an enzymatic reaction; the oxidation of 1-, d-, and epi-inositol by *Acetobacter suboxydans*. J Biol Chem 174:173–188
108. Magasanik B, Chargaff E (1948) The oxidation of d-quercitol by *Acetobacter suboxydans*. J Biol Chem 175:939–943
109. Magasanik B, Franzl RE, Chargaff E (1952) The stereochemical specificity of the oxidation of cyclitols by *Acetobacter suboxydans*. J Am Chem Soc 74:2618–2621. doi: 10.1021/ja01130a045
110. Anderson L, Takeda R, Angyal S, McHugh D (1958) Cyclitol oxidation by *Acetobacter suboxydans*. II. Additional cyclitols and the "Third Specificity Rule". Arch Biochem Biophys 78:518–531
111. Criddle W, Fry J, Keaney M (1974) Myo-inositol dehydrogenase(s) from *Acetomonas oxydans*. Optimization of conditions for solubilization of membrane-bound enzyme. Biochem J 137:449–452
112. Criddle W, Fry J, Keaney M, Lucas C, Tovey J (1977) Myo-inositol dehydrogenase(s) from *Acetomonas oxydans*. Mol Cell Biochem 16:3–8
113. Wissler J, Freivogel K, Wiesner W (1995) Cyclitol. WO9704101
114. Yagi J, Yamashita T, Kato A, Takaki Y, Sakai H (1967) Studies on erythorbic acid production by fermentation. Part I. Erythorbic acid-producing strain and cultural condition. Agric Biol Chem 31:340–345
115. Takahashi T (1969) Erythorbic acid fermentation. Biotechnol Bioeng 11:1157–1171. doi: 10.1002/bit.260110611
116. Takahashi T, Mitsumoto M, Kayamori H (1960) Production of D-araboascorbic acid by penicillium. Nature 188:411–412
117. Takahashi T, Yamashita H, Kato E, Mitsumoto M, Murakawa S (1976) Purification and some properties of D-glucono-γ-lactone dehydrogenase D-erythorbic acid producing enzyme of *Penicillium cyaneo-fulvum*. Agric Biol Chem 40:121–129
118. Salusjärvi T, Kalkkinen N, Miasnikov AN (2004) Cloning and characterization of gluconolactone oxidase of *Penicillium cyaneo-fulvum* ATCC 10431 and evaluation of its use for production of D-erythorbic acid in recombinant *Pichia pastoris*. Appl Environ Microbiol 70:5503–5510. doi:10.1128/aem.70.9.5503-5510.2004
119. Murakawa S, Takahashi T (1977) Biosynthesis of a new ascorbic acid analogue by D-gluconolactone dehydrogenase of *Penicillium cyaneo-fulvum*. Agric Biol Chem 41:2103–2104
120. Takahashi T, Mitsumoto M (1963) Transformation and hydrolysis of D-gulono-gamma and delta-lactone. Nature 199:765–767
121. Neidelman S, Amon W, Geigert J (1980) Production of 2-keto-D-gluconic acid and hydrogen peroxide. US 4351902
122. Shao Y, Seib P, Kramer K, Van Galen D (1993) Synthesis and properties of D-erythroascorbic acid and its vitamin C activity in the tobacco hornworm (*Manduca sexta*). J Agric Food Chem 41:1391–1396
123. Kim S-T, Huh W-K, Lee B-H, Kang S-O (1998) D-Arabinose dehydrogenase and its gene from *Saccharomyces cerevisiae*. Biochimica et Biophysica Acta (BBA) Protein Structure and Molecular Enzymology 1429:29–39. doi:10.1016/S0167-4838(98)00217-9
124. Amako K, Fujita K, Shimohata T, Hasegawa E, Kishimoto R, Goda K (2006) NAD+-specific D-arabinose dehydrogenase and its contribution to erythroascorbic acid production in *Saccharomyces cerevisiae*. FEBS Lett 580:6428–6434

125. Huh W-K, Lee B-H, Kim S-T, Kim Y-R, Rhie G-E, Baek Y-W, Hwang C-S, Lee J-S, Kang S-O (1998) D-Erythroascorbic acid is an important antioxidant molecule in *Saccharomyces cerevisiae*. Mol Microbiol 30:895–903. doi:10.1046/j.1365-2958.1998.01133.x
126. Hancock R, Galpin J, Viola R (2000) Biosynthesis of L-ascorbic acid (vitamin C) by *Saccharomyces cerevisiae*. FEMS Microbiol Lett 186:245–250
127. Shimizu K, Nishiyama K, Inoue T, Takano N, Mikata M, Masataka Y, Azuma T, Osawa S (1967) Studies on erythorbic acid production by fermentation. Part II. Erythorbic acid production by Jar Fermentor. Agric Biol Chem 31:346–352
128. Sun W-J, Zhou Y-Z, Zhou Q, Cui F-J, Yu S-L, Sun L (2012) Semi-continuous production of 2-keto-gluconic acid by *Pseudomonas fluorescens* AR4 from rice starch hydrolysate. Biores Technol 110:546–551. doi:10.1016/j.biortech.2012.01.040
129. Matsushita K, Ameyama M (1982) D-Glucose dehydrogenase from *Pseudomonas fluorescens*, membrane-bound. Methods Enzymol 89:149–154
130. Ramakrishnan T, Cambell J (1955) Gluconic dehydrogenase of *Pseudomonas aeruginosa*. Biochim Biophys Acta 17:122–127
131. Matsushita K, Shinagawa E, Ameyama M (1982) D-Gluconate dehydrogenase from bacteria, 2-keto-D-gluconate-yielding, membrane-bound. Methods Enzymol 89:187–193
132. Chundawat S, Beckham G, Himmel M, Dale B (2011) Deconstruction of lignocellulosic biomass to fuels and chemicals. Annu Rev Chem Biomol Eng 2:121–145
133. Silveira MM, Jonas R (2002) The biotechnological production of sorbitol. Appl Microbiol Biotechnol 59:400–408

Adv Biochem Eng Biotechnol (2014) 143: 189–228
DOI: 10.1007/10_2014_269

Published Online: 28 March 2014

Amino Acids in Human and Animal Nutrition

Andreas Karau and Ian Grayson

Abstract Amino acids are key components of human and animal nutrition, both as part of a protein-containing diet, and as supplemented individual products. In the last 10 years there has been a marked move away from the extraction of amino acids from natural products, which has been replaced by efficient fermentation processes using nonanimal carbon sources. Today several amino acids are produced in fermentation plants with capacities of more than 100,000 tonnes to serve the requirements of animal feed and human nutrition. The main fermentative amino acids for animal nutrition are L-lysine, L-threonine, and L-tryptophan. DL-Methionine continues to be manufactured for animal feed use principally by chemical synthesis, and a pharmaceutical grade is manufactured by enzymatic resolution. Amino acids play an important role in medical nutrition, particularly in parenteral nutrition, where there are high purity requirements for infusion grade products. Amino acids are also appearing more often in dietary supplements, initially for performance athletes, but increasingly for the general population. As the understanding of the effects of the individual amino acids on the human metabolism is deepened, more specialized product mixtures are being offered to improve athletic performance and for body-building.

Keywords Amino acid · Animal feed · Biotransformation · Corynebacterium glutamicum · Dietary requirements · Fermentation · Flavorings · Isolation · Medical nutrition · Parenteral nutrition · Sports nutrition · Supplements · Sweeteners · Synthesis

Contents

A. Karau (✉) · I. Grayson
Evonik Industries AG, Business Line Health Care, Rodenbacher Chaussee 4,
63457 Hanau-Wolfgang, Germany
e-mail: andreas.karau@evonik.com

1 Introduction

Proteins are composed of the 20 naturally occurring α-amino acids, and are an essential component of both animal and human nutrition. Although a balanced diet can supply sufficient amino acids to satisfy nutritional demand, animal feed in particular can be deficient in essential amino acids to maintain optimal growth and health of the animal. In addition, amino acids are used to supplement or fortify human food, to add flavor (e.g., monosodium glutamate, MSG), or to supply completely the body's amino acid requirements in the case of medical nutrition. This review discusses the key processes for the industrial manufacture of the proteinogenic amino acids, together with their key derivatives and some nonproteinogenic amino acids and the application of these for both human and animal nutrition.

The importance of amino acids in nutrition, and the equivalence between protein and protein hydrolysates in providing nitrogen balance in the diet was recognized at the start of the twentieth century [1]. The use of amino acids as supplements increased through the first half of the twentieth century mainly due to usage in animal feeding but also for human nutrition. More recently, the application of amino acids in medical nutritional supplements and in products directed towards athletes and the general public has become much more widespread. The general chemistry and application of amino acids in industry have been presented in several reviews [2, 3] and are not covered here.

2 Production of Amino Acids

2.1 Historical Production of Amino Acids

The industrial production of amino acids began early in the last century with the extraction of individual amino acids from natural sources. These included hydrolysis of the nonsugar portion of molasses (from sugar beet or sugar cane), [4]

Table 1 Typical amino acid composition of commercial molasses hydrolysate [4]

Amino acid	Approximate composition (%)
L-Glutamic acid	45
L-Aspartic acid	10
L-Alanine	5
Glycine	5
Taurine	5
L-Tyrosine	5
L-Isoleucine	4
L-Leucine	4
L-Serine	4
L-Valine	3
Others	10

Table 2 Typical amino acid composition of animal protein hydrolysate [5]

Typical composition from ox-hide gelatin (%)	Amino Acid
23.2	Glycine
13.9	L-Proline
11.3	L-4-*trans*-Hydroxyproline
9.8	L-Glutamic acid
9.3	L-Alanine
7.3	L-Arginine
5.6	L-Aspartic acid
3.6	L-Lysine
3.4	L-Serine
2.9	L-Leucine
2.2	L-Valine
1.9	L-Phenylalanine
1.9	L-Threonine
1.4	L-Isoleucine
2.3	Others

and acid hydrolysis of animal products such as beef gelatin or keratin from chicken feathers or animal hair [4, 5]. Table 1 shows the typical amino acid composition of molasses, and Table 2, that of animal-derived protein.

The protein hydrolysis method provides ready access to up to 17 of the proteinogenic amino acids. Following filtration of the insoluble cystine and tyrosine fraction, the amino acid solution is separated into acidic, neutral, and basic fractions by extraction, and the individual amino acids are separated and purified by ion exchange chromatography followed by crystallization (Fig. 1). It can be seen that cysteine is converted into the dimer, cystine, whereas asparagine and glutamine are hydrolyzed to the respective acids, and methionine and tryptophan are largely destroyed by the hydrolysis process.

The discovery in 1908 of monosodium glutamate, extracted from kelp, as a flavoring agent [6] led to the development of processes for its extraction from acid

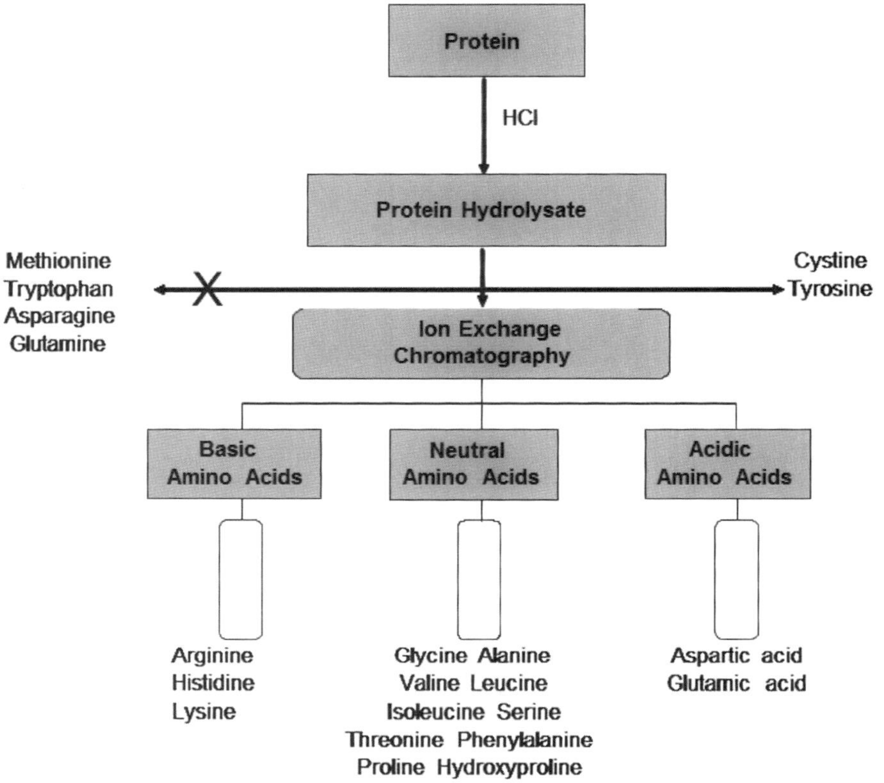

Fig. 1 Flow diagram for the isolation of amino acids from animal protein hydrolysis (Evonik Rexim)

hydrolysis of wheat gluten or from defatted soya bean, and the establishment of an industrial process by the Ajinomoto Company for commercial manufacture of MSG. Today, although some amino acids are still produced in part by extraction from natural sources, modern production methods have overtaken these traditional routes in terms of quality, efficiency, and price competitiveness.

2.2 Modern Production of Amino Acids

The processes applied today depend mainly on the volumes produced and whether the pure L-enantiomer or the racemate is the desired product for the particular application.

All the proteinogenic amino acids have been prepared by chemical synthesis, both in the racemic and enantiomerically pure forms, but only two amino acids are routinely manufactured by chemical synthesis on a large scale. These are glycine and DL-methionine. Glycine is a simple synthetic task as it is the only achiral natural amino acid. Methionine is unusual in that both the D- and L-forms are nutritionally equivalent, so for the major use, animal feed, the DL-form is routinely applied. L-Methionine is manufactured from the racemate by an enzymatic resolution process. When L-phenylalanine was required in large amounts as a raw material for the sweetener aspartame in the 1980s, a number of synthetic routes were developed, both enzymatic processes and those using homogeneous asymmetric catalysis. A number of the chemical processes were run at the pilot scale, but none have been developed as a full-scale production process [7]. These chemical methods were later displaced as fermentation processes became the method of choice for the manufacture of most amino acids.

The large-scale fermentation of amino acids began in the 1950s when Kinoshita, Udaka, and coworkers at Kyowa Hakko in Japan discovered the glutamate-producing bacterium now known as *Corynebacterium glutamicum* [8]. This glutamate fermentation method has been developed into a low-cost process that today produces more than 1,000,000 tons of L-glutamic acid for the food industry. A few years later a mutant of *C. glutamicum* was developed that produced a high yield of L-lysine, one of the major amino acids required for animal nutrition. Large-scale biotransformation processes have also been implemented, for example, the Degussa acylase process for the production of L-methionine, which has been in use since the 1980s [9].

Over the last 60 years, the development and optimization of bacterial strains used in amino acid production have changed dramatically [10]. In the past, the optimization was done based on random mutation and selection. Mutagenesis was performed by ultraviolet radiation or by chemical mutagens (e.g., using nitrosoguanidine). The strains were selected either in the presence of an analogue of the amino acid or of an important intermediate in the biosynthetic pathway of the relevant amino acid. These strains were selected based on growth, product titer, and also for low by-product formation. Based on this approach, feedback inhibition of the amino acid itself or relevant intermediates could be eliminated. Several practical industrial production strains have been obtained using this approach. Its advantage was that no detailed genetic information on the microbial strains was required. However, the increase in performance is limited, as specific properties cannot be changed using the mutagenesis approach, and in addition negative mutations can also accumulate (e.g., slow growth rates, by-product formation, etc.). Furthermore, it was not possible to maximize the yield based on carbon using this method.

During the 1990s, more targeted strain optimization approaches were implemented based on the development of genetic engineering tools. This included the amplification of relevant genes in the product pathway, the knock-out of specific genes that caused side-product formation and the modification of specific enzyme properties by protein evolution methods. In addition, improvements were made by modifications of the importers and exporters of the product and relevant substrates. Based on this approach, it was possible to reduce selectively the formation of critical side products, to eliminate bottlenecks in the biosynthetic pathway, and also to increase the carbon yield.

Table 3 Key manufacturing processes used today for the main amino acids

Amino acid	Production processes used
L-Alanine	Biotransformation
L-Arginine	Fermentation
L-Aspartic acid	Biotransformation
L-Cysteine	Fermentation, Extraction
L-Glutamic acid	Fermentation
L-Glutamine	Fermentation
Glycine	Chemical synthesis
L-Histidine	Fermentation
L-4-*trans*-Hydroxyproline	Biotransformation, Fermentation, Extraction
L-Isoleucine	Fermentation
L-Leucine	Fermentation
L-Lysine	Fermentation
DL-Methionine	Chemical synthesis
L-Methionine	Biotransformation
L-Phenylalanine	Fermentation
L-Proline	Fermentation
L-Serine	Biotransformation, Fermentation
L-Threonine	Fermentation
L-Tryptophan	Fermentation
L-Tyrosine	Fermentation
L-Valine	Fermentation

More recently, optimization based on system biology approaches has become increasingly relevant. In this approach the strains are characterized at four different levels: those of the genome, transcriptome, proteome, and metabolome. Supported by the appropriate analytical technologies, modeling and simulation tools, the strains are analyzed for current limitations in product formation, side-product formation, and the formation of central metabolites. Based on these results, targets for further genetic engineering work are then identified. The strains improved by genetic engineering are then tested in fermentation and downstream processing, and again characterized by the four different levels. This is carried out in several cycles until the desired strain performance is achieved [11, 12].

Today fermentation or enzymatic routes have been developed for almost all the major L-amino acids, and these have largely displaced the previous extraction methods. This has happened at the same time that customers have become increasingly concerned about the safety of animal-derived products, because of the possibility of infectious agents such as prions surviving the isolation process. In 2006, major amino acid manufacturers, including Degussa and Kyowa Hakko, announced the cessation of their protein extraction processes, and the establishment of fermentation methods for the major amino acids. However, extraction of cysteine from hair or feathers is still widely used, particularly in China, and extraction of 4-hydroxyproline from gelatin is still common. Table 3 shows the chief manufacturing processes for the main proteinogenic amino acids.

Table 4 Biosynthetic pathways for the main amino acids

Precursor	1st formed acid	2nd formed acid	3rd formed acid
Oxaloacetic acid	L-Aspartic acid	L-Asparagine	
		L-Lysine	
		L-Methionine	
		L-Threonine	L-Isoleucine
Glucose	L-Phenylalanine		
	L-Tyrosine		
	L-Tryptophan		
	L-Histidine		
Pyruvic acid	L-Alanine		
	L-Valine		
	L-Leucine		
3-Phosphoglyceric acid	L-Serine	L-Cysteine	
		Glycine	
2-Ketoglutaric acid	L-Glutamic acid	L-Glutamine	
		L-Proline	L-4-Hydroxyproline
		L-Arginine	
		L-Ornithine	

2.3 Production of Specific Amino Acids

This section describes the chief industrial processes used for manufacture of the main proteinogenic amino acids. The biotechnological processes can be grouped according to the key metabolic pathways involved in the biosynthesis of the amino acids (Table 4).

2.3.1 Glycine

Glycine is manufactured exclusively by chemical synthesis, and two main processes are practiced today (Scheme 1). The direct amination of chloroacetic acid with a large excess of ammonia gives good yields of glycine without producing large amounts of di- and trialkylated products [13]. This process is widely used in China, where the main application of the glycine is as a raw material for the herbicide glyphosate.

The other main process is the Strecker synthesis. The direct Strecker reaction of formaldehyde and ammonium cyanide produces methylene amino acetonitrile, which must be hydrolyzed in two stages to produce glycine [14]. A more efficient approach is to aminate the intermediate glycolonitrile, followed by hydrolysis [15]. An alternative method, which is more often applied for the homologous amino acids, is the Bucherer–Bergs reaction. Reaction of formaldehyde and ammonium carbonate or bicarbonate gives the intermediate hydantoin, which can be hydrolyzed to glycine in a separate step [16].

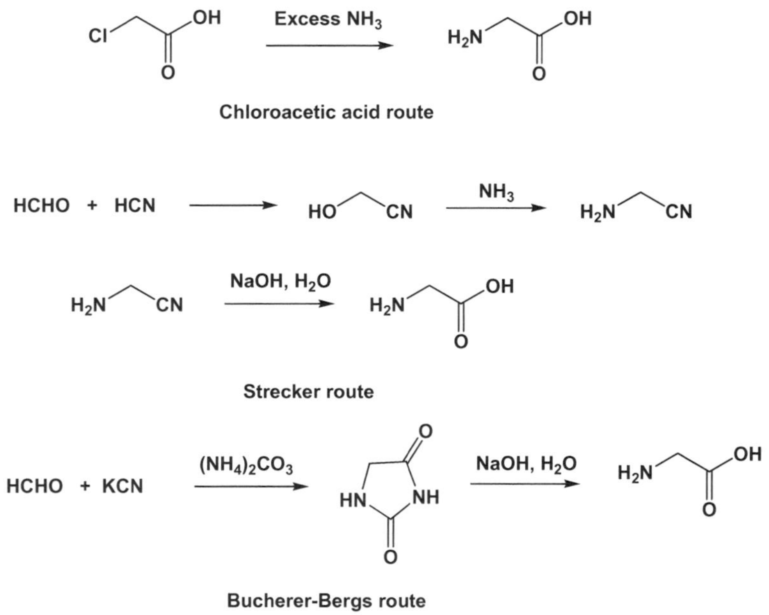

Scheme 1 Industrial processes for the manufacture of glycine

2.3.2 DL-Methionine

DL-Methionine is the second amino acid that is almost exclusively manufactured by chemical synthesis. The process used today was originally developed by Werner Schwarze at Degussa in the 1940s, and has been continually improved and refined since. Today DL-methionine is manufactured in several plants, each with a capacity of more than 100,000 tonnes. To operate the process on an industrial scale also requires back-integration into the key hazardous raw materials acrolein, methyl mercaptan, and hydrogen cyanide (Scheme 2). After the formation of the hydantoin, the key step is alkaline hydrolysis of the hydantoin, to produce methionine directly in up to 95 % yield based on acrolein [17, 18].

2.3.3 L-Lysine

C. glutamicum and, to a lesser extent, *E. coli* are the main organisms used today for industrial L-lysine production. The first L-producing strains based on *C. glutamicum* were reported in 1961 [19], and those based on *E. coli* in 1995 [20]. The advantages of using *E. coli* versus *C. glutamicum* include the achievement of higher growth rates at higher fermentation temperatures. The formation of lysine is highly influenced by two enzymes, aspartate kinase (AK) and homoserine dehydrogenase (HDH). AK converts aspartate into aspartate semialdehyde, and is

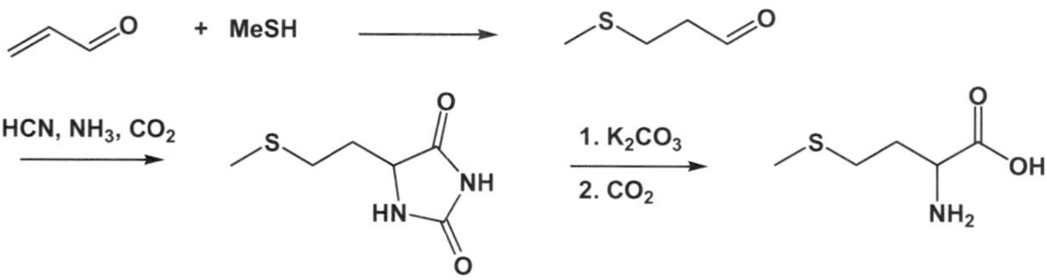

Scheme 2 Process for the manufacture of DL-Methionine

highly feedback-inhibited by lysine and threonine. HDH converts aspartate semialdehyde into homoserine, which is an intermediate for the biosynthesis of threonine, methionine, and isoleucine. L-Lysine–producing strains therefore often contain a deregulated AK and/or a reduced activity HDH [21]. Despite the improvement of the flux from aspartate towards lysine, the availability of key metabolites from the central metabolic pathways is also essential. Here the formation of oxaloacetate directly from phosphoenol pyruvate or via pyruvate is essential for the carbon yield as some unnecessary cycles are included. For example, inactivation of the enzyme phosphoenol pyruvate carboxykinase, which catalyzes the reverse reaction from oxaloacetate to phosphoenol pyruvate gave an improvement in lysine formation [22]. By overexpression of pyruvate carboxylase, the conversion yield of glucose to lysine could be increased by 50 % [23]. With a synthetic lysine hyperproducing strain, containing 12 defined modifications from the wild type, a carbon yield of 0.55 g/g and a product titer of 120 g/L over 30 h fermentation could be obtained [24].

Today, however, the main commercial process for L-lysine remains the fermentation of *C. glutamicum*. This is performed in fed-batch mode in large-scale fermenters of up to 500 m^3 volume, with production capacities in excess of 100,000 tonnes. The commercial manufacturing process has been comprehensively described by Pfefferle [25].

2.3.4 L-Threonine

L-Threonine can be produced using strains of *E. coli* or *C. glutamicum*. As threonine is also an amino acid of the aspartate family, aspartate semialdehyde is a common intermediate with the biosynthesis of L-lysine. In order to optimize a high-yielding L-threonine–producing strain, the following strategy is applied: the pathway towards L-lysine is minimized by reducing the activity of dihydrodipicolinate synthase (*dapA*) and at the same time the pathway towards L-threonine is favored by overexpression of the genes of the threonine operon, which consists of the genes for homoserine dehydratase (*thrA*), homoserine kinase (*thrB*), and threonine synthase (*thrC*). As L-threonine is also a precursor for L-isoleucine, further conversion of L-threonine into L-isoleucine has to be minimized by deactivation of the threonine

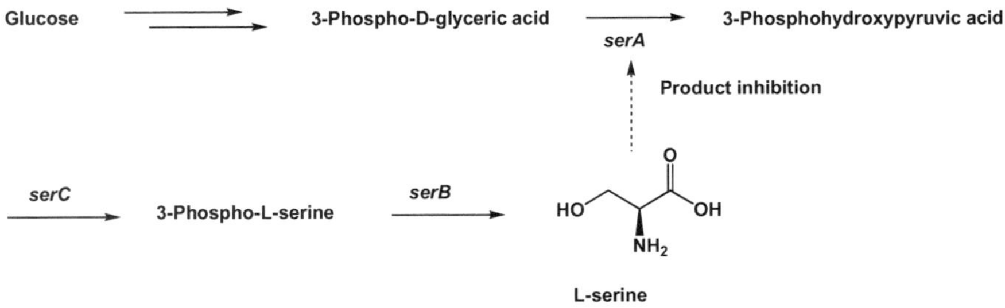

Fig. 2 Biosynthetic pathway for the production of L-serine by fermentation

dehydratase gene (*ilvA*). In the meantime, *E. coli* based strains have also been developed by the application of systems biology, not only by deletion or downregulation of the competing pathways such as L-lysine, L-methionine, and L-isoleucine, but also by optimization of the supply of key precursors such as oxaloacetate. The *E. coli* strain has been reported to produce 82 g/L L-threonine in 48 h with a carbon yield of 39 % [26]. A more detailed description of the development of a commercial L-threonine process has been given by Debabov [27]. Today L-threonine is manufactured on a commercial scale of several thousand tonnes using the *E. coli* fermentation process.

2.3.5 L-Serine

Serine is the first amino acid produced in the 3-phosphoglycerate pathway. It is further converted to glycine and L-cysteine. Industrially L-serine can be produced by direct fermentation or by an enzymatic process from glycine. The enzymatic route developed by Mitsui reacts glycine with formaldehyde using serine hydroxymethyltransferase (SHMT). With an overexpression of SHMT in *E. coli*, concentrations of serine of over 300 g/L in 35 h reaction time have been reported, with a glycine conversion of >98 % [28]. This process requires the addition of tetrahydrofolic acid to the system as a cofactor.

An alternative to enzymatic production is a direct fermentation to give L-serine. Strains based on *Brevibacterium flavum* [29] and *C. glutamicum* [30] have been described. In both strains, the enzymes phosphoglycerate dehydrogenase (*serA*), phosphoserine phosphatase (*serB*), and phosphoserine transaminase (*serC*) have been overexpressed. These enzymes are involved in the biosynthesis pathway from 3-phosphoglycerate (Fig. 2). Note that because of product inhibition by L-serine of *serA*, feedback-resistant mutants have been developed to increase yields. A mutant strain of *B. flavum* with a feedback-resistant *serA** and overexpression of the *serA**, *serB*, and *serC* has been reported to accumulate 35.2 g/L L-serine with a carbon yield of 32 % based on glucose [29]. In addition it has been shown that increased yields in *C. glutamicum* can be obtained by deleting the L-serine degrading enzyme L-serine dehydratase (sdaA).

2.3.6 L-Proline and L-4-Hydroxyproline

L-Proline is produced via the 2-ketoglutarate pathway from L-glutamic acid. In 1973 it was shown that an L-glutamic acid producer based on *C. glutamicum* could produce L-proline in the presence of an excessive amount of biotin and high concentrations of ammonium chloride [31]. Under these conditions around 40 g/L proline was obtained over 100 h with a conversion yield of 18 % based on glucose.

More detailed investigations with a proline overproducing strain of *Serratia marcescens* have demonstrated that the key bottleneck is product inhibition by L-proline of the first enzyme in the conversion of L-glutamate, γ-glutamyl kinase (GK; Fig. 3) [32]. With overexpression of a deregulated GK and some media optimization steps, a titer of 100 g/l in 4 days with a yield of 32 % based on sucrose was obtained [33]. In addition, elimination of the proline degradation pathway via proline oxidase is essential for obtaining an improved proline yield [34]. In addition to *S. marcescens*, *E. coli* strains have also been described for L-proline production.

With a route to L-proline established, access is now possible to L-*trans*-4-hydroxyproline by fermentation. This is achieved by use of the enzyme L-proline-4-hydroxylase, which belongs to the group of α-ketoglutarate–dependent dioxygenases and requires α-ketoglutaric acid as a cofactor (Fig. 4). Researchers at Kyowa Hakko have described the conversion of L-proline into L-4-hydroxyproline using an isolated enzyme, or alternatively by integration of the L-proline-4-hydroxylase enzyme into a proline-producing strain [35]. If the L-proline degradation pathway is inactivated, a complete conversion of L-proline into L-4-hydroxyproline can be achieved. Using resting *E. coli* cells carrying a plasmid containing an L-proline-4-hydroxylase gene from *Dactylosporangiumn sp.*, approximately 41 g/L of L-hydroxyproline could be obtained from proline in 100 h. The required α-ketoglutarate was produced in situ from glucose [36].

2.3.7 L-Glutamic Acid and L-Glutamine

As the first amino acid to be produced by fermentation, the development of the process to manufacture L-glutamic acid has been the subject of much research and process optimization. The overproduction of L-glutamic acid by strains of *Corynebacterium glutamicum* is induced by the addition of detergents such as Tween 60. In addition, extensive metabolic engineering of *C. glutamicum* has been performed, particularly by Ajinomoto, to generate the strains used today in commercial production, which can give production titers of more than 100 g/L [37]. All stages in the pathway from glucose to L-glutamic acid have been subject to study and optimization, including the uprating of the citric acid cycle, and deletion of genes for further metabolism of L-glutamic acid [38].

L-Glutamine is also produced from a strain of *C. glutamicum*, at volumes of only a few thousand tonnes. In this case, the ATP-dependent enzyme glutamine synthetase is upregulated, and the process is performed near the isoelectric point of

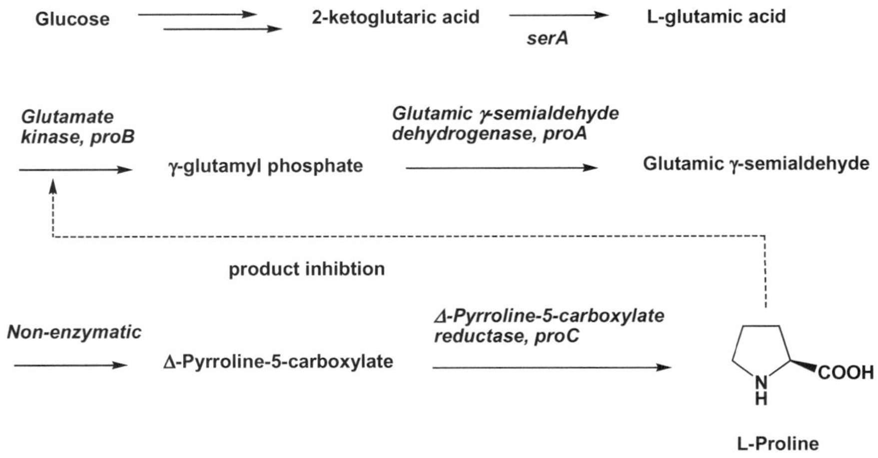

Fig. 3 Biosynthetic pathway for the production of L-proline by fermentation

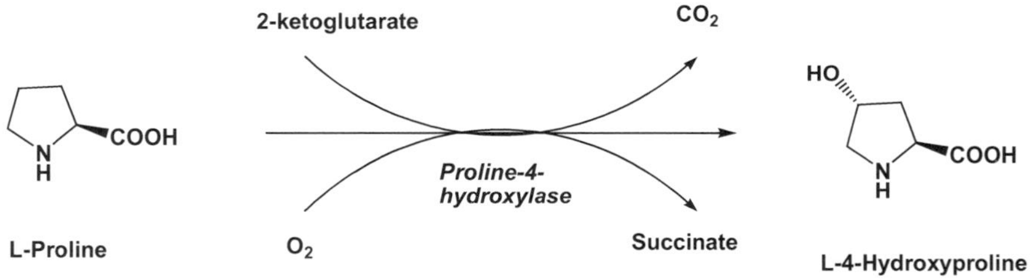

Fig. 4 Biosynthetic pathway for the production of L-*trans*-4-hydroxyproline by fermentation

L-glutamine, pH 5.6, as the product is readily hydrolyzed under more acid or alkaline conditions [39].

2.3.8 L-Tryptophan, L-Phenylalanine, and L-Tyrosine

L-Tryptophan, L-phenylalanine, and L-tyrosine are known as the aromatic amino acids, and are all derived from phosphoenol pyruvate via chorismate. A complex series of product inhibition feedback mechanisms controls the synthesis of the three aromatic amino acids (Fig. 5). Today, L-tryptophan is manufactured by fermentation on a scale of several thousand tonnes, using *E. coli* or *C. glutamicum*. Metabolic engineering has improved the yield of each of the aromatic amino acids, but the product titers are still low compared with other amino acids such as lysine or glutamic acid [40]. The fermentation route to L-tryptophan has today completely replaced the previous enzymatic process from indole and serine, using tryptophan synthetase from *E. coli* [41].

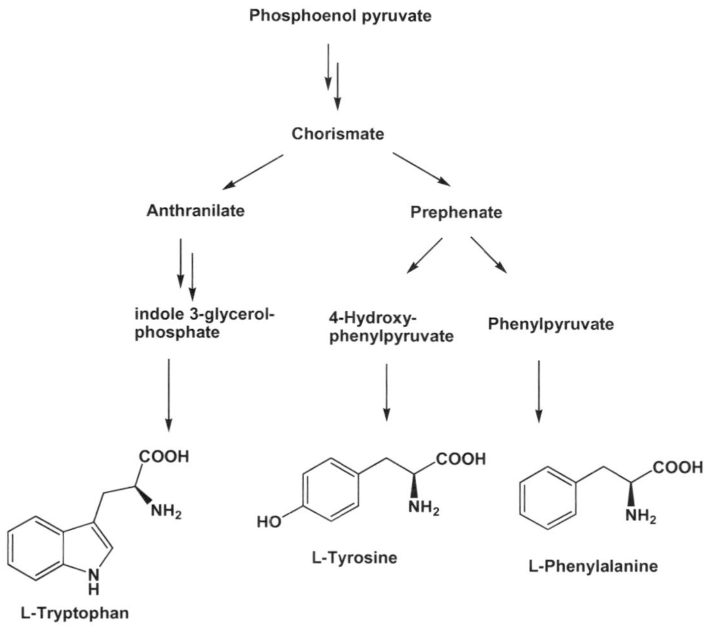

Fig. 5 Simplified metabolic pathway for the manufacture of the aromatic amino acids in *E. coli*

L-Phenylalanine is produced by fermentation on a large scale as the precursor for aspartame. It is generally produced in *E. coli* [40, 42], and the fermentation approach has completely replaced the enzymatic process, which involved the acylase-catalyzed hydrolysis of racemic N-acetylphenylalanine [43]. Fermentation methods for L-tyrosine using *C. glutamicum* have also been developed [44]; however, the market for this amino acid is smaller than that for phenylalanine. In 2006 Kyowa Hakko announced that they were introducing an industrial fermentation process for the manufacture of L-tyrosine.

2.3.9 L-Valine, L-Leucine, and L-Isoleucine

L-Isoleucine is a member of the aspartic acid pathway, and is produced by further downstream metabolism of L-threonine. It is manufactured on an industrial scale by fermentation using strains of *C. glutamicum* [45]. Optimization of the process requires an efficient method for the separation of L-isoleucine from the by-product L-valine [46].

L-Valine and L-leucine are both members of the pyruvic acid pathway. Fermentation methods have been reported using both *E. coli* and *C. glutamicum* strains. For example, patents have been filed by Ajinomoto for the manufacture of L-valine [47] and L-leucine [48] using *E. coli*. An *E. coli* strain giving increased L-valine production and high carbon yield has been constructed using a rational metabolic engineering approach [49].

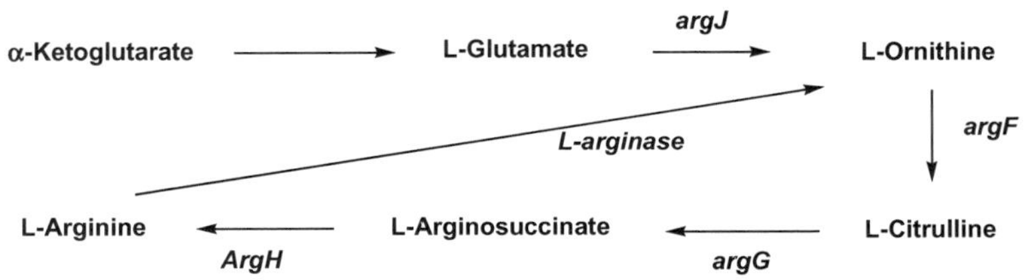

Fig. 6 Metabolic pathway for the manufacture of L-arginine and L-ornithine in *C. glutamicum*

2.3.10 L-Arginine and L-Ornithine

L-Arginine and L-ornithine are important amino acids as both are constituents of the urea cycle. They are members of the α-ketoglutarate family, being downstream products from L-glutamic acid (Fig. 6). L-Ornithine and L-citrulline are intermediates in the metabolic pathway, and the process can be engineered to produce either L-arginine, L-citrulline, or L-ornithine preferentially. L-Arginine is usually manufactured on a commercial scale using engineered strains of *C. glutamicum* [50, 51], although production using *Corynebacterium crenatum* [52] and *E. coli* [53] have also been reported.

L-Ornithine has been made in the past by the action of L-arginase on either L-arginine or DL-arginine [54]. Today, however, a modified *C. glutamicum* is employed for the direct fermentation of L-ornithine [55].

2.3.11 L-Histidine

L-Histidine is grouped with the aromatic amino acids, but the metabolic route diverges at an early stage from the other members of the group. It is produced in *C. glutamicum* in a 10-step sequence starting from phosphoribosyl pyrophosphate [56]. Originally production titers of up to 10.5 g/L were reported [57], but this has since been increased by workers at Kyowa Hakko to 22.5 g/L [58]. In a parallel development, the fermentation of L-histidine using *E. coli* has been reported by Ajinomoto, with titers up to 19.1 g/L [59]. Both the titer and the carbon yield for L-histidine are lower than those reported for L-phenylalanine and L-tryptophan, and L-histidine remains one of the more challenging amino acids to produce on an industrial scale.

2.3.12 L-Cysteine and L-Methionine

L-Cysteine and L-methionine are linked in their metabolic pathways in that both require a sulfuration step in their biosynthesis. L-Cysteine is derived from L-serine and L-methionine from L-homoserine (Fig. 7) [60]. Modification of *E. coli* for L-cysteine production requires removing the feedback inhibition of the enzyme

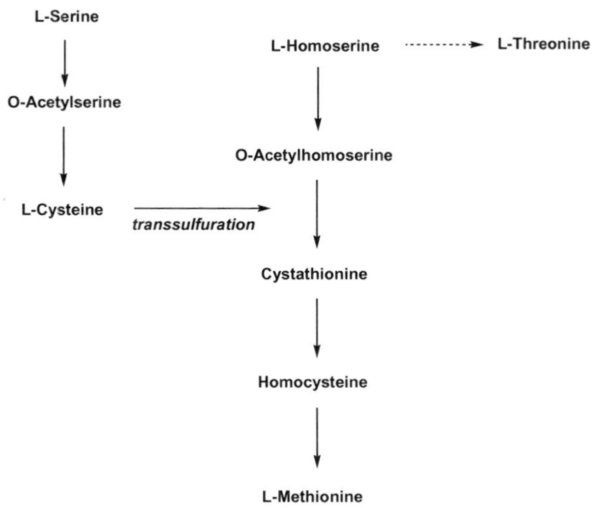

Fig. 7 The linked pathways for the production of L-cysteine and L-methionine in bacteria [60]

L-serine *O*-acetyltransferase. Early work in this area produced strains with an L-cysteine production rate of 1–2 g/L [60]. Wacker announced in 2001 that they were beginning commercial production of L-cysteine by fermentation; a relevant patent gives the production in *E. coli* as about 12 g/L [61]. Ajinomoto had previously utilized an enzymatic production process for L-cysteine involving the asymmetric hydrolysis of DL-2-amino-Δ^2-thiazoline-4-carboxylic acid [60]. However Ajinomoto has also filed patents on the fermentative production of L-cysteine in modified strains of *E. coli* [62].

There is as yet no industrial-scale fermentation method for the manufacture of L-methionine. L-Methionine is a member of the aspartate family, and is produced in *C. glutamicum* by suppressing the formation of L-threonine, and removing the feedback inhibition of homoserine dehydrogenase by L-threonine (Fig. 7). Metabolic engineering of *C. glutamicum* has produced a strain giving 0.16 g/L L-methionine along with 5.6 g/L L-threonine [63]. Another set of modifications to *C. glutamicum* has been shown to produce 2.9 g/L L-methionine together with 23.8 g/L L-lysine and small amounts of homoserine and valine [64]. Both the titer of L-methionine and the carbon yield of this fermentation process are still far below those of other commercial amino acid processes [12].

Today L-methionine is manufactured on an industrial scale by enzymatic resolution of the readily available DL-methionine via its *N*-acetyl derivative. The main process employed is the acylase resolution method [9, 10, 43]. The aminoacylase from *Aspergillus oryzae* gives excellent results, and is applied in production using enzyme membrane reactors [65, 66]. The D-*N*-acetylmethionine is recycled to the racemate either thermally or enzymatically using an *N*-acylamino acid racemase from *Amycolatopsis orientalis* [67] (Scheme 3).

As the industrial process for the manufacture of DL-methionine proceeds via the hydantoin intermediate, a direct enzymatic hydrolysis of the hydantoin would be advantageous in reducing the number of process steps required for L-methionine.

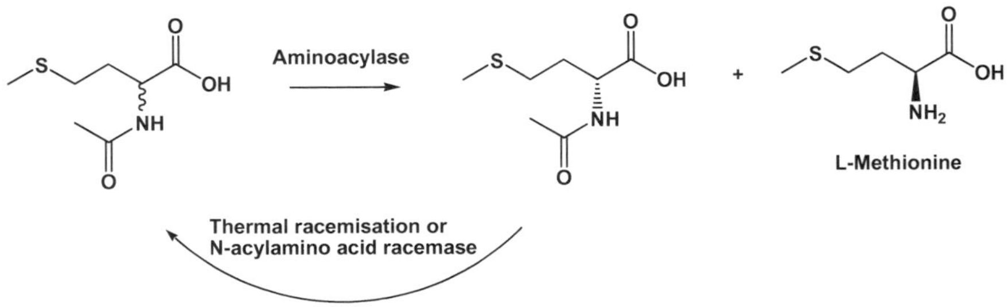

Scheme 3 The acylase process for the industrial manufacture of L-methionine

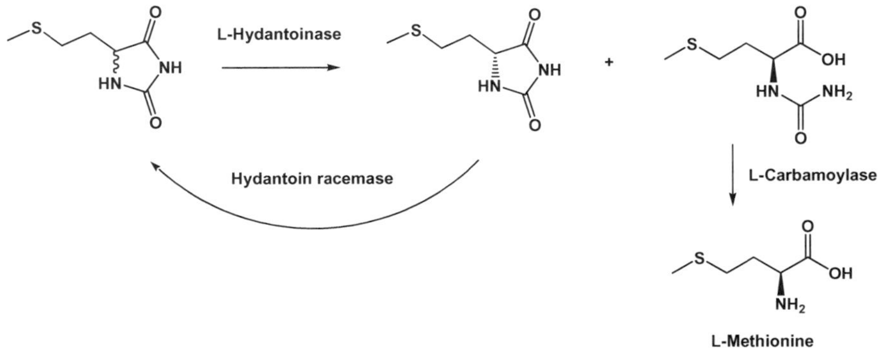

Scheme 4 The hydantoinase process for the industrial manufacture of L-methionine

Naturally occurring hydantoinases are used to manufacture D-amino acids, such D-phenylglycine. By applying directed evolution techniques, it was possible to engineer a hydantoinase from *Arthrobacter sp.* to hydrolyze only the L-enantiomer of the methionine hydantoin [68]. This was combined with a carbamoylase and a hydantoin racemase from *Arthrobacter sp.* and overexpressed in *E. coli* to give an efficient whole-cell biocatalyst for the manufacture of L-methionine by dynamic kinetic resolution [67] (Scheme 4).

2.3.13 L-Aspartic acid and L-Alanine

These two amino acids are manufactured by tandem enzymatic processes starting from readily available fumaric acid. The manufacture of L-aspartic acid, required as a raw material for the sweetener aspartame, is one of the few large-scale applications of an ammonia lyase enzyme [9, 69]. The process is performed with an immobilized L-aspartate ammonia lyase from *E. coli*, as well as with whole cell catalysts from *Brevibacterium flavum*, and requires an excess of ammonia to drive the equilibrium towards aspartic acid. The optimization of this process with substrate concentrations up to 2 M and high space–time yields makes it more attractive for industrial application than a direct fermentation route.

Scheme 5 Enzymatic processes for the manufacture of L-aspartic acid and L-alanine

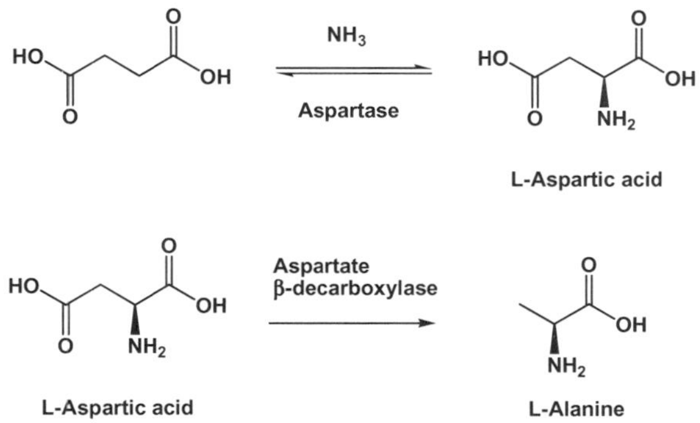

L-Alanine is derived from L-aspartic acid by irreversible decarboxylation using L-aspartate β-decarboxylase [69]. On an industrial scale, this is performed as a continuous whole-cell biotransformation process using *Pseudomonas dacunhae* (e.g., by Tanabe). Although the two stages from fumaric acid can be combined, the different optimum reaction conditions for each enzymatic step mean that separate manufacturing steps for L-aspartic acid and L-alanine are more efficient (Scheme 5).

2.4 Industrial Production of Amino Acids by Fermentation

2.4.1 The Industrial Fermentation Process

Commercial fermentation processes for the main proteinogenic amino acids are performed under industrial conditions at large scale, the production volume depending on the market demand for the individual amino acids. Some minor amino acids are produced in smaller quantities of around 100 tonnes, mainly for the pharmaceutical market, whereas more than 500,000 tonnes L-lysine is produced for animal feed applications, and more than 1,500,000 tonnes L-glutamic acid is manufactured as a food flavoring agent. The processes used are mainly aerobic batch processes or fed-batch fermentation processes, operated under sterile conditions [10, 70]. Repetitive fed-batch processes have been reported, which are claimed to maximize productivity and equipment utilization. However, these processes require a high genetic and metabolic stability of the production strains as the strain performance can be reduced with an increasing number of strain generations.

Glucose, sucrose, or molasses is generally used as the carbon source, depending on regional availability. As the carbon sources have a significant impact on production costs, particularly for the large-volume amino acids such as L-glutamic acid, L-lysine, and L-threonine, there is a clear trend towards back-integration into starch production and the evaluation of alternative carbon sources such as lignocellulose hydrolysates or glycerol. Gaseous ammonia or its aqueous solution is generally used as the nitrogen source. Ammonia is used to regulate the pH during the fermentation process and to provide the required nitrogen for the synthesis of

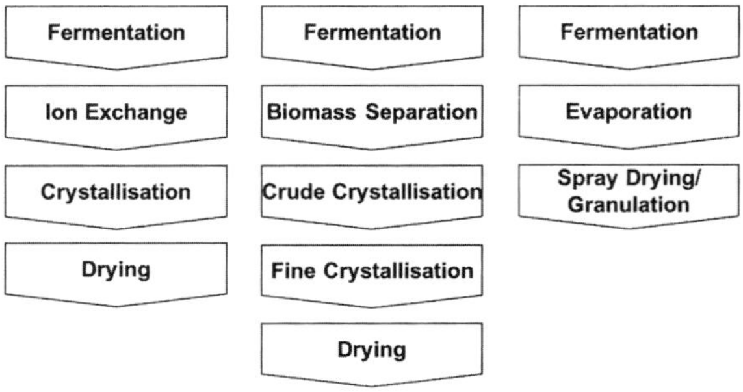

Fig. 8 Different purification processes used for feed-grade amino acids

the amino acids. Typically the pH is maintained in the range of 6.5–7.5. The optimal fermentation temperature depends on the strain and the amino acid being produced; for example, for L-lysine 31–35 C has been reported to be optimal. The optimal fermenter volume depends on the production volume of the amino acid. For large-volume products such as L-lysine, fermenter volumes up to 500 m^3 have been reported [71].

2.4.2 Industrial-Scale Downstream Processing of Amino Acids

For feed-grade amino acids there are three main purification routes (Fig. 8). L-Lysine, for example, is purified via an ion exchange process followed by a crystallization step. The ion exchange process is designed so that the biomass from the fermenter can pass the typically strong cationic resin, and only the lysine is bound to the resin. L-Lysine is eluted with ammonia and crystallized as the hydrochloride salt. The ion exchange step can be executed as a batch process or using simulating moving-bed chromatography. The latter process reduces the amount of ammonia and sulfuric acid used to regenerate the resins.

The second option uses a double crystallization process for purification, preceded by a centrifugation or filtration step to remove the biomass. This process is applied, for example, for the manufacture of L-threonine [72].

A third alternative is to apply spray drying or granulation to the crude fermentation product. In this method, the entire contents of the fermenter are processed and dried, to give a crude product suitable for direct application in animal feed, for example, Biolys®. There is no purification of the amino acid, but other valuable products from the fermentation broth, such as salts, proteins, sugars, and other amino acids, are present in the product, giving it added nutraceutical benefits.

For amino acids used in the food and pharmaceutical markets, the purity requirements are higher than those required for animal feed applications. Typical purity requirements are >99 % based on dry matter with additional limits for specific impurities, based on HPLC, salt content, heavy metals, microbial counts, and endotoxins. This explains why the purification processes are more complex compared with the feed-grade products. The number of purification steps depends on the purity and on the amino acid profile of the fermentation broth. If the starting

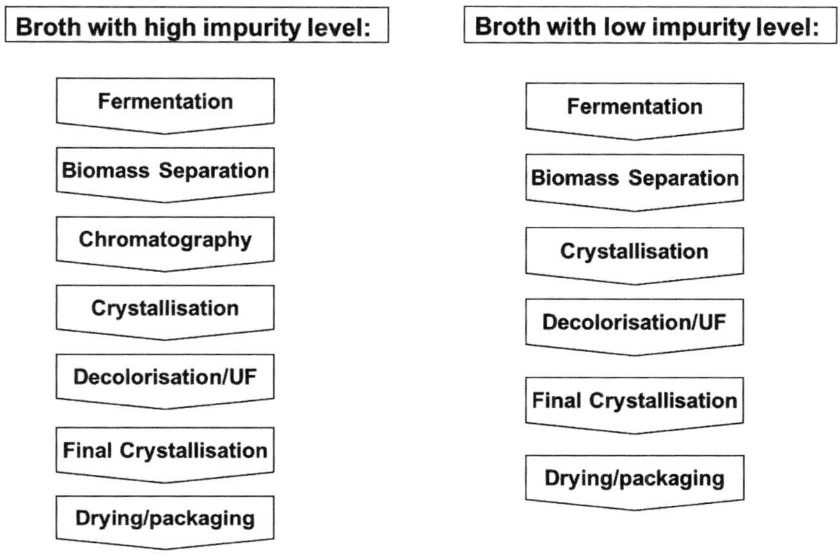

Fig. 9 Purification processes used for food and pharmaceutical-grade amino acids

broth has a low purity, a chromatographic step is typically included to increase the purity followed by a crude crystallization step. After the crude crystallization, the product is redissolved and decolorization and ultrafiltration are performed. The ultrafiltration step is designed to reduce the bio-burden and endotoxin levels. However, in recent years development of the bacteria used has meant that significant progress has been made with respect to the reduction of impurities such as unwanted by-product amino acids, and in increasing the overall purity at the end of the fermentation process. Today it is sometimes possible to eliminate the chromatographic purification step and achieve an amino acid of the desired purity using only two crystallizations and a decolorization step (Fig. 9). For the decolorization step, charcoal or resin treatment or nanofiltration membranes are generally used.

3 Application in Animal Nutrition

3.1 Dietary Requirements of the Amino Acids

There is a trend in most countries towards increased meat consumption, and this has led to the intensification of livestock production. Intensively reared animals are fed on a variety of feedstuffs, depending on the availability of feed and on the geographical location. The main protein sources are soybean meal, wheat, corn, barley, meat and bone meal, and fish meal. Most feed ingredients, even mixtures of

Table 5 Limiting amino acids in the diets of pigs and poultry

Animal	1st limiting Amino acid	2nd limiting Amino acid	3rd limiting Amino acid	4th limiting Amino acid
Pig	Lysine	Threonine	Methionine	Tryptophan
Poultry	Methionine	Lysine	Threonine	Valine

these feed sources, do not provide balanced protein for animal nutrition. The amino acid profile of the feed does not match the physiological demands of the animal. Different animal species require a different balance of amino acids, and the requirements of a particular animal may vary with age and sex. The main essential amino acids that are deficient in most animal feeds are methionine, lysine, threonine, and tryptophan. Supplementation of feed with a specific mixture of the limiting amino acids is essential to produce healthy animals without wastage of feed protein.

If the dietary protein is deficient in one or more essential amino acids, then the animal has to consume more protein to receive all the required amino acids. This leads to excretion of the excess of other amino acids, and to an increased cost of feed materials. It is estimated that 1 % reduction in protein in the feed will lead to a reduction of 10 % nitrogen in manure, 10 % ammonia emissions to the air, 3 % lower water consumption, and 5 % manure volume [73]. The amino acid that is most lacking is termed the first limiting amino acid, followed by the second limiting amino acid, and so on. Table 5 shows the generally accepted limiting amino acids for pigs and for poultry, although these may vary slightly depending on the protein feed available in a particular geographical area.

Further data are available for other species, for example, farmed fish, and the limiting amino acids, the amino acid content of each feedstuff, and the specific amino acid requirements for each animal (subdivided by age and sex) are presented in reference sources [2, 74]. Manufacturers of feed amino acids also provide extensive tables with recommendations for supplementation for a large number of different animals and conditions [75, 76]. There are also software packages available, for example, AMINODat® [77]. These present the analyses of many thousands of different feedstuffs, and permit the calculation of amino acid supplementation for different animals, depending on the particular conditions of the livestock farmer. It is also possible to calculate the environmental benefit from the reduction in animal feed by correct supplementation with amino acids. The amino acid requirements tables are not based on the total amino acids in the feeds, but are corrected to standardized ileal digestible amino acids (SID), which differ between animal species. The SID method has been shown to lead to formulations that better meet the amino acid requirements for the animal than those formulated on a total amino basis.

For many years only the first three limiting amino acids were considered in feed calculations. However, increasing accuracy in feed analysis and supplementation has led to the identification of the fourth and fifth limiting amino acids. For example, in broiler chickens valine and isoleucine are considered as colimiting fourth and fifth amino acids [78]. Similarly tryptophan is widely considered to be the fourth limiting amino acid in pigs. As well as the essential amino acids, animals also receive the nonessential amino acids as part of their feed. The composition of different feed products with respect to their relative content of nonessential amino acids has also been assessed [79].

As well as the traditional areas of pigs and poultry, amino acid supplementation has recently gained importance in fish farming. The most important group is the carp family *Cyprinidae*, with a production of approximately 20 million tonnes in 2008, mainly in China. Studies have shown that about 40 % of the fish meal used as protein feed in carp aquaculture can be replaced by supplementation with methionine as the first limiting amino acid and lysine as the second [80].

3.2 Products and Formulation

All amino acids for feed purposes are manufactured either as granulated solids or as liquid concentrates. It is easier for small producers to formulate solid amino acids with animal feed to make suitable feed pellets. Pelleted feeds are used for the rearing of pigs and poultry and generally result in a more rapid weight gain and better utilization of the feed than other forms such as meal or mash. Of the main feed amino acids, L-tryptophan, L-threonine, and the branched-chain amino acids such as L-valine are manufactured as essentially pure amino acids for incorporation into animal feed (Table 6).

L-Lysine, manufactured by fermentation, is formulated in different ways, depending on the purification steps performed after fermentation [25]. L-Lysine hydrochloride has a lysine content of 78 % and is a commonly used supplement form. The hydrochloride is formed from the crude fermentation product by ion exchange. Lysine is also sold as a liquid concentrate of the free base (ca. 50 %), and as an appproximately 50 % granulated lysine sulfate (Biolys) which is isolated directly from the fermentation process by evaporation and granulation. This form also contains small amounts of other amino acids, proteins, and sugars produced during the fermentation process.

Methionine is used in animal feed as the racemate, as both isomers can be metabolized by animals and humans. DL-Methionine is usually sold as a powder (e.g., MetAMINO® from Evonik). As an alternative to methionine itself, the product methionine hydroxy acid (2-hydroxy-4-methylthiobutanoic acid; Fig. 10) is offered by some manufacturers. It is available as an 88 % liquid concentrate in water (Rhodimet® AT88 from Adisseo and Alimet® from Novus International) or as an 84 % preparation of the calcium salt (MHA® from Novus International). Care must be taken when handling the free acid form of MHA because of its low pH. Studies have shown that methionine hydroxy acid has a lower efficacy in feed than DL-methionine. In one example in pigs, it was calculated that the bioequivalence of the hydroxy acid to DL-methionine was 64 % on a weight/weight basis or 73 % on a molar basis [81].

Guanidinoacetic acid is a natural precursor of creatine, lacking the N-methyl group (Fig. 10). It is formed in the animal from arginine and glycine. When meat

Table 6 Properties of the key animal feed amino acids (manufacturer's data)

Amino acid or salt	Min. assay as free amino acid (%)	Sol. water (g/l at 20°C)	Crude protein (N *x* 6.25) (%)	pH value
L-Lysine free base	50	Miscible	61.9	10–11
L-Lysine HCl	78	642	93.4	5.6–5.9
L-Lysine sulfate (Biolys®)	50	360	75.0	5–7
DL-Methionine	99	30	58.1	5–6
Methionine hydroxy acid solution	88	Miscible	0	<1
Methionine hydroxy acid calcium salt	84	74	0	11
L-Threonine	98	90	72.4	5.0–6.5
L-Tryptophan	98	11	84.0	4.5–7.0
L-Valine	96.5	57	72.1	5–6
Guanidinoacetic acid	96	4	221	

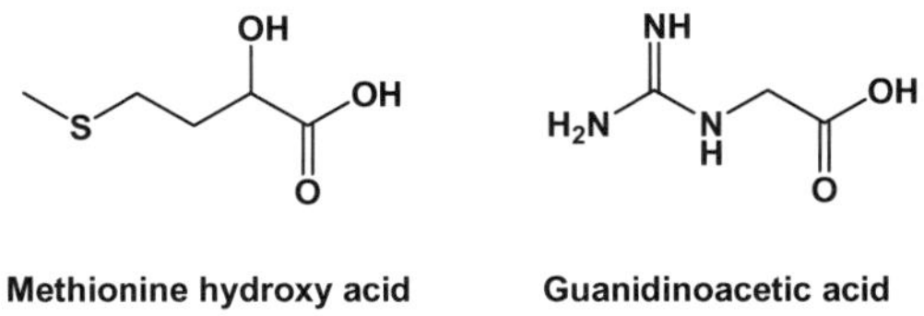

Fig. 10 Alternatives to amino acids for feed additives

and bone meal were banned in Europe for poultry feed in 2001, a drop in growth rates for chickens was observed. Studies have shown that supplementation with creatine or with guanidinoacetic acid (CreAMINO®) increases growth rates in chickens [82, 83].

After glutamic acid and monosodium glutamate for food flavoring, the amino acids for feed are those amino acids manufactured in the highest amounts. With increasing technology in agriculture and a growth in demand for meat products, particularly in Asia, the demand for feed amino acids is expected to increase strongly over the next five years. The estimated world production for 2011 is shown in Table 7 [84].

4 Application in Human Nutrition

4.1 Dietary Requirements of the Amino Acids

For nutritional purposes, the proteinogenic amino acids are divided into essential (indispensable) and nonessential (dispensable) amino acids. The nonessential amino acids can normally be produced by the human body in sufficient quantities to maintain health, although supplementation may be necessary in cases of illness. The essential amino acids, on the other hand, must be obtained through diet in sufficient quantities. Earlier studies by Rose [85] and Hegsted [86] listed eight

Table 7 World manufacture of feed amino acids, 2011 [84]

Amino acid	Production (tonnes)	Expected annual growth 2011–2016 (%)
DL-Methionine (includes methionine hydroxy acid)	700,000	3.6
L-Lysine (includes hydrochloride and sulfate salts)	600,000	4.5
L-Threonine	120,000	5.8
L-Tryptophan	4,000	11.4

Table 8 Recommended minimum daily requirements for the essential amino acids [87]

Amino acid	mg/kg per day	mg/g protein
L-Histidine	10	15
L-Isoleucine	20	30
L-Leucine	39	59
L-Lysine	30	45
L-Methionine and L-Cysteine	15	22
L-Phenylalanine and L-Tyrosine	25	38
L-Threonine	15	23
L-Tryptophan	4	6
L-Valine	26	39
Total essential amino acids	184	277

essential amino acids that had to be obtained from diet, and estimated the daily adult requirement for these amino acids. More recently L-histidine has been added to the list of essential amino acids, as hemoglobin production is affected if it is not present. In addition, there are some amino acids termed semiessential, as they are generally produced in insufficient quantity by the body, particularly in infants or by patients suffering from infectious diseases, such as tuberculosis or HIV/AIDS. These include the arginine group (arginine, ornithine, citrulline) and cysteine. The latest WHO report [87] gives the recommended daily requirements for the essential amino acids (Table 8). For dietary purposes the sulfur amino acids (methionine and cysteine) are grouped together, as are the aromatic amino acids (phenylalanine and tyrosine). The requirements in terms of g protein are based on a typical protein intake of 0.66 g/kg body weight per day. However, the recommended protein consumption for adults is between 0.8 and 0.9 g/kg body weight/day in most western countries. Athletes and body-builders may have considerably higher protein consumption than these figures, often by using supplements. Note that the recommended minimum dietary amounts of amino acids have been increased since the previous WHO report in 1985 [88].

For infant nutrition there is a requirement to supply amino acids for growth as well as to maintain metabolism. The recommended minimum essential amino acid requirements for babies, infants, and school-age children are given in Table 9. Above 10 years of age, the juvenile requirements are increasingly approximate to the adult requirements. The recommended protein intake, expressed in mg/kg per

Table 9 Recommended requirements of the essential amino acids for infants [87]

Amino Acid	mg/kg per Day		
Age	6 Months	1–2 Years	3–10 Years
L-Histidine	22	15	12
L-Isoleucine	36	27	23
L-Leucine	73	54	44
L-Lysine	64	45	35
L-Methionine and L-Cysteine	31	22	18
L-Phenylalanine and L-Tyrosine	59	40	30
L-Threonine	34	23	18
L-Tryptophan	9.5	6.4	4.8
L-Valine	49	36	29

day, is considerably higher for babies and infants. A summary of the requirements at different ages, based on the WHO report, has recently been published [89].

Different foodstuffs contain varying amounts of the individual amino acids, and there are wide variations within a specific foodstuff depending on the growing conditions and the method of preparation for consumption. The FAO has collected data for a wide range of food products [90] and a selection is given in Table 10 for some common foods. The average amino acid content (in mg/100 g foodstuff) is given together with the total protein percentage in the food.

The methods used to determine the composition group glutamine and asparagine together with their respective acids. Cysteine is presented as cystine. A more recent analysis of a range of food products has produced percentage compositions of the amino acid components including glutamine and asparagine, which is broadly in line with the previous publication [91]. These percentage compositions are given in Table 11.

Although the recommended amino acid intake may be completely achieved by consumption of animal protein, other diets, particularly cereal-rich diets, do not provide the full amounts of all the essential amino acids for a typical total consumption of protein. For each foodstuff, the first limiting amino acid can be defined, which is the amino acid most deficient in the food product. For example, in most cereals, lysine is the first limiting amino acid. The first and second limiting amino acids for a selection of foodstuffs are given in Table 12 [2].

WHO has produced an amino acid scoring pattern for the essential amino acids [87] that reflects the figures given in Table 8 for adults and Table 9 for infants and children. The numbers given, particularly for pre-school–age children, have been criticized. A protein digestibility-corrected amino acid score (PDCAAS) has been adopted to measure the amount of dietary amino acids actually taken up by the body [92]. The concentration of the first limiting amino acid in the test protein is compared with the concentration of the amino acid in the reference scoring pattern, and then corrected for the digestibility of the test protein. For example, the amount of animal or milk protein required to be added to a child's cereal diet to provide the reference amount of lysine can then be calculated. The calculation of the

Table 10 Average amino acid compositions of selected foodstuffs (mg/100 g product) [90]

Amino Acid	Milk	Cheese	Egg	Beef	Chicken	Fish	Wheat	Rice	Bean	Potato	Tomato	Apple
L-Alanine	119	566	773	1033	682	1126	472	474	927	89	27	17
L-Arginine	113	651	754	1118	1114	1066	602	650	1257	100	24	10
L-Aspartic acid	264	1277	1190	1590	1834	1947	644	808	2648	248	129	78
L-Cysteine	28	76	301	226	262	220	332	84	188	12	7	5
L-Glutamic acid	764	3691	1576	2703	3002	2655	3900	1622	3721	204	457	42
Glycine	68	344	410	860	1059	906	512	393	839	76	20	14
L-Histidine	92	556	301	603	525	665	299	197	627	30	17	7
L-Isoleucine	162	956	778	852	1069	900	426	300	927	76	20	13
L-Leucine	328	1864	1091	1435	1472	1445	871	648	1685	121	30	23
L-Lysine	268	1559	863	1573	1590	1713	374	299	1593	96	32	22
L-Methionine	86	530	416	478	502	539	196	183	234	26	7	3
L-Phenylalanine	185	950	709	778	800	737	589	406	1154	80	20	10
L-Proline	314	2324	515	668	829	692	1298	369	789	75	18	13
L-Serine	199	970	946	713	781	816	600	427	1228	83	28	16
L-Threonine	153	725	634	812	794	861	382	307	878	75	25	14
L-Tryptophan	48	217	184	198	205	211	142	98	223	83	9	3
L-Tyrosine	163	973	515	637	669	689	391	275	559	55	14	6
L-Valine	189	1393	847	886	1018	1150	577	433	1016	93	24	15
Total	**3543**	**19622**	**12803**	**17163**	**18207**	**18338**	**12607**	**7973**	**20493**	**1622**	**908**	**311**
% Protein	3.5 %	18.0 %	12.4 %	17.7 %	20.0 %	18.8 %	12.2 %	7.5 %	22.1 %	2.0 %	1.1 %	0.4 %

Table 11 Relative amounts of amino acids in selected foodstuffs [91]

Amino Acid	Milk (%)	Cheese (%)	Egg (%)	Beef (%)	Chicken (%)	Fish (%)	Wheat (%)	Rice (%)	Bean (%)	Potato (%)	Tomato (%)	Apple (%)
L-Alanine	3.3	3.0	5.1	6.1	4.7	6.6	3.8	5.8	3.9	4.4	2.4	5.6
L-Ariginine	3.6	2.6	7.4	6.7	6.7	6.0	3.9	8.2	6.6	4.8	4.3	4.1
L-Asparagine	4.5	5.5	5.1	3.9	3.5	4.4	3.8	5.5	9.0	19.5	6.2	10.5
L-Aspartic acid	3.3	3.8	4.2	5.8	6.9	7.9	1.4	3.7	3.3	1.1	14.0	6.9
L-Cysteine	0.7	0.8	2.0	1.2	1.1	1.0	2.5	2.2	1.3	1.1	1.5	1.7
L-Glutamic acid	8.8	7.5	5.9	9.6	10.4	9.5	8.1	7.5	4.8	1.0	28.4	4.3
L-Glutamine	12.0	11.1	7.2	6.4	5.2	5.3	21.7	11.0	13.2	17.2	12.5	6.6
Glycine	2.0	2.0	3.1	4.6	7.0	5.3	4.1	4.5	3.8	3.5	2.5	4.4
L-Histidine	2.9	2.4	2.4	3.7	2.8	3.1	2.4	2.7	3.2	2.2	1.1	3.3
L-Isoleucine	4.4	4.4	6.3	5.4	5.8	4.6	4.5	3.9	4.8	4.3	2.2	3.9
L-Leucine	9.2	8.5	7.7	8.0	7.6	7.9	7.0	8.2	8.3	5.7	3.1	7.6
L-Lysine	7.4	6.2	8.0	8.9	7.7	8.3	2.9	4.2	7.1	5.1	4.0	8.1
L-Methionine	2.3	3.7	3.2	2.6	2.6	2.4	1.4	2.2	1.4	1.3	0.8	1.0
L-Phenylalanine	5.2	4.6	4.6	4.5	4.5	4.9	4.9	6.2	6.0	4.6	3.3	6.4
L-Proline	8.7	14.5	4.1	4.0	4.3	3.0	11.1	4.3	4.3	3.6	2.3	4.7
L-Serine	5.2	4.8	6.8	4.8	4.0	4.5	4.9	4.3	5.8	4.6	2.7	4.5
L-Threonine	4.2	3.0	4.3	4.3	4.7	4.4	3.0	3.9	3.9	5.6	2.7	3.6
L-Tryptophan	1.4	1.4	2.1	1.2	1.2	1.4	0.8	1.0	1.4	1.2	1.1	1.0
L-Tyrosine	4.5	4.7	4.2	3.3	3.7	3.4	3.5	3.4	3.1	2.7	2.0	4.5
L-Valine	6.5	5.6	6.1	5.2	5.5	6.2	4.4	7.3	5.0	6.7	3.0	7.6

Table 12 Limiting amino acids in a selection of foods

Food product	1st limiting amino acid	2nd limiting amino acid
Peanut	Threonine	Lysine, Methionine
Fish	Methionine	Lysine
Milk (Casein)	Methionine	Tryptophan
Beans	Methionine	
Soy protein	Methionine	Lysine
Wheat	Lysine	Threonine
Rice	Lysine	Threonine, Tryptophan
Maize	Lysine	Tryptophan, Threonine

Table 13 Comparison of the master amino acid pattern [94] with the WHO scoring pattern

Amino acid	MAP lower limit (%)	MAP upper limit (%)	WHO scoring pattern (%)
L-Isoleucine	12.2	16.5	11.4
L-Leucine	18.2	27.4	22.5
L-Lysine	12.6	23.6	17.2
L-Methionine	2.3	7.8	8.4
L-Phenylalanine	8.4	13.1	14.5
L-Threonine	9.7	12.9	8.8
L-Tryptophan	2.1	4.7	2.3
L-Valine	12.6	19.0	14.9

recommended amino acid intake for children's diets based on the adult amino acid pattern has recently been criticized [93].

A mixture of amino acids has been patented [94] and put on the market as the Master Amino Acid Pattern (MAP). This mixture contains the original eight essential amino acids (without histidine) and is claimed to provide a readily absorbed mixture of the essential amino acids in the correct ratios. A number of trials have been performed showing that the MAP is equivalent to a protein supplement for athletes, and that it provides a low calorie source of amino acids to aid in a weight reduction program [95]. The levels of amino acids claimed in the patent are similar to those in the latest WHO scoring pattern [87], but are lower than the recommended WHO levels for the sulfur-containing amino acids (as methionine) and for the aromatic amino acids (as phenylalanine; Table 13).

The minimum dietary amounts of amino acids have been defined, however, there has been limited work on the maximum tolerable dose for individual amino acids. Athletes in training, weightlifters, and body-builders consume large amounts of amino acids, either as dietary protein or as readily absorbable protein hydrolysates. A study has reviewed the available literature of both human and animal studies where doses of amino acids far above the normal dietary amounts have been consumed [96]. Most amino acids do not cause problems in adults, even in large doses, but some side effects have been reported with individual compounds.

Cysteine: Nausea, insomnia, and dizziness occurred at doses of 20 g/day.

Glutamic acid: This amino acid is neurotoxic in young animals, but a comparable effect has not been observed in humans. The symptoms of "Chinese restaurant syndrome," connected with the consumption of large quantities of MSG, have not been confirmed in rigorously controlled trials. However, it is acknowledged that a small proportion of the population may be sensitive to MSG at a dose level of 3 g. Asthmatics may also be sensitive to glutamic acid.

Histidine: A dose of 24–64 g/day in overweight subjects led to a number of side effects. No problems were observed in another trial at 4.5 g/day.

Methionine: A dose of 5 g/day is considered to be acceptable, but higher levels have produced side effects including high plasma homocysteine, which is correlated with cardiovascular diseases.

Phenylalanine: This amino acid is only problematic in patients with phenylketonuria, which results in a build-up of phenylalanine and its metabolites in the blood.

Tryptophan: There was an outbreak of eosinophilia-myalgia syndrome in the 1980s which was linked to the use of L-tryptophan in supplements. It is now believed that the cause was unrelated to the tryptophan content. In human trials, large doses do not appear to cause problems and the product is sold as a sleeping aid.

Tyrosine: There are usually no problems except in premature babies, who can develop neonatal tyrosinemia, due to a deficiency of hepatic tyrosine aminotransferase.

4.2 Application in Medical Nutrition

Medical nutrition is defined as a mixture of food components given under controlled clinical conditions by medical personnel. The food components can be taken orally or by gastric tube (enteral feeding). Parenteral feeding is given in hospitals by direct infusion to patients who are unable to eat, digest, or absorb nutrients through the digestive system. Parenteral nutrition provides supplementary nutrients by infusion, and should be differentiated from total parenteral nutrition, where the patient's complete nutritional requirements are met by infusion.

Medical nutrition products were introduced in the 1950s, initially for patients with genetic diseases where specific nutrients could not be metabolized. Nutritionally complete, ready-to-use oral and enteral products were introduced in the 1970s, and since then specialized products have been developed for specific conditions of the liver, kidney, or intestinal tract. Parenteral nutrition was developed with animal studies in the 1960s, and the first human trial results were published in 1968. Since then its efficacy in providing adequate nutrition under sterile conditions in a clinical setting has been demonstrated worldwide [97].

Oral or enteral nutritional compositions usually supply their amino acid requirement in the form of protein or protein hydrolysate, homogenized for easy administration and digestion. Individual amino acids may be added to provide a balanced composition of both essential and nonessential amino acids. However, there are conditions where oral or enteral feeding is possible, but the patient cannot tolerate protein or protein hydrolysate, and a balanced mixture of amino acids is added to the formula. For example, the product EleCare® from Abbott Nutrition is indicated in infants and children for the dietary management of protein maldigestion, malabsorption, short-bowel syndrome, eosinophilic GI disorders, GI-tract impairment, or other conditions where an amino-acid based diet is required. It contains glutamine as the main amino acid, together with all essential and most nonessential amino acids, together with carbohydrates, lipids, vitamins, and minerals [98].

A dietary product that contains an elemental protein source composed of individual amino acids, together with other easily digestible nutrients, minerals, and vitamins is termed an elemental diet. This type of diet was originally developed for astronauts, but is now used for patients with particular conditions, such as Crohn's disease [99] and inflammatory bowel disorders, as well as after colonic or rectal surgery. It contains no indigestible bulk or fibrous material, and the nutrients are absorbed in the upper small intestine [100]. Specialized products are also available; for example, an amino acid mixture lacking phenylalanine is indicated for infants with phenylketonuria. Dipeptides based on glutamine (glycyl-glutamine and alanyl-glutamine) have also been used as additives to elemental diets, similar to their application in parenteral solutions, as they are an effective way of delivering glutamine to the digestive system, and the dipeptides are rapidly absorbed from the upper small intestine. There is evidence that using alanyl-glutamine in an enteral feeding solution is more effective in the synthesis of arginine, than when the peptide is administered parenterally [101]. The human H^+/peptide cotransporters PEPT1 and PEPT2 have been studied extensively for their affinity for different oligopeptides, as well as for peptidomimetics, which are often found as pharmaceuticals [102, 103].

Amino acids for parenteral nutrition are provided as a balanced solution of amino acids; as a component of a nutritional product containing carbohydrates, vitamins, and minerals; or as specialized solutions lacking certain amino acids or enriched in specific products. There are specific mixtures indicated for particular diseases. For example, solutions rich in the branched-chain amino acids leucine, isoleucine, and valine, and poor in methionine are available for patients with liver disease, whereas solutions containing only the essential amino acids are indicated for other patients, such as those with kidney disease. Typical compositions from major manufacturers are given in Table 14. In some cases dipeptides are used in place of individual amino acids. For example, glycyl-glutamine and glycyl-tyrosine are used in the product GLAMIN®. The product NEPHROTECT®, indicated for kidney disease and for dialysis patients, contains the dipeptide glycyl-tyrosine and the amino acid derivative N-acetylcysteine. Other dipeptides have also been employed. Alanyl-glutamine is available as an aqueous solution to be added to standard amino acid mixtures for use in intensive care units, and in burns and other trauma cases (e.g., DIPEPTIVEN®, Fresenius Kabi). Dipeptides of tyrosine are more soluble than the parent amino acid, and dipeptides based on glutamine are more stable in solution than the parent amino acid. The role of dipeptides and other derivatives of amino acids in parenteral nutrition has been reviewed [104]. There are also special parenteral solutions for infant nutrition, particularly for neonates, which contain amino acids considered semiessential for

Table 14 Composition of typical amino acid mixtures for infusion solutions [106]

Amino acid	CLINIMIX® 5/15 (Baxter)	SYNTHAMIN® 5.5 % (Baxter)	GLAMIN® (Fresenius Kabi)	AMINOVEN® 10 % (Fresenius Kabi)
Essential:				
L-Leucine	3.65	8.04	7.90	7.40
L-Isoleucine	3.00	6.60	5.60	5.00
L-Valine	2.90	4.40	7.30	6.30
L-Lysine[1]	2.90	6.38	9.00	6.60
L-Phenylalanine	2.80	6.16	5.85	5.10
L-Histidine	2.40	5.18	6.80	3.00
L-Threonine	2.10	4.62	5.60	4.40
L-Methionine	2.00	4.40	5.60	4.30
L-Tryptophan	0.90	1.98	3.80	2.00
Nonessential:				
L-Alanine	10.35	22.80	16.00	14.00
L-Arginine	5.75	12.64	11.30	12.00
L-Aspartic acid			3.40	
L-Glutamic acid			5.60	
L-Glutamine[2]			20.00[2]	
Glycine	5.15	11.32	12.21[2]	11.00
L-Proline	3.40	7.58	6.80	11.20
L-Serine	2.50	5.50	4.50	6.50
L-Tyrosine[2]	0.20	0.44	2.28[2]	0.40
Taurine				1.00

Values are in g/1000 ml of admixed product

[1] Lysine is often administered as the hydrochloride or acetate salt

[2] Glutamine and tyrosine are administered as the dipeptides glycyl-glutamine and glycyl-tyrosine, respectively

premature neonates (arginine, cysteine, and taurine), for example, TrophAmine® from B. Braun. The evidence for the addition of different amino acids to parenteral mixtures has recently been reviewed [105]. Apart from the evident clinical effect of glutamine supplementation, and the use of additional amino acids for premature neonates, there is little evidence to support the supplementation of standard parenteral essential amino acid mixtures with other amino acids. There is some evidence to support increased dosing of branched-chain amino acids in cases of liver failure, but further trials are needed.

Amino acids and dipeptides intended for oral or enteral nutrition have to meet the usual standards for nutraceutical additives. It is not normally necessary for them to be manufactured to cGMP. Amino acids and dipeptides for infusion solutions must be manufactured to cGMP and they must be pyrogen-free. Certification as halal or kosher is required for products to be sold into these markets, both for enteral and parenteral products.

The key western manufacturers of clinical nutrition products are Abbott Nutrition, Nestlé, Fresenius Kabi, Nutricia (Danone), and Mead Johnson. In addition, B. Braun and Baxter concentrate on the parenteral nutrition market. The market size in the European Union and United States was estimated at $5.4 billion (2008 figures), of which the largest share ($3.3 billion) was for enteral nutrition products. The market for parenteral nutrition products is $1.2 billion and for infant nutrition products, $883 million. The annual growth rates for 2008–2013 have been estimated at 13 % for the US market and 10 % for the EU market [107].

4.3 Sports Nutrition and Supplements

The use of amino acid supplementation for sports nutrition began with the use of protein supplements or protein hydrolysates such as from milk (whey or casein) or soy protein. These products are designed to replace amino acids lost during exercise. Many commercial mixtures are available, many containing additional ingredients such as vitamins and minerals, and also flavorings. Commercial products advertise their contents of key amino acids for muscle building, notably glutamine and the branched-chain amino acids, valine, leucine, and isoleucine.

Supplementation with individual amino acids or with mixtures of amino acids in tablet or capsule form is also recommended by supplement producers for sports nutrition as well as for general nutrition. The doses for these amino acid mixtures are generally lower (1–5 g) than the recommended doses for protein supplements (20–50 g). Unlike the Master Amino Acid Pattern [94] these mixtures contain both essential and nonessential amino acids, and are rich in glycine and proline (Table 15). The intended dose is several capsules with meals or after exercise.

In addition to this use of general amino acid supplementation, a great deal of research has been done on the effects of specific amino acids on sports nutrition. This has concentrated on the branched-chain amino acids, glutamine and arginine. The effects of other proteinogenic amino acids have been summarized in reviews, but are not considered here [108, 109].

The branched-chain amino acids, valine, leucine, and isoleucine, are of particular interest in sports nutrition, as they are not metabolized in the liver. These amino acids are transported directly to the muscle, where they are metabolized into the corresponding α-keto-acids. The nitrogen is taken up by α-ketoglutarate and converted into glutamate by the enzyme branched-chain amino transferase (Fig. 11). The glutamate is transported to the liver, where the ammonia is transferred into the urea cycle. The branched-chain keto-acid is then irreversibly decarboxylated by the branched-chain keto-acid–dehydrogenase complex, and converted to energy through the citric acid cycle. The regulation of the branched-chain keto-acid–dehydrogenase determines whether the branched-chain amino acid is used to provide energy, or is converted into protein to build up the muscle. In addition, leucine is believed to play a role not only as a building block for protein, but also as a modulator for protein metabolism [110–112].

A lot of research has focused on the causes of muscle fatigue after exercise. Early work in the field found free ammonia from the metabolism of branched-chain amino acids at high levels in blood plasma, and it was suggested that free

Table 15 Composition of typical amino acid supplement mixtures (manufacturers' data)

Amino acid	Amino 1500 mg tablets	Health aid multi amino acid tablets (mg)
Essential:		
L-Leucine	36	23
L-Isoleucine	16	12
L-Valine	30	18
L-Lysine	45	32
L-Phenylalanine	22	16
L-Histidine	12	6
L-Threonine	26	15
L-Methionine	12	6
L-Tryptophan	–	–
Nonessential:		
L-Alanine	115	70
L-Arginine	105	63
L-Aspartic acid	61	46
L-Carnitine	–	4
L-Cysteine	–	4
L-Citrulline	–	3
L-Glutamic acid	136	80
Glycine	318	167
L-Hydroxylysine	21	–
L-Hydroxyproline	168	–
L-Ornithine	22	4
L-Proline	187	110
L-Serine	46	28
L-Tyrosine	8	2
Total (per capsule)	1,386	709

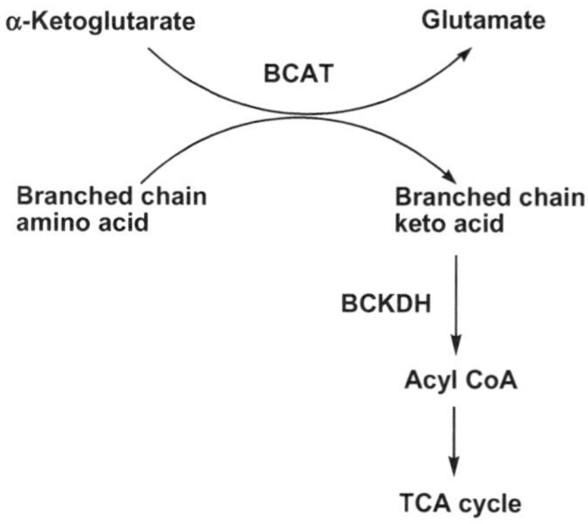

Fig. 11 Simplified metabolic pathway for the branched-chain amino acids. *BCAT* branched-chain aminotransferase; *BCKDH* branched-chain keto-acid dehydrogenase

ammonia could also cross the blood–brain barrier [113]. Additional evidence came from the study of patients with McArdle's disease, who cannot use muscle glycogen and therefore metabolize more branched-chain amino acids. An imbalance in the keto-glutarate/glutamate transamination leads to hyperammonemia [114].

Although several studies claim that supplementation with branched-chain amino acids after exercise reduces the feeling of muscle soreness and aids recovery from exercise [115, 116], there is no firm evidence that supplements help to build muscle, and scientific literature reviews are skeptical of their effects [112]. Nevertheless there are many commercial supplements on the market containing branched-chain amino acids, usually in the ratio leucine:valine:isoleucine = 2:1:1, and there is considerable anecdotal evidence of positive effects on trained athletes.

In order to improve nitrogen metabolism and prevent hyperammonemia, attempts have been made to replace one or more of the branched-chain amino acids with the corresponding α-keto–acid. In an early example, the supplementation of leucine was replaced with sodium α-ketoisocaproate in a group of postoperative patients. The keto-acid supplement increased blood acetoacetate concentration and reduced nitrogen wastage [117]. Supplements for patients with renal insufficiency containing keto-acids have been marketed for some time. For example, the product Ketosteril® consists of an amino acid mixture where five of the essential amino acids have been replaced by their keto-acid or hydroxy-acid equivalents. The effect of replacing the branched-chain amino acids with their keto-equivalents has also been tested in sports nutrition [118]. The keto-acid supplementation improved exercise tolerance and training effects, along with a better stress-recovery rate. Supplements containing some α-keto–acids are already on the market. One example is the product Vantage VO_2 Max Endurance, which contains the keto-acid salts L-arginine, α-keto-isocaproate, and L-ornithine α-ketoglutarate.

L-Glutamine is also considered a key amino acid for sports nutrition. It can be converted into the body to α-ketoglutarate and to L-glutamic acid, and participates in both the citric acid and transamination cycles. Glutamine supplementation is claimed to increase the amount of the amino acid in skeletal muscles, and to prevent impairment of the immune system through a decrease in free glutamine levels during exercise. Glutamine is also a source of glucose and a precursor of glutathione in the body [108, 119]. As glutamine is unstable in solution, it is often marketed as a more stable dipeptide. The dipeptide L-alanyl- L-glutamine, which is marketed by Kyowa Hakko under the trade name Sustamine™, has been shown to provide a significant ergogenic effect during hydration stress in endurance exercise [120].

L-Arginine is considered as a semiessential amino acid required for normal growth. Arginine is a key component of the urea cycle, which operates in the liver, and is recycled via ornithine and citrulline (Fig. 12). Arginine metabolism is thus the main route for the excretion of ammonia from breakdown of the amino acids. In principle, arginine supplementation may be able to improve performance by reducing the concentration of free ammonia in the blood. Some studies have shown that arginine supplements reduce muscle fatigue. In addition arginine can have an ergogenic effect by stimulating the production of growth hormone, and as an intermediate in the synthesis of creatine in the body. Finally arginine is a precursor of nitric oxide under the action of nitric oxide synthase (Fig. 13). As a vasodilator, nitric oxide can improve performance by increasing blood flow to the muscles [109, 121].

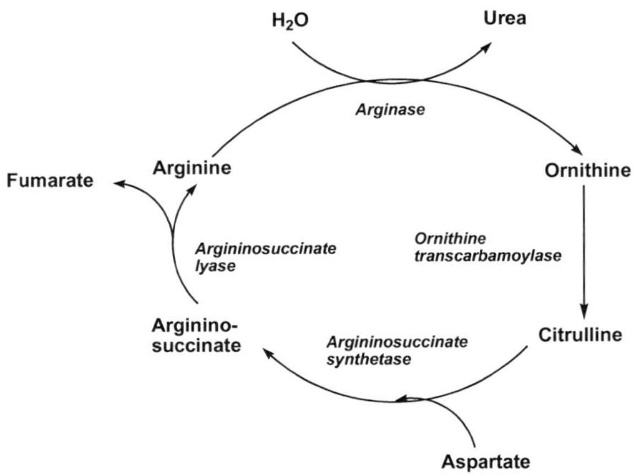

Fig. 12 Simplified metabolic pathway for the metabolism of arginine

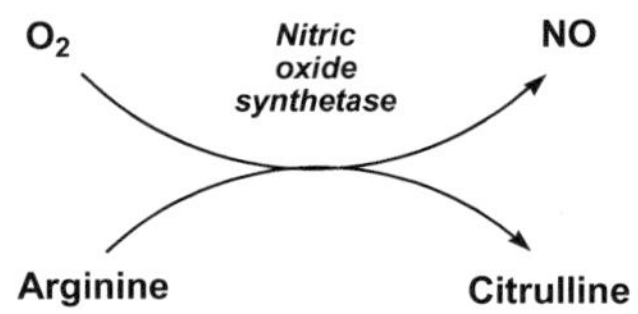

Fig. 13 Formation of nitric oxide from arginine

A number of other nonproteinogenic amino acids have been promoted as supplements and for improving athletic performance. Some of the best-known examples are shown in Fig. 14. L-Carnitine is synthesized in the body from lysine and methionine. It has a physiological role in the transport of fatty acids into the mitochondria and thus promoting oxidative metabolism. It is claimed that improving fat metabolism, carnitine reduces body fat and increases the amount of muscle. L-Carnitine has been widely promoted as an ergogenic and as an aid to recovery from exercise. However, reviews of the scientific evidence suggest there is no clear performance from carnitine supplementation [122, 123].

Carnosine (β-alanyl-L-histidine) is a naturally occurring dipeptide found mainly in skeletal muscle. Carnosine is believed to contribute to pH buffering in the muscle. Both carnosine and β-alanine have been studied for their ergogenic effects, and are sold as sports nutrition supplements. β-Alanine is sold as a precursor to carnosine. There is some evidence that supplementation with β-alanine or with carnosine can have an ergogenic effect on high-intensity exhaustive exercise [124]. There is also evidence that β-alanine supplementation increased muscle carnitine levels, but all authors agree that more research is needed [125, 126].

Creatine is synthesized in the body from glycine, arginine, and methionine, and is also found in meat and fish. It is mainly concentrated in skeletal muscle. Creatine is continually degraded to creatinine and excreted, so supplies need to be

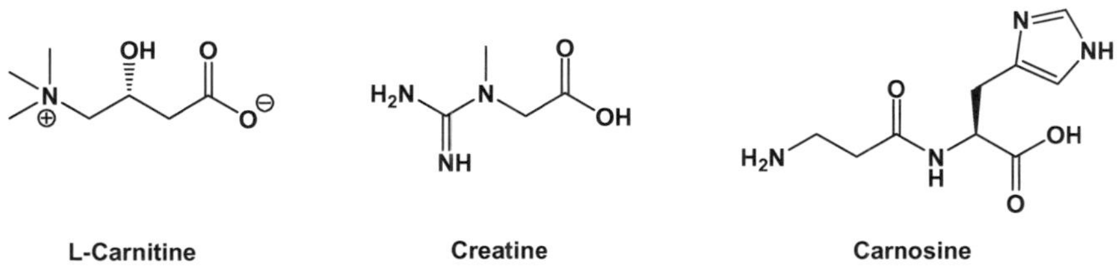

Fig. 14 Nonproteinogenic amino acids used in sports supplements

replaced. Creatine is commonly sold as a supplement to increase exercise capacity and to build lean body mass. There is considerable evidence that there is a clear ergogenic effect with a dose of 3–5 g/day, and the general scientific opinion is that creatine is safe and has a positive effect on strength and lean body mass [127].

4.4 Sweeteners and Flavorings

The flavoring agent monosodium glutamate and the sweetener aspartame (L-aspartyl-L-phenylalanine methyl ester) are covered by other chapters in this book. This section deals with the taste and flavor properties of the other amino acids and some short-chain peptides.

Amino acids can give flavor to food products on their own, but mainly by reaction with other food components during cooking. The chief processes are the Maillard reaction between amino acids and reducing sugars, leading to browning and the formation of flavor compounds, and the Strecker reaction between amino acids and α-dicarbonyl compounds, leading to transamination and the production of an aldehyde with one carbon atom less. The aldehydes formed are often aroma compounds. Cysteine is used as an additive to meat as it produces meat-flavor compounds by the Maillard reaction, for example, on roasting. Cysteine is also used as a flavor additive in bread dough.

Several surveys have been made of the taste properties of individual amino acids, ranking them in terms of the quality of the taste (sweet, sour, bitter, salty, or umami) and intensity [128–130]. Not all sources agree on the exact classification, as some amino acids are placed in more than one class. It is generally agreed, however, that glycine and alanine are sweet, and increasing side-chain length imparts a bitter taste (Table 16).

Some authors detect a salty taste as well, but others attribute this to the counter-ion or salt of the amino acid, when studies have been done at different pH values. The D-isomers of the common amino acids have also been tested, and many of these are considered to be sweet, whereas the L-isomers are bitter [128]. This is particularly true for the aromatic amino acids and the branched-chain amino acids.

Short-chain peptides have also been tested for their flavor characteristics. Most di- and tripeptides have a bitter taste, particularly when they contain amino acids with hydrophobic alkyl or aromatic side chains [131, 132]. Dipeptides containing

Table 16 Flavor properties of the common amino acids

Amino acid	Taste
L-Alanine	Sweet
Glycine	Sweet
L-Lysine	Sweet and bitter
L-Proline	Sweet and bitter
L-Serine	Sweet and sour
L-Threonine	Sweet and sour
L-Cysteine	Sulfurous, meaty
L-Methionine	Sulfurous and bitter
L-Arginine	Bitter
L-Histidine	Bitter
L-Isoleucine	Bitter
L-Leucine	Bitter
L-Phenylalanine	Bitter
L-Tyrosine	Bitter
L-Tryptophan	Bitter
L-Valine	Bitter
L-Aspartic acid	Sour
L-Glutamic acid	Sour and umami
L-Aparagine	Sour
Sodium L-glutamate	Umami
Sodium L-aspartate	Umami

glycine and alanine residues were judged to be flat in taste. However, tripeptides of alanine and glycine containing at least one alanine residue were all found to be sweet [133]. A sixth taste, *kokumi*, has recently been proposed. This is not in itself a specific taste, but a taste enhancer, which increases the intensity of salt, sweet, and umami tastes. *Kokumi* is produced in food by calcium ions and particularly by the tripeptide glutathione (γ-L-glutamyl-L-cysteinyl-glycine), and acts by inducing a response in calcium-sensing receptor cells [134].

The sweet taste and general acceptability of glycine as an additive has led to the use of glycine as a carrier for metal salts. These glycine chelates are often added to food supplement mixtures to provide essential trace metals. The metals are usually the divalent forms of calcium, copper, iron, magnesium, manganese, and zinc, and the trivalent form of chromium [135]. Chelates of these metals with a number of other amino acids have also been patented [136]. The amino acid chelates provide the metal in a more soluble and more readily absorbable form, without any effect on the taste of the food product, as demonstrated in trials with bread products [137].

5 Conclusions and Outlook

Amino acids are today indispensable components of animal nutrition products, and also in human medical nutrition. They are becoming increasingly common as supplements for athletic performance and also for improving general health. This trend is likely to continue with the increase in a desire for a healthy lifestyle, coupled with the rise of a middle class in countries such as China. The future will not consist of new individual products, but of more imaginative combinations and formulations, and the incorporation of amino acid supplements into common foodstuffs. In animal feed, the increased desire for meat consumption in emerging markets will also lead to an increased requirement for the main feed amino acids, and new plants will continue to be constructed near these new markets. There may be developments in the marketing of short-chain peptides as an alternative to the individual amino acids, and in the combination of amino acids with other products such as keto-acids to achieve optimum results in sports nutrition.

In manufacturing, the outlook is for a greater reliance on optimized fermentation processes as a source for manufactured amino acids, even for products such as methionine, which today is largely produced by chemical processes. High-yielding fermentation processes will be achieved by a greater use of the techniques of genetic engineering and synthetic biology, so that yields for many amino acids will approach those of L-glutamic acid and L-lysine. Efficient fermentation methods will be developed for the two amino acids that are still produced by extraction in many parts of the world, L-cysteine and L-4-hydroxyproline, making these amino acids more acceptable in the market.

References

1. Greenstein JP, Winitz M (1961) Chemistry of the amino acids. Wiley, New York, pp 245–432
2. Drauz K, Grayson I, Kleemann A, Krimmer HP, Leuchtenberger W, Weckbecker C (2007) Amino Acids. In Ullmann's encyclopedia of industrial chemistry. Wiley, Weinheim
3. Araki K, Ozeki T (2003) Amino Acids. In Kirk-Othmer encyclopedia of chemical technology. Wiley, Hoboken
4. Evonik Rexim, internal company data
5. Eastoe JE (1955) Biochem J 61:589
6. Ikeda K (1908) J Tokyo Chem Soc 30:820
7. Blaser HU, Spindler F, Studer M (2001) Appl Catal A 221:119
8. Kinoshita S, Udaka S, Shimono M (1957) J Gen Appl Microbiol 3:193
9. Gröger H, Drauz K (2004) Methods for the enantioselective biocatalytic production of L-amino acids on an industrial scale. In: Blaser H-U, Schmidt E (eds) Asymmetric catalysis on industrial scale. Wiley, Weinheim, pp 131–147
10. Leuchtenberger W, Huthmacher K, Drauz K (2005) Appl Microbiol Biotechnol 69:1
11. Park JH, Lee SY (2008) Curr Opin Biotechnol 19:454
12. Becker J, Wittmann C (2012) Curr Opin Biotechnol 23:718
13. Orten JM, Hill RM (1941) Org Synth Coll 1:299
14. Anslow WK, King H (1941) Org Synth Coll 1:298
15. Mihara N, Furuya O, Wada K (1975) US Patent 3,875,221
16. White HC, Wysong DV (1952) US Patent 2,663,713
17. Mannsfeld SP, Pfeiffer A, Tanner H, Wagner H, Liebetanz E (1978) US Patent 4,069,251
18. Lussling T, Müller KP, Schreyer G, Theissen F (1981) US Patent 4,303,621
19. Nakayama K, Kitada S, Kinoshita S (1961) J Gen Appl Microbiol 7:145
20. Kojima M, Ogawa Y, Kawamura K, Sano K (1995) WO Patent 9516042

21. Shiio I, Sano K (1969) J Gen Appl Microbiol 15:267
22. Riedel C, Rittmann D, Dangel P, Moeckel B, Petersen S, Sahm H, Eikmanns B (2001) J Mol Microbiol Biotechnol 3:573
23. Peters-Wendisch PG, Schiel B, Wendisch VF, Katsoulidis E, Sahm H, Eikmanns B (2001) J Mol Microbiol Biotechnol 3:295
24. Becker J, Zelder O, Haefner S, Schroeder H, Wittmann C (2004) J Bacteriol 13:159
25. Pfefferle W, Möckel B, Bathe B, Marx A (2003) Adv Biochem Eng Biotechnol 79:60
26. Lee KH, Park JH, Kim TY, Kim HU, Lee SY (2007) Mol Syst Biol 3:149
27. Debabov VG (2003) Adv Biochem Eng Biotechnol 79:114
28. Ura D, Hashimukai T, Matsumoto T, Fukuhara N (1990) EP Patent 421477
29. Sugimoto M, Osumi T, Nakamatsu T, Hibino W, Ito M (1999) EP Patent 931,833
30. Ziegler P, Eggeling L, Sahm H (2001) WO Patent 01/64899
31. Nakanishi T, Yokote Y, Taketsugu Y (1973) Hakko Kogaku Zasshi 51:742
32. Omori K, Suzuki I, Yuji K (1992) J Gen Microbiol 138:693
33. Masuda M, Takamatu S, Nishimura N, Komatsubara S, Tosa T (1993) Appl Biochem Biotechnol 43:189
34. Suguira M, Takagi T, Kisumi M (1985) Appl Microbiol Biotechnol 21:213
35. Ozaki A, Shibasaki T, Mori H, Maruyama A, Motoyama H (1998) EP Patent 826,773
36. Shibasaki T, Mori A, Ozaki A (2000) Biosci Biotechnol Biochem 64:746
37. Kimura E (2003) Adv Biochem Eng Biotechnol 79:38
38. Wittmann C (2010) Adv Biochem Eng Biotechnol 120:21
39. Kusumoto I (2001) J Nutr 131:2552S
40. Ikeda M (2006) Appl Microbiol Biotechnol 69:615
41. Bang W, Lang S, Sahm H, Wagner F (1983) Biotechnol Bioeng 25:999
42. Sprenger GA (2007) Aromatic amino acids. In: Wendisch VF (ed) Amino acid biosynthesis. Springer, Heidelberg, pp 93–128
43. Bonmarius AS, Drauz K, Klenk H, Wandrey C (1992) Ann NY Acad Sci 672:126
44. Katsumoto R, Ikeda M (1987) EP Patent 263,515
45. Tsuchida T, Miwa K, Nakamori S (1984) US Patent 4,442,208
46. Weuster-Bolz D, Karutz M, Joksch B, Schärtges D, Wandrey C (1996) Appl Microbiol Biotechnol 46:209
47. Matsui H, Tsuchida T, Nakamori S (1983) US Patent 4,391,907
48. Ono Y, Sato K (1998) US Patent 5,763,231
49. Park JH, Lee KH, Kim TY, Lee SY (2007) PNAS 104:7797
50. Ikeda M, Mitsuhashi S, Tanaka K, Hayashi M (2009) Appl Environ Microbiol 75:1635
51. Utagawa T (2004) J Nutr 134:2854S
52. Xu M, Rao Z, Lan C, Dou W, Zhang X, Xu H, Jin J, Xu Z (2011) Appl Biochem Biotechnol 163:707
53. Caldara M, Dupont G, Leroy F, Goldbeter A, De Vuyst L (2008) J Biol Chem 283:6347
54. Makrealyas K, Drauz K (1995) US Patent 5,405,761
55. Claes W, Gerstmeier R (2011) US Patent Appl 2011/0244529
56. Kulis-Horn RK, Persicke M, Kalinowski J (2014) Microbial Biotechnol, 7:5
57. Araki K, Kuga T (1985) US Patent 4,495,283
58. Mizukami T, Hamu A, Ikeda M, Oka T, Katsumata R (1994) Biosci Biotech Biochem 58:635
59. Klyachko EV, Shakulov RS, Kozlov YI (2007) US Patent 7,399,618
60. Wada M, Takagi H (2006) Appl Microbiol Biotechnol 73:48
61. Winterhalter C, Leinfelder W (1999) US Patent 5,972,663
62. Nonaka G, Takagi H, Ohtsu I (2011) US Patent Appl 2011/0033902
63. Park YH, Cho KM, Shin YU, Um H (2006) WO Patent Appl 2006/001616
64. Park SD, Lee JY, Sim SY, Kim Y, Lee HS (2007) Metab Eng 9:327
65. Wöltinger J, Drauz K, Bonmarius AS (2001) Appl Catal A 221:171
66. Wöltinger J, Karau A, Leuchtenberger W, Drauz K (2005) Adv Biochem Eng Biotechnol 92:289

67. May O, Verseck S, Bonmarius A, Drauz K (2002) Org Process Res Dev 6:452
68. May O, Nguyen PT, Arnold FH (2000) Nat Biotechnol 18:317
69. Sonke T, Kaptein B, Schoemaker HE (2009) In amino acids, peptides and proteins in organic chemistry, vol 1. Wiley, Weinheim, pp 79–117
70. Ikeda M (2003) Adv Biochem Eng Biotechnol 79:1
71. Nagai Y, Ito H, Yasueda H (2009) In encyclopedia of industrial biotechnology. Wiley, Weinheim, pp 1–10
72. Hermann T (2003) J Biotechnol 104:155
73. Müller A, Redshaw MS (2011) Evonik Amino News 15:1. Available via https://www.aminoacidsandmore.com
74. D'Mello JPF (2003) Amino acids in animal nutrition. CABI, Wallingford
75. Lemme A, Htoo J, Helmbrecht A (2010) Evonik Amino News 14:1. Available from https://www.aminoacidsandmore.com
76. Relandeau C, Eudaimon M (2008) Ajinomoto lysine formulator's handbook. Available via http://ajinomoto-eurolysine.com
77. Evonik Industries AG (2013) AMINODat®. http://feed-additives.evonik.com
78. Dozier WA III, Corzo A, Kidd MT, Tillman PB, Branton SL (2011) Br Poul Sci 52:238
79. Li X, Rezaei R, Li P, Wu G (2011) Amino acids 40:1159
80. Nwanna LC, Lemme A, Metwally A, Schwarz FJ (2012) Aquaculture 356–357:365
81. Kim BG, Lindemann MD, Rademacher M, Brennan JJ, Cromwell GL (2006) J Anim Sci 84:104
82. Lemme A et al (2007) Paper presented at 16th European symposium on poultry nutrition, Strasbourg, 26–30 Aug
83. Michiels J, Maerten L, Buyse J, Lemme A, Rademacher M, Dierick NA, De Smet S (2012) Poult Sci 91:402
84. Malveda MP (2012) CEH marketing research report: major amino acids. IHS, Englewood
85. Rose WC (1957) Nutr Abstr Rev 27:531
86. Irwin MI, Hegsted DM (1971) J Nutr 101:539
87. World Health Organization (2007) Protein and amino acid requirements in human nutrition. WHO technical report series 935
88. World Health Organization (1985) Energy and protein requirements. WHO technical report series 724
89. Millward DJ (2012) Brit J Nutr 108:S3
90. FAO (1970) Nutritional studies no 24, Amino acid content of foods and biological data on proteins. Interprint, Rome. http://www.fao.org/docrep/005/AC854T/AC854T00.htm
91. Sosulski FW, Imafidon GI (1990) J Agric Food Chem 38:1351
92. Schaafsma G (2000) J Nutr 130:1856S
93. Millward DJ (2012) Brit J Nutr 108:S31
94. Luca M (1992) US Patent 5,132,113
95. Luca-Moretti M, Grandi A, Luca E (2003) Adv Ther 20:282
96. Garlick PJ (2004) J Nutr 134:1633S
97. Dudrick SJ (2009) J Am Coll Nutr 28:243
98. Abbott Nutrition (2012) Company product information on EleCare. http://abbottnutrition.com/brands/products/elecare
99. O'Morain C, Segal AW, Levi AJ (1984) Brit Med J 288:1859
100. Russell RI (1975) Gut 16:68
101. Lighart-Melis GC et al (2009) Am J Clin Nutr 90:95
102. Adibi SA (1997) Gastroenterology 113:332
103. Biegel A, Brandsch M et al (2006) Amino Acids 31:137
104. Fürst P (2003) Clin Nutr Suppl 2:S23
105. Yarandi SS, Zhao VM, Hebbar G, Ziegler TR (2011) Curr Opin Clin Nutr Metab Care 14:75
106. Manufacturers' data
107. Sahoo A (2009) Opportunities and key players in clinical nutrition. Business Insights, London

108. Williams M (2005) J Int Soc Sports Nutr 2:63
109. Spruce NF, Titchenal AC (2009) Other individual amino acids. In: Driskell JA (ed) Sports nutrition fats and proteins. CRC Press, Boca Raton, pp 279–353
110. Massey KA, Blakeslee CH, Pikow HS (1998) Amino Acids 14:271
111. Shimomura Y et al (2006) J Nutr 136:529S
112. Gleeson M (2009) Branched-Chain Amino Acids. In Ref 109, 243
113. Banister EW, Cameron BJC (1990) Int J Sports Med 11:S129
114. Wagenmakers AJM, Coakley JH, Edwards RHT (1990) Int J Sports Med 11:S101
115. Ohtani M, Sugita M, Maruyama K (2006) J Nutr 136:538S
116. Blomstrand E (2006) J Nutr 136:544S
117. Sapir DG et al (1983) Lancet 321:1010
118. Liu Y, Lange R, Langanky J, Hamma T, Yang B, Steinacker JM (2012) J Int Soc Sports Nutr 9:37
119. Jonnalagadda SS (2009) Glutamine. In Ref 109, 261
120. Hoffman JR et al (2010) J Int Soc Sports Nutr 7:8
121. Campbell BI, La Bounty PM, Roberts M (2004) J Int Soc Sports Nutr 1:35
122. Karlic H, Lohringer A (2004) Nutrition 20:709
123. Brass EP (2004) Ann New York Acad Sci 1033:67
124. Hobson RM, Saunders B, Ball G, Harris RC, Sale C (2012) Amino Acids 43:25
125. Caruso J, Charles J, Unruh K, Giebel R, Learmonth L, Potter W (2012) Nutrients 4:585
126. Derave W, Sale C (2012) Amino Acids 43:1
127. Buford TW, Kreider RB, Stout JR, Greenwood M, Campbell B, Spano M, Ziegenfuss T, Lopez H, Landis J, Antonio J (2007) J Int Soc Sports Nutr 4:6
128. Belitz HD, Grosch W, Schieberle P (2009) Food chemistry. Springer, Berlin, pp 8–92
129. Kirimura J, Shimizu A, Kimizuka A, Ninomiya T, Karsuya N (1969) J Agric Food Chem 17:689
130. Schallenberger RS (1993) Taste chemistry. Chapman and Hall, London, pp 213–252
131. Ishibashi N, Arita Y, Kanehisa H, Kouge K, Okai H, Fukui S (1987) Agric Biol Chem 51:2389
132. Ishibashi N, Sadamori K, Yamamoto O, Kanehisa H, Kouge K, Kikuchi E, Okai H, Fukui S (1987) Agric Biol Chem 51:3309
133. Ishibashi N, Ono I, Kato K, Shigenaga T, Shinoda I, Okai H, Fukui S (1988) Agric Biol Chem 52:91
134. Maruyama Y, Yasuda R, Kuroda M, Eto Y (2012) PLoS One 7:e34489
135. Ashmead HH (1989) US Patent 4,830,716
136. Ashmead SD, Wheelwright DC, Ericson C, Pedersen M (2002) US Patent 6,458,981
137. Layrisse M, Garcia-Casal MN, Solano L, Baron MA, Arguello F, Llovera D, Ramirez J, Leets I, Tropper E (2000) J Nutr 130:2196

Adv Biochem Eng Biotechnol (2014) 143: 229–256
DOI: 10.1007/10_2013_235

Published Online: 19 February 2014

Food and Feed Enzymes

Marco Alexander Fraatz, Martin Rühl and Holger Zorn

Abstract Humans have benefited from the unique catalytic properties of enzymes, in particular for food production, for thousands of years. Prominent examples include the production of fermented alcoholic beverages, such as beer and wine, as well as bakery and dairy products. The chapter reviews the historic background of the development of modern enzyme technology and provides an overview of the industrial food and feed enzymes currently available on the world market. The chapter highlights enzyme applications for the improvement of resource efficiency, the biopreservation of food, and the treatment of food intolerances. Further topics address the improvement of food safety and food quality.

Keywords Biotechnology · Enzyme · Feed · Food · Phytase

Abbreviations

AMFEP	Association of Manufacturers and Formulators of Enzyme Products
BCWH	Bacterial cell wall hydrolase
BHA	2-*tert*-butyl-4-hydroxyanisole and 3-*tert*-butyl-4-hydroxyanisole
BHT	2,6-*bis*(1,1-dimethylethyl)-4-methylphenol
COT	British Committee on Toxicology
DCP	Dichloropropanol
EC	European Commission or Enzyme Commission
EFSA	European Food Safety Authority
GM	Genetically modified
JECFA	Joint FAO/WHO Expert Committee on Food Additives
MCPD	Monochloropropanediol
SCF	EU Scientific Committee on Food
StEP	Staggered extension process
TFA	Trans fatty acid

M. A. Fraatz · M. Rühl · H. Zorn (✉)
Institute of Food Chemistry and Food Biotechnology, Justus Liebig University Giessen, Heinrich-Buff-Ring 58, 35392 Giessen, Germany
e-mail: holger.zorn@lcb.chemie.uni-giessen.de

Contents

1 A Brief History of Enzymes

Humans have used the unique catalytic properties of enzymes, in particular for food production, for thousands of years. One of the oldest surviving records of mankind, written approximately 4,500 years ago by the Sumerians and known as the *Epic of Gilgamesh* (Tablet II), refers to the positive aspects of the biotechnological products bread and beer:

Enkidu knew nothing about eating bread for food,
and of drinking beer he had not been taught.
The harlot spoke to Enkidu, saying:

"Eat the food, Enkidu, it is the way one lives.
Drink the beer, as is the custom of the land."

Enkidu ate the food until he was sated,
he drank the beer-seven jugs!—and became expansive and sang with joy!
He was elated and his face glowed.
He splashed his shaggy body with water,
and rubbed himself with oil, and turned into a human [1].

The oldest wine law can be found as a part of the Code of Hammurabi, eternalized on the famous stele of Hammurabi,[1] king of Babylon (1792–1750 BC). In addition, the ancient Greeks and Romans worshiped the gods Dionysus and Bacchus as being responsible for the success of fermentation processes. However, the actual reason for a reliable fermentation was not disclosed for a few hundred years.

The first direct observation and description of microorganisms was made by the Dutchman Antoni van Leeuwenhoek (1632–1723) in 1675 [2]. By using a simple

[1] The original stele is on exhibition in the Louvre in Paris.

homemade microscope, he was able to visualize microorganisms. Then 150 years later, in 1814, an important biotechnological reaction was described by Gottlieb Sigismund Kirchhoff (1764–1833). He noted that germinating grain contains a compound that converts starch into sugar [3]. Erhard Friedrich Leuchs (1800–1837) observed the same phenomenon for human saliva in 1831 [4].

Two years later, the French researchers Anselme Payen (1795–1871) and Jean-François Persoz (1805–1868) separated a substance from malt extract by means of alcohol precipitation that was capable of starch hydrolysis [5]. They called it diastase (amylase), derived from the Greek word *diastasis*, which means "separation". The French scientist Émile Duclaux (1840–1904) proposed to honor Payen and Persoz by introducing the suffix *-ase* as a naming convention for enzymes [6].

The Swedish chemist Jöns Jacob Berzelius (1779–1848) demonstrated that malt extract can break down starch more efficiently than sulphuric acid. Additionally, he coined the term "catalysis" [7]. The expression was derived from Greek words *kata,* which means "down", and *lyein,* which means "loosen".

Further contributing to our understanding of the fermentation process were the studies of Theodor Schwann (1810–1882), Friedrich Traugott Kützing (1807–1893), and especially of Louis Pasteur (1822–1895). Among other things, Schwann isolated animal pepsin to study it, and Pasteur provided the first experimental evidence for a microbial cause of fermentation [8]. In 1874, the Danish pharmacist Christian Ditlev Ammentorp Hansen (1843–1916) brought a standardized enzyme preparation (rennet) for the production of cheese on the market. The term *enzyme*—derived from Greek meaning "in yeast"—was suggested in 1876 by Wilhelm Friedrich Kühne (1837–1900) [9]. Friedrich Wilhelm Ostwald (1853–1932) discovered that enzymes are extremely important for chemical processes within living organisms and act as catalysts. In 1909, Ostwald was awarded the Nobel Prize in Chemistry "in recognition of his work on catalysis", among other contributions [10].

In 1894, the Japanese chemist Jōkichi Takamine (1854–1922) applied presumably for the first patent on a microbial enzyme in the United States [11]. A fungal enzyme was isolated from koji (*Aspergillus oryzae*) and was called takadiastase. Later, Takamine licensed the exclusive production rights for the enzyme to the Parke-Davis company. He became the "father of commercial enzymology"[2] and a millionaire. Takadiastase was sold as a digestive aid for the treatment of dyspepsia. Emil Fischer (1852–1919), a German chemist and Nobel Prize laureate of 1902 [12], published the Lock and Key Model for enzymes to visualize the substrate and enzyme interaction in 1894 [13].

Eduard Buchner's (1860–1917) investigations of the cell-free fermentation laid the cornerstone for modern enzymology [14]. He was able to prove that it was not the living yeast cells that were essential for fermentation, but rather certain

[2] Dr. Clifford W. Hesseltine (1991) said, "Dr. Jokichi Takamine was the father of commercial enzymology."

enzymes produced by the cells (no *vis vitalis* required). Buchner's results were in clear contradiction to the view of Pasteur, who presumed that the complex apparatus of intact yeast cells would be required. In 1907, Buchner was awarded the Nobel Prize in Chemistry "for his biochemical researches and his discovery of cell-free fermentation" [15].

The origin of modern enzyme kinetics dates back to 1903. Victor Henri (1872–1940) described the initial formation of an enzyme-substrate complex as an essential step in enzymatic reactions [16]. Inspired by his findings, Leonor Michaelis (1875–1949) and Maud Leonora Menten (1879–1960) developed the Michaelis–Menten equation 10 years later [17, 18]. In 1907, Otto Karl Julius Röhm (1876–1939) patented the first enzymes for large-scale industrial application: proteolytic enzymes extracted from cow stomachs, which brought significant technical advantages to the bating of hides for the production of leather. To produce and market the new product called Oropon, he established the company Röhm & Haas (Esslingen, Germany) together with his friend and businessman Otto Haas (1872–1960). In 1914, Otto Röhm applied for a patent for the first enzymatic detergent and named it Burnus. To distribute Burnus, Röhm & Haas acquired the soap factory August Jacobi and Son (Darmstadt, Germany) in 1916.

Sir Alexander Fleming (1881–1955), the discoverer of penicillin and Nobel Prize laureate of 1945 [19], coined the name *lysozyme* in 1922 for antibacterial enzymes isolated from hen egg white. James Batcheller Sumner (1887–1955) isolated the enzyme urease in 1926 and concluded that enzymes are proteins [20]. Twenty years later, he was awarded one-half of the Nobel Prize in Chemistry (1946) "for his discovery that enzymes can be crystallized" [21]. At the same time, Kaj Ulrik Linderstrøm-Lang (1896–1959) investigated chemical properties of proteins and laid down basic formalisms for their purification [22]. In 1929, Arthur Harden (1865–1940) and Hans von Euler-Chelpin (1873–1964) were awarded jointly the Nobel Prize in Chemistry "for their investigations on the fermentation of sugar and fermentative enzymes" [23]. Amongst others, they discovered the first coenzyme and called it "coferment".

Interestingly, it was the discovery of a class of enzymes, namely the DNA-cleaving enzymes (restriction endonucleases), that paved the way for modern molecular biology. For this pioneering work, the Swiss microbiologist and geneticist Werner Arber (born 1929) along with the American researchers Daniel Nathans (1928–1999) and Hamilton Othanel Smith (born 1931) were awarded the Nobel Prize in Physiology or Medicine in 1978 [24]. With the rapid development of modern molecular biotechnology, it was then possible to modify enzymes and to produce them recombinantly. The first commercial recombinant fat-splitting enzyme, a lipase from a genetically engineered fungal microorganism, was introduced in 1987 by the company Novo [25]. The fungal enzyme was called Lipolase and was directly used in the Japanese detergent Hi-Top made by the Lion Corporation.

The enormous technological advances in recent decades have led to the development of novel tools for molecular biotechnology and recombinant production systems. They gave new opportunities for the alteration of enzymatic properties, which started almost 50 years ago by "chemical mutation" [26, 27]. At the end of the twentieth century, directed evolution by recombining DNA sequences of enzymes via DNA shuffling, error-prone polymerase chain reaction, or staggered extension process (StEP) resulted in enzymes with improved properties [28]. In addition to this more random mutational approach, the rational enzyme design was used for optimization of enzyme properties. On the basis of known amino acid sequences and crystallographic structures of enzymes, site-directed mutations have been made to delete or replace one or more amino acids to obtain improved catalytic properties [29].

Parallel to the molecular improvement of enzymes, new sources for novel and interesting enzymes were put into focus: the metagenomes [30]. Here, the genomic information of different habitats (e.g. soil, salt lakes, deep sea, and tree tops) is analyzed and transferred into host organisms for recombinant production of novel enzymes. Since then, the list of available enzymes for the production of food and animal feed has been growing constantly.

2 Legal Situation

Depending on the intended use of food enzymes, they can be divided into the following categories: food ingredients, food additives, and processing aids. Enzymes added for nutritional reasons are regarded as food ingredients, but this is rarely the case. Most enzymes are added to food for technological reasons. If these enzymes are still present in an active form in the end products, they fall under the definition of food additives. Otherwise, they commonly belong to the category of processing aids.

Prior to 2009 and partly still valid, the European regulations concerning enzymes were based mainly on four legislative acts: Directive 89/107/EEC, regulating food additives; Directive 83/417/EEC, regulating caseins and caseinates; Directive 2001/112/EC, regulating fruit juices and similar products; and Regulation (EC) 1493/1999, regulating the common organization of the wine market. Within the group of enzymes used for technological applications, only those enzymes considered to be food additives had to be assessed for safety before they were placed on the market. Local exceptions existed in Denmark and France, where national regulations for enzymes used as processing aids applied.

In 2008, the European Commission (EC) published a regulation that defined all food enzymes with technological purposes as a separate group and regulates them harmonized (Regulation (EC) No. 1332/2008). Together with regulations on food additives (Regulation (EC) No. 1333/2008), flavorings (Regulation (EC) No. 1334/2008), and a common authorization procedure (Regulation (EC) No. 1331/2008), these regulations are known as the so-called Food Improvement Agents Package. For the first time in the European history, the Food Improvement Agents Package provided a mandatory and harmonized authorization procedure for all food enzymes with technological purposes.[3] A list of food enzymes was created and "only food enzymes included in the Community list may be placed on the market as such and used in foods […]".[4] By controlling the authorization procedure in terms of regulations instead of directives, the regulations are universally binding and directly applied. Modifications as part of the implementation process of directives by the Member States are not possible, which ensures legal certainty and should facilitate international trade.

Since the late 1980s, the use of genetic engineering in particular has revolutionized commercial enzyme production and massively accelerated the rate of innovation. New enzyme activities, new applications, and improved performance of existing enzymes were made possible. In this context, the increasing use of enzymes from extremophilic microorganisms played an important role, too. Additionally, genetic engineering permitted the production of enzymes from previously noncultivable microorganisms.

A possible downside of this new era of enzyme usage is the lack of long-term experience in matters of human consumption for many of these enzymes. Therefore, an important concern of the European legislature was to establish a universal procedure for the safety assessment of enzymes. In the past, risk assessments of food enzymes were carried out mainly by the Joint FAO/WHO Expert Committee on Food Additives (JECFA) or were controlled by national authorization procedures. Very few reviews were also executed by the EU Scientific Committee on Food (SCF). Guidelines for the risk assessment of food enzymes have been published by the JECFA, the SCF, the Association of Manufacturers and Formulators of Enzyme Products (AMFEP) and the British Committee on Toxicology (COT).

This has changed after the entry of Regulation (EC) No. 1331/2008 into force. Only enzymes that were already authorized as food additives (E 1103 invertase and E 1105 lysozyme) or for the production of wine according to Regulation (EC) No. 1493/1999 (urease, β-glucanase, and lysozyme) will be automatically added to the positive list.[5] All other enzymes have to go through the newly established regulatory process. Therefore, the need for approval also includes those enzymes that were already approved under national laws (e.g. in France or in Denmark) or for which a positive evaluation of the JECFA exists. The approval process includes an assessment of health risks posed by the European Food Safety Authority (EFSA). The requirements for risk assessment are expressed as EFSA guidelines. To assess the need for technological benefit or misleading, however, no guidelines are provided.

The community list of enzymes shall include the name of the enzyme, its specifications (including origin, purity criteria, etc.), the foods to which the enzyme may be added, the conditions under which the enzyme may be used,

---

[3] Enzymes that are exclusively for the production of additives or the production of processing aids are, however, excluded from the new enzyme regulation.

[4] Regulation (EC) No. 1332/2008, Article 4.

[5] Regulation (EC) No. 1332/2008, Article 18(1).

selling restrictions, and specific requirements with respect to the labeling.[6] However, the positive list is not yet completed. Until then, the national provisions in force will continue to apply in the Member States.[7] It is assumed that the list will be completed by 2021. Additionally, certain transitional periods apply.[8]

3 Market Overview

3.1 Food Enzymes

For centuries, enzymes are used traditionally for dairy, baking, brewing, and winemaking, although not in isolated form. Enzymes are needed for cheese production and used for a wide variety of other dairy goods. Enzymes increase dough volume, lead to crispy crusts, and keep bread soft and fresh longer. They can be used to compensate for variations of flour and malt quality. In addition, breweries use enzymes to lower calories and alcohol concentration of beer. In winemaking, enzymes are used to maintain wine color and clarity or to reduce the sulfur content. Additionally, enzymes can enhance the filterability and improve the flavor of wine. They also help improving the quality, stability, and yield of fruit juices. Last but not least, the application of enzymes revolutionized the starch and sugar industry by making the hydrolysis of starch and rearrangement of glucose to fructose much more cost-effective.

The large number of applications makes enzymes highly valuable for the food and feed industry. According to the Novozymes Report 2011 [31], the global industrial enzyme market had a volume of approximately €2.7 billion. In the enzyme business,[9] food and beverage enzymes accounted for 29 % of sales, second only to household care enzymes (31 %). Feed and other technical enzymes accounted for 13 % of sales. With a market share of 47 % in industrial use enzymes, Novozymes was the global market leader, followed by DuPont (21 %), which recently acquired Danisco, and DSM (6 %). In the European market, AB Enzymes, Christian Hansen, and Henkel are prominent companies.

Of the six different existing enzyme classes (EC 1: oxidoreductases, EC 2: transferases, EC 3: hydrolases, EC 4: lyases, EC 5: isomerases, and EC 6: ligases) all but ligases are sold commercially for food and feed production. Currently, approximately 260 different enzymes are available in the European Union (Table 1). They can be isolated from plant[10] (3 %) and animal (6 %) materials but the majority are produced by means of fungi (filamentous ascomycetes and

[6] Regulation (EC) No. 1332/2008, Article 7.

[7] Regulation (EC) No. 1332/2008, Article 24.

[8] Regulation (EC) No. 1332/2008, Article 18(2).

[9] Including biobusiness.

[10] In relation to the total number of enzymes.

Table 1 List of enzymes used in the food and feed industry, together with their production strains and application areas [25, 32–36]

Accepted name	Systematic name	EC number	Production strains	Application areas
Oxidoreductases		*EC 1.x.x.x*		
Glucose oxidase	β-D-glucose:oxygen 1-oxidoreductase	EC 1.1.3.4	*Aspergillus niger, A. oryzae* (GM), *Penicillium chrysogenum*	Bakery, eggs, starch processing
Hexose oxidase	D-hexose:oxygen 1-oxidoreductase	EC 1.1.3.5	*Hansenula polymorpha* (GM)	Bakery, starch processing
Cellobiose dehydrogenase (acceptor)	Cellobiose:acceptor 1-oxidoreductase	EC 1.1.99.18	*Fusarium venenatum* (GM)	
Laccase	Benzenediol:oxygen oxidoreductase	EC 1.10.3.2	*Aspergillus niger, A. oryzae* (GM), *Trametes hirsuta, T. versicolor, Trichoderma longibrachiatum*	Beverages, cork treatment
L-ascorbate oxidase	L-ascorbate:oxygen oxidoreductase	EC 1.10.3.3	*Cucurbita pepo*	Clinical tests
Catalase	Hydrogen-peroxide:hydrogen-peroxide oxidoreductase	EC 1.11.1.6	*Aspergillus niger, A. oryzae* (GM), *Micrococcus luteus*	Beverages, eggs, starch processing, others
Peroxidase	Phenolic donor:hydrogen-peroxide oxidoreductase	EC 1.11.1.7	Lactoserum, soy	Starch processing, others
Linoleate 13*S*-lipoxygenase	Linoleate:oxygen 13-oxidoreductase	EC 1.13.11.12	soy, *Escherichia coli* (GM)	Bakery, flavor production
Transferases		*EC 2.x.x.x*		
Phosphatidylcholine-sterol *O*-acyltransferase	Phosphatidylcholine:sterol *O*-acyltransferase	EC 2.3.1.43	*Bacillus licheniformis* (GM)	Bakery, eggs, meat and fish, milk and cheese, others
Protein-glutamine γ-glutamyltransferase (transglutaminase)	Protein-glutamine γ-glutamyltransferase	EC 2.3.2.13	*Streptomyces mobaraensis*	Meat and fish, bakery, milk and cheese
Dextransucrase	Sucrose:(1→6)-α-D-glucan 6-α-D-glucosyltransferase	EC 2.4.1.5	*Leuconostoc mesenteroides*	Dextran production
1,4-α-glucan branching enzyme	(1→4)-α-D-glucan: (1→4)-α-D-glucan 6-α-D-[(1→4)-α-D-glucano]-transferase	EC 2.4.1.18	*Bacillus amyloliquefaciens* (GM)	Starch processing

(continued)

Table 1 (continued)

Accepted name	Systematic name	EC number	Production strains	Application areas
Cyclomaltodextrin glucanotrans-ferase	(1→4)-α-D-glucan 4-α-D-[(1→4)-α-D-glucano]-transferase (cyclizing)	EC 2.4.1.19	*Bacillus licheniformis* (GM)	Production of cyclodextrins
1,4-α-glucan 6-α-glucosyltrans-ferase	(1→4)-α-D-glucan:(1→4)-α-D-glucan(D-glucose) 6-α-D-glucosyltransferase	EC 2.4.1.24	*Aspergillus niger, Trichoderma longibrachiatum* (GM)	Starch processing
Hydrolases		*3.x.x.x*		
Carboxylesterase	Carboxylic-ester hydrolase	EC 3.1.1.1	*Rhizomucor miehei*	Hydrolysis of various carboxylic esters
Triacylglycerol lipase	Triacylglycerol acylhydrolase	EC 3.1.1.3	*Aspergillus niger* (GM), *A. oryzae* (GM), *Bacillus licheniformis* (GM), calf, *Candida lipolytica, C. rugosa,* goat, *Hansenula polymorpha* (GM), lamb, *Mucor javanicus, Penicillium roqueforti, Pichia angusta* (GM), *Rhizomucor miehei, Rhizopus niveus* (non-GM & GM), *R. oryzae*	Bakery, milk and cheese, starch processing, others
Phospholipase A_2	Phosphatidylcholine 2-acylhydrolase	EC 3.1.1.4	*Aspergillus oryzae* (GM), *A. niger* (GM), ox, pig, *Streptomyces violaceoruber* (non-GM and GM), *Trichoderma longibrachiatum* (GM)	Bakery, eggs, starch processing, others
Lysophospholipase	2-lysophosphatidyl-choline acylhydrolase	EC 3.1.1.5	*Aspergillus niger* (non-GM and GM), *Trichoderma longibrachiatum* (GM)	Beverages, starch processing
Pectinesterase	Pectin pectylhydrolase	EC 3.1.1.11	*Aspergillus niger* (non-GM and GM), *A. oryzae* (GM), *A. sojae, Penicillium funiculosum, Rhizopus oryzae, Trichoderma longibrachiatum* (GM)	Beverages, fruits and vegetables
Tannase	Tannin acylhydrolase	EC 3.1.1.20	*Aspergillus oryzae, A. niger*	Beverages

(continued)

Table 1 (continued)

Accepted name	Systematic name	EC number	Production strains	Application areas
Acylglycerol lipase	Glycerol-ester acylhydrolase	EC 3.1.1.23	*Penicillium camembertii*	Lipids
Phospholipase A_1	Phosphatidylcholine 1-acylhydrolase	EC 3.1.1.32	*Aspergillus oryzae* (GM)	Milk and cheese
Feruloyl esterase	4-hydroxy-3-methoxycinnamoyl-sugar hydrolase	EC 3.1.1.73	*Aspergillus niger, Streptomyces werraensis*	Biomass degradation
Hydroxyacylglutathione hydrolase	*S*-(2-hydroxyacyl)-glutathione hydrolase	EC 3.1.2.6	*Bacillus amyloliquefaciens* (GM)	Bakery
Phytase (3-, 4-)	*Myo*-inositol-hexakisphosphate x-phosphohydrolase	EC 3.1.3.8/26	*Aspergillus niger* (non-GM and GM), *A. oryzae (GM), Schizosaccharomyces pombe* (GM), *Trichoderma longibrachiatum* (GM)	Feed, bakery
Phosphodiesterase I	Oligonucleotide 5′-nucleotidohydrolase	EC 3.1.4.1	*Leptographium procerum,* malt, *Penicillium citrinum*	Yeast extract production
Phospholipase C	Phosphatidylcholine cholinephosphohydrolase	EC 3.1.4.3	*Pichia pastoris* (GM)	Emulsifiers
Ribonuclease P		EC 3.1.26.5	*Penicillium citrinum*	Yeast extract production
α-Amylase	4-α-D-glucan glucanohydrolase	EC 3.2.1.1	*Aspergillus niger, A. oryzae, Bacillus amyloliquefaciens* (non-GM and GM), *B. licheniformis* (non-GM and GM), *B. stearothermophilus, Microbacterium imperiale, Pseudomonas fluorescens* (GM), *Trichoderma longibrachiatum* (GM)	Beverages, bakery, starch processing
β-Amylase	4-α-D-glucan maltohydrolase	EC 3.2.1.2	Barley, *Penicillium multicolor,* wheat, soy	Beverages, starch processing

(continued)

Table 1 (continued)

Accepted name	Systematic name	EC number	Production strains	Application areas
Glucan 1,4-α-glucosidase	4-α-D-glucan glucohydrolase	EC 3.2.1.3	*Aspergillus niger* (non-GM and GM), *A. oryzae, Hypocrea jecorina* (GM), *Rhizopus niveus R. oryzae, Trichoderma longibrachiatum* (GM)	Bakery, beverages, starch processing
Cellulase	4-(1,3;1,4)-β-D-glucan 4-glucanohydrolase	EC 3.2.1.4	*Aspergillus niger, Humicola insolens, Hypocrea jecorina* (GM), *Penicillium funiculosum, Talaromyces emersonii, Trichoderma longibrachiatum* (non-GM and GM), *T. viride*	Bakery, beverages, starch processing
Endo-1,3(4)-β-glucanase	3-(1→3;1→4)-β-D-glucan 3(4)-glucanohydrolase	EC 3.2.1.6	*Aspergillus niger, A. oryzae* (GM), *Bacillus amyloliquefaciens* (non-GM and GM), *Cellulosimicrobium cellulans, Disporotrichum dimorphosporum, Humicola insolens, Penicillium funiculosum, Talaromyces emersonii, Trichoderma longibrachiatum* (non-GM and GM)	Beverages
Inulinase	1-β-D-fructan fructanohydrolase	EC 3.2.1.7	*Aspergillus niger, A. oryzae* (GM)	Starch processing
Endo-1,4-β-xylanase	4-β-D-xylan xylanohydrolase	EC 3.2.1.8	*Aspergillus niger* (non-GM and GM), *A. oryzae* (GM), *Bacillus amyloliquefaciens* (GM), *B. licheniformis* (GM), *Disporotrichum dimorphosporum, Humicola insolens, Penicillium funiculosum, Talaromyces emersonii, Trichoderma longibrachiatum* (non-GM and GM), *T. viride*	Bakery, beverages, starch processing, others

(continued)

Table 1 (continued)

Accepted name	Systematic name	EC number	Production strains	Application areas
Dextranase	6-α-D-glucan 6-glucanohydrolase	EC 3.2.1.11	*Chaetomium erraticum, C. gracile, Penicillum lilacinum*	Sugar refinery
Chitinase	(1→4)-2-acetamido-2-deoxy-β-D-glucan glycanohydrolase	EC 3.2.1.14	*Streptomyces violaceoruber* (GM)	Production of *N*-acetyl-glucosamine
Polygalacturonase	(1→4)-α-D-galacturonan glycanohydrolase	EC 3.2.1.15	*Aspergillus niger* (non-GM and GM), *A. wentii, Rhizopus oryzae, Trichoderma longibrachiatum*	Beverages
Lysozyme	Peptidoglycan *N*-acetylmuramoylhydrolase	EC 3.2.1.17	Chicken egg	Preservation
α-Glucosidase	α-D-glucoside glucohydrolase	EC 3.2.1.20	*Aspergillus niger*	Starch processing
β-Glucosidase	β-D-glucoside glucohydrolase	EC 3.2.1.21	*Aspergillus niger*	Sugar specialties
α-Galactosidase	α-D-galactoside galactohydrolase	EC 3.2.1.22	*Aspegillus niger, A. oryzae* (GM), *Saccharomyces cerevisiae* (GM)	Medical applications
β-Galactosidase (lactase)	β-D-galactoside galactohydrolase	EC 3.2.1.23	*Aspergillus niger, A. oryzae* (non-GM and GM), *Bacillus circulans, Kluyveromyces fragilis, K. lactis* (non-GM and GM)	Milk processing
β-Fructofuranosidase (invertase)	β-D-fructofuranoside fructohydrolase	EC 3.2.1.26	*Aspergillus niger, Saccharomyces cerevisiae*	Sucrose processing
Pullulanase	Pullulan 6-α-glucanohydrolase	EC 3.2.1.41	*Bacillus acido-pullulyticus, B. amyloliquefaciens* (GM), *B. brevis, B. licheniformis* (GM), *B. subtilis* (GM), *Klebsiella planticola, Pullulanibacillus sp., Trichoderma longibrachiatum* (GM)	Beverages, starch processing
β-L-*N*-acetylhexos-amini-dase	β-*N*-acetyl-D-hexosaminide *N*-acetylhexosaminohydrolase	EC 3.2.1.52	*Streptomyces violaceoruber* (GM)	

(continued)

Table 1 (continued)

Accepted name	Systematic name	EC number	Production strains	Application areas
α-*N*-arabinofuranosidase	α-L-arabinofuranoside arabinofuranohydrolase	EC 3.2.1.55	*Aspergillus niger* (non-GM and GM)	Bakery
Glucan 1,3-β-glucosidase	3-β-D-glucan glucohydrolase	EC 3.2.1.58	*Penicillium funiculosum, Trichoderma harzianum*	Starch processing
Glucan 1,4-α-maltotetrao-hydrolase	4-α-D-glucan maltotetraohydrolase	EC 3.2.1.60	*Bacillus licheniformis* (GM), *B. subtilis* (GM)	Flour processing
Mycodextranase	(1→3)-(1→4)-α-D-glucan 4-glucanohydrolase	EC 3.2.1.61	*Bacillus licheniformis* (GM)	Bakery
Isoamylase	Glycogen α-1,6-glucano-hydrolase	EC 3.2.1.68	*Pseudomonas amyloderamosa*	Starch processing
Mannan *endo*-1,4-β-mannosidase	1,4-β-D-mannan mannanohydrolase	EC 3.2.1.78	*Aspergillus niger, Bacillus halodurans, Trichoderma longibrachiatum* (GM)	Processing of mannans and galacto-mannans
Arabinan *endo*-1,5-α-L-arabinanase	5-α-L-arabinan 5-α-L-arabinanohydrolase	EC 3.2.1.99	*Aspergillus niger*	Processing of L-arabinan
Glucan 1,4-α-malto-hydrolase	4-α-D-glucan α-maltohydrolase	EC 3.2.1.133	*Bacillus amyloliquefaciens* (GM), *Microbacterium imperiale*	Bakery, starch processing
Aminopeptidase		EC 3.4.11.x	*Aspergillus niger, A. oryzae, Rhizopus oryzae*	Milk processing
Leucyl aminopeptidase		EC 3.4.11.1	*Aspergillus oryzae*	Beverages, production of soy sauce
Serine-type carboxypeptidase		EC 3.4.16.x	*Aspergillus niger* (GM)	Meat and fish, milk processing, flavoring preparations
Serine endopeptidase		EC 3.4.21.x	*Aspergillus oryzae, A. wentii, Bacillus amyloliquefaciens, B. licheniformis, Cryphonectria parasitica, Fusarium venenatum* (GM), *Rhizomucor miehei*	Protein hydrolysis
Chymotrypsin		EC 3.4.21.1	*Bacillus licheniformis* (GM), beef pancreas	Protein hydrolysis

(continued)

Table 1 (continued)

Accepted name	Systematic name	EC number	Production strains	Application areas
Trypsin		EC 3.4.21.4	*Fusarium venenatum* (GM)	Milk and cheese
Thrombin		EC 3.4.21.5	Cattle, pig	Meat and fish
Prolyl oligopeptidase		EC 3.4.21.26	*Aspergillus niger* (GM)	Beverages, others
Subtilisin		EC 3.4.21.62	*Bacillus licheniformis*	Protein hydrolysis
Oryzin		EC 3.4.21.63	*Aspergillus oryzae*	Bakery
Aqualysin 1		EC 3.4.21.111	*Bacillus amyloliquefaciens* (GM)	Bakery
Papain		EC 3.4.22.2	*Carica papaya*	Beverages, others
Aspartic endopeptidases		EC 3.4.23.x	Animal origin, *Aspergillus niger* (GM), *A. oryzae, A. wentii, Bacillus licheniformis, Micrococcus caseolyticus*	Bakery, beverages, milk and cheese, protein hydrolysis
Pepsin		EC 3.4.23.1/2	Cattle, pig	Protein hydrolysis
Pepsin A		EC 3.4.23.1	Bovine rennet	Milk and cheese
Chymosin		EC 3.4.23.4	*Aspergillus niger* (GM), calf, *Escherichia coli* (GM), *Kluyveromyces lactis* (GM)	Milk and cheese, others
Aspergillopepsin I		EC 3.4.23.18	*Aspergillus oryzae* (non-GM and GM), *A. wentii*	Milk and cheese, others
Endothiapepsin		EC 3.4.23.22	*Cryphonectria parasitica*	Others
Mucorpepsin		EC 3.4.23.23	*Aspergillus oryzae* (GM), *Mucor pusillus, Rhizomucor miehei*	Meat and fish, protein hydrolysis
Thermolysin		EC 3.4.24.27	*Geobacillus caldoproteolyticus*	Protein hydrolysis
Bacillolysin		EC 3.4.24.28	*Bacillus amyloliquefaciens* (non-GM and GM)	Bakery, beverages, meat and fish, flour processing, protein hydrolysis
Deuterolysin		EC 3.4.24.39	*Aspergillus oryzae, A. wentii*	Beverages, protein hydrolysis

(continued)

Table 1 (continued)

Accepted name	Systematic name	EC number	Production strains	Application areas
Asparaginase	L-asparagine amidohydrolase	EC 3.5.1.1	*Aspergillus niger* (GM), *A. oryzae* (GM)	Potato products, bakery, fruits and vegetables, starch processing, others
Glutaminase	L-glutamine amidohydrolase	EC 3.5.1.2	*Aspergillus niger, Bacillus amyloliquefaciens, B. subtilis*	Protein hydrolysis
Urease	Urea amidohydrolase	EC 3.5.1.5	*Lactobaillus fermentum*	Others
Protein-glutamine glutaminase	Protein-L-glutamine amidohydrolase	EC 3.5.1.44	*Chryseobacterium proteolyticum*	Production of seasonings
AMP deaminase	AMP aminohydrolase	EC 3.5.4.6	*Aspergillus melleus, A. oryzae*	Yeast extract
Lyases		*EC 4.x.x.x*		
Pectate lyase	(1→4)-α-D-galacturonan lyase	EC 4.2.2.2	*Bacillus subtilis*	Beverages
Poly(β-D-mannuronate) lyase	Poly[(1→4)-β-D-mannuronide] lyase	EC 4.2.2.3	*Sphingobacterium multivorum*	Processing of alginates
Acetolactate decarboxylase	(2*S*)-2-hydroxy-2-methyl-3-oxobutanoate carboxy-lyase [(3*R*)-3-hydroxybutan-2-one-forming]	EC 4.1.1.5	*Bacillus amyloliquefaciens* (GM), *Saccharomyces cerevisiae* (GM)	Beverages
Pectin lyase	(1→4)-6-*O*-methyl-α-D-galacturonan lyase	EC 4.2.2.10	*Aspergillus niger* (non-GM and GM), *A. sojae, Penicillium funiculosum, Rhizopus oryzae, Trichoderma longibrachiatum* (GM)	Beverages
Isomerases		*EC 5.x.x.x*		
Glucose isomerase		EC 5.3.1.x	*Actinoplanes missouriensis, Bacillus coagulans, Streptomyces murinus, S. olivochromogenes*	Sugar production
Xylose isomerase	D-xylose aldose-ketose-isomerase	EC 5.3.1.5	*Streptomyces murinus, S. olivochromogenes, S. rubiginosus* (non-GM and GM)	Sugar production
Glucose-6-phosphate isomerase	D-glucose-6-phosphate aldose-ketose-isomerase	EC 5.3.1.9	*Streptomyces violaceoniger*	Sugar production

GM genetically modified

basidiomycetes: 58 %, yeasts: 5 %) and bacteria (28 %). One-third of the offered enzymes originate from genetically modified organisms.[11]

3.2 *Feed Enzymes*

In addition to a variety of applications in the food industry, enzymes may also be used during the manufacture of feed. In this capacity, they are used primarily to increase the availability of essential nutrients, complementing the spectra of activities of enzymes already present in feed and animals' own digestive enzymes. Thus, the animal feedstuff can be used more efficiently and the use of enzymes helps to conserve resources and avoid waste. The most important enzyme in the feed sector is phytase (see 4.1.3). In addition, xylanase and β-glucanase are of importance.

4 Applications

The manifold industrial applications of enzymes have been reviewed in great detail [e.g. 37, 38]. Very good overviews on enzymes used in food and feed technology were published by Whitehurst and van Oort [39] and Bedford and Partridge [40]. Current trends have been reviewed by Son and Ravindran [41].

Due to the existing literature and the enormous variety of enzymes used in food and animal feed (Table 1), it is not necessary or even possible to cover all applications in the context of this book chapter. Therefore, a number of prominent examples in the areas of resource efficiency, biopreservation, health, safety, and food quality are highlighted here.

4.1 *Resource Efficiency*

4.1.1 Amylase

Due to the increasing availability of efficient and stable biocatalysts, chemical processes are more and more replaced by biotechnological methods. The latter often show superior economic and ecological operating figures. Therefore, the acid-catalyzed saccharification and the production of sugar specialties have been replaced worldwide by enzymatic procedures. In these systems, amylases are

[11] Counting self-cloned organisms, although legal self-cloning of nonpathogenic naturally occurring microorganisms is excluded from Directive 90/219/EEC.

essential and thus have become one of the most important industrial enzymes [42]. Besides the conversion of starch to sugar syrups, they can be used in other areas of the food industry, as well as in the detergent, textile, paper, and pharmaceutical industries [37, 43, 44].

α-Amylases (4-α-D-glucan glucanohydrolase, EC 3.2.1.1) degrade starch and similar carbohydrates by endohydrolysis of their (1→4)-α-D-glucosidic bonds. The majority of α-amylases belong to the group of metalloenzymes and require calcium ions (Ca^{2+}). By means of direct evolution, the performance of several amylases was further maximized and adapted to the needs of the starch processing industry [45]. The enhancement of the thermostability of amylase was achieved by DNA-shuffling techniques [46]. The baking industry and consumers might benefit from genetically optimized starch-modifying enzymes. Based on the improved thermal stability of an α-amylase in the acidic pH range, the retrogradation of sourdough breads ("staling") can be delayed [47].

4.1.2 Peptidase

Seasonings are biotechnologically obtained by hydrolysis of vegetable proteins in multistage enzymatic processes. Unlike the previously common catalysis with the aid of hydrochloric acid, which leads to the presence of small remainders of fat in the formation of monochloropropanediols (mainly 3-MCPD) and dichloropropanols (1,3- and 1,2-DCP) [48], no toxic byproducts are produced in the enzymatic process. Peptidases (EC 3.4) from edible mushrooms with new catalytic properties allow for an even more efficient protein hydrolysis [49].

4.1.3 Phytase

Phytases are phosphatases able to hydrolyze O–P bonds in phytic acid liberating inorganic phosphate. Phytases can be grouped according to the attack on the hexaphosphoric ester into 3-phytase (*myo*-inositol-hexakisphosphate 3-phosphohydrolase, EC 3.1.3.8) as well as 4-phytase (*myo*-inositol-hexakisphosphate 4-phosphohydrolase, EC 3.1.3.26), releasing the phosphate at the corresponding position at the inositol ring (Fig. 1).

Phytic acid is used by plants to store different type of anions (Cu^{2+}, $Fe^{2+/3+}$, Ca^{2+}, Mg^{2+}, etc.). The resulting salts are known as phytates. Nonruminant animals do not have the enzymatic ability to hydrolyze phytates; therefore, phosphate and minerals are not absorbed, but rather they pass through the intestinal tract undigested. The addition of phytases to food and feed does therefore enhance the availability of phosphate and minerals bound in phytates.

More than 20 years ago, the first commercial phytase product for feed, Natuphos, was released [50]. Nowadays, several other phytase products are available on the market for the improvement of animal feed, such as Ronozyme from Novozymes and Finase from AB Enzymes. The range of applications in nonruminant

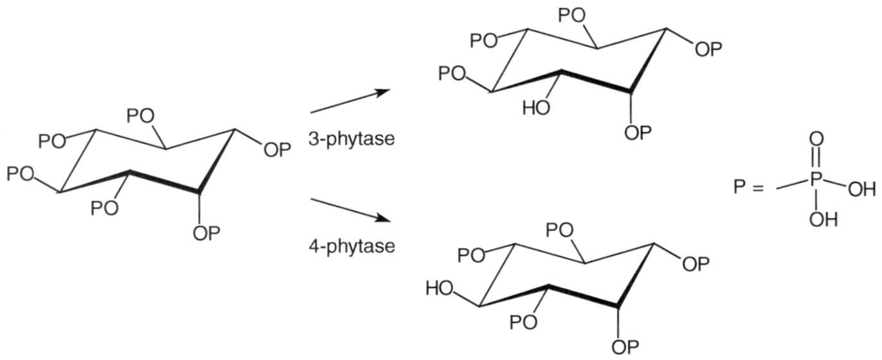

Fig. 1 Hydrolysis of *myo*-inositol-1,2,3,4,5,6-hexakisphosphate by 3- and 4-phytase

feed is large, as discussed in several reviews on the usage of phytases in the fodder of pigs [51, 52], poultry [53], and fish [54, 55]. Additionally, phytase might also be used in food for the improvement of the nutritional value of cereal food products by degrading phytate [56].

Commercial recombinant production of phytases occurs generally, if not exclusively, in ascomycetes [32, 50], whereas the phytase genes originate from different phyla: bacteria, ascomycetes, and basidiomycetes (*Peniophora lycii* [57]). Industrial production takes place in liquid media in stirred bioreactors on a $6 \times 120\ m^3$ scale. In addition, the cultivation of phytase producing filamentous fungi in solid-state [58] or solid-substrate [59] fermentation systems has also been studied, but it lacks industrial adaptability. Another opportunity for the production of phytases is the usage of transgenic plants, such as maize, rice, soybean, and wheat [60–63].

4.2 Biopreservation

Foods that spoil during manufacturing or storage endanger human life, waste important resources, and cost the food industry vast sums of money. Therefore, foods have been preserved since the dawn of mankind. In recent decades, this has often been achieved by adding antimicrobial preservatives (e.g. sulfites, sorbic acid, benzoic acid, and their salts) and/or antioxidants (e.g. 2-*tert*-butyl-4-hydroxyanisole and 3-*tert*-butyl-4-hydroxyanisole (BHA), 2,6-*bis*(1,1-dimethyl-ethyl)-4-methylphenol (BHT)). In particular, synthetic food additives are often disliked by consumers because they are occasionally considered critically [64].

Alternatively, natural or controlled microbiota or antimicrobials can be used to enhance the safety and to extend the shelf life of food [65]. Lactic acid bacteria are commonly used. They often produce lactic and acetic acid, hydrogen peroxide, and under certain conditions peptide bacteriocins, which prevent the development of pathogens and spoilage microorganisms [66, 67]. The most important bacteriocin is the peptide nisin [68], as discussed in detail in Chap. 2

4.2.1 Lysozyme

The cell wall of bacteria is composed of the peptidoglycan murein. Murein is a cross-linked heteropolymer consisting of sugars and amino acids. Its task is to maintain the turgor pressure and shape of the bacterial cell. Peptidoglycan can be cleaved by bacterial cell wall hydrolases (BCWHs), which leads to bacteriolysis. BCWHs are found ubiquitously in nature in animals, plants, protozoa, bacteria, and bacterial viruses [69]. Due to their high specificity, they only attack bacteria.

The most important and best characterized BCWH is lysozyme (peptidoglycan *N*-acetylmuramoylhydrolase, EC 3.2.1.17) from hen egg white. Lysozyme was discovered by Sir Alexander Fleming in 1922 and has been extensively studied since then. It became one of the few approved natural antimicrobials for use in food in the European Union (E 1105). The main application of lysozyme is to prevent the growth of gram-positive bacteria in semi-hard cheeses (e.g. to prevent late blowing caused by *Clostridium tyrobutyricum*) [70]. Further applications include spoilage control in wine [71], beer [72], fish [73], and meat [74].

4.3 Health, Safety and Quality

4.3.1 Lactase

β-D-Galactosidase (*β*-D-galactoside galactohydrolase, EC 3.2.1.23) hydrolyses the glycosidic bond of the disaccharide lactose into its monomers glucose and galactose. Therefore, *β*-D-galactosidase is also known as lactase. *β*-D-Galactosidases can be found throughout nature. Commercial enzymes are mainly of fungal origin of the genera *Kluyveromyces* (yeast) and *Aspergillus* (filamentous ascomycete) [75]. Although several *β*-D-galactosidases are available on the market, new efforts are still conducted to improve the enzymatic properties by site-directed mutagenesis [76] or to find new *β*-D-galactosidase genes by screening metagenome databases [77].

From a technological point of view, the milk sugar lactose is hydrolyzed because of the resulting increase of sweetness and the reduced susceptibility to crystallization during spray drying of milk and whey [78, 79]. Nevertheless, the most apparent application is the production of "lactose-free" milk products, which enables the consumption of dairy products by lactose-intolerant people.

For industrial processes, *β*-D-galactosidases can be applied by immobilization of the enzyme on carriers, such as cellulose, alginate, or other polymers, for hydrolysis of milk or whey products [80]. Conversion rates of lactose in batch and continuous operation mode by immobilized *β*-D-galactosidase might reach 95 %

[81]. To reach even lower lactose concentrations of less than 0.01 %, a combination of β-D-galactosidase activity and ultrafiltration as well as nanofiltration methods can be used [82]. Besides the degradation of substances that lead to food intolerances, the elimination of food allergens by means of a specific degradation of allergenic epitopes is another promising application of enzymes [83].

4.3.2 Asparaginase

Using the enzyme asparaginase (L-asparagine amidohydrolase, EC 3.5.1.1), it is possible to significantly reduce the formation of the cooking carcinogen acrylamide during roasting, deep-frying, or baking of foods [84]. The enzyme hydrolyzes free asparagine to aspartic acid, thereby preventing the formation of acrylamide by reaction of asparagine with reducing sugars at elevated temperatures during the Maillard reaction (Fig. 2) [85].

The mitigation of acrylamide formation is especially important for a number of cereal- and potato-based products, including crackers, crispbread, gingerbread, biscuits, French fries, and potato chips. After asparaginase pretreatment, the acrylamide concentration of certain foods could be reduced by up to 97 % [84, 86]. By means of in vitro directed evolution, the properties of asparaginase were optimized. For example, an Asp133Leu mutation of a wild-type enzyme showed a significantly improved thermal stability. The enzyme's half-life at 50 °C increased from 3 to 160 h, and the half-inactivation temperature of the mutant was 9 °C higher.

4.3.3 Lipase

Trans fatty acids (TFAs) are fatty acids with at least one double bond in *(E)*-configuration. The consumption of TFAs increases the risk of coronary heart diseases. Thus, their concentrations in lipid-containing products should be reduced [87]. Naturally, TFAs occur in small amounts in meat and milk of ruminants, but the most significant concentrations of TFAs develop during partial hydrogenation and deodorization of fats [88]. The formation of TFAs during fat hardening can be avoided by lipase catalyzed transesterification to increase the slip melting points of fats. [89].

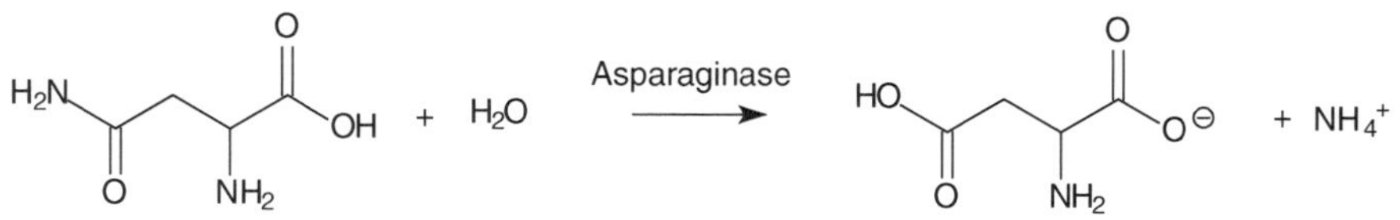

Fig. 2 Asparaginase catalyzed hydrolysis of asparagine to aspartic acid

For enzymatic transesterification between different lipids, triacylglycerol lipases (triacylglycerol acylhydrolase, EC 3.1.1.3) acting on the SN1 and SN3 positions of the triglyceride are used. Various lipases have been applied for the production of table margarine out of fat-oil blends. Lipases of the ascomycetes *Thermomyces lanuginosa* and *Rhizomucor miehei*, as well as a lipase of the proteobacterium *Pseudomonas* sp., were used for transesterification of fat blends consisting of palm stearin and vegetable oil [90, 91]. Fully hydrogenated oils in blends with vegetable oils also have been used [92]. In all studies, an increase of the slip melting points and the solid fat content was achieved in the fat-oil blend, thus indicating an alternative method for fat hardening via hydrogenation.

4.4 Further Applications

4.4.1 Laccase

Laccases are multicopper-oxidases (benzenediol:oxygen oxidoreductase, EC 1.10.3.2) that are able to oxidize phenolic substrates (e.g. 2,6-dimethoxy-phenol), aromatic amines (e.g. 1-hydroxybenzotriazole), or polycyclic aromatic hydrocarbons (e.g. anthracene) [93]. The oxidation of the substrate occurs via a one-electron reduction and is accompanied by a reduction of molecular oxygen to water.

Most laccases are of fungal origin [94], but they also occur in bacteria, insects, and plants [95–97]. Due to the broad substrate range of laccases, their possible industrial usage is widespread. Nevertheless, only few applications have been commercialized up to now, mostly in the textile industry [98]. In the food and feed sector, laccases have been evaluated for different applications, such as the stabilization of beverages, the reduction of off-flavors, the improvement of wheat dough, and the usage of laccases as biosensors in the food processing industry [37, 98–100].

Off-flavors in wine may occur due to microbial conversion of phenolic compounds present in the wine itself or in the cork stoppers. In a commercial product from Novozymes called Suberase, laccase is used for polymerization of phenolic compounds in the cork, which act as precursors for malodors, such as pentachlorophenol or 2,4,6-trichloroanisol [101].

In apple juice, addition of laccase reduced the amount of the phenolic off-flavors 2,6-dibromophenol, guaiacol, and α-terpineol [102]. Several studies used laccases to increase the stability of apple juices by polymerization of phenols and their subsequent removal by ultrafiltration methods [103]. A positive side effect is the decrease of molecular oxygen in the juice due to its consumption by laccases. Nevertheless, desired phenols might also be oxidized; thus, the sensory attributes and nutritional value might be altered.

The polymerization reactions catalyzed by laccases can also be used to improve the shade of food, such as the coloration of tea-based products [104]. In the bakery industry, laccase might be used together with proteases or xylanases to improve the dough quality. It was proposed that laccases oxidize ferulic acid attached to the arabinoxylan present in cereal flour. The obtained phenolic radicals can undergo a nonenzymatic reaction, resulting in cross-linked feruloylated arabinoxylans [105]. In oat flour-based bread, the usage of laccase increased the loaf-specific volume and decreased the crumb hardness [106]. Contradictory, in another study, laccase alone decreased the specific volume and increased the crumb hardness [105]. A combined usage of laccase with xylanase improved again the oat flour bread properties [105, 107].

4.4.2 Peroxidase

Peroxidases (EC 1.11.1.x) are a diverse group of oxidoreductases using peroxide as an electron acceptor. Their substrate spectrum ranges from hydrogen peroxide decomposed by catalase (hydrogen-peroxide:hydrogen-peroxide oxidoreductase, EC 1.11.1.6), phenolic compounds degraded by Mn-dependent and lignin peroxidases (Mn(II):hydrogen-peroxide oxidoreductase, EC 1.11.1.13 and 1,2-*bis* (3,4-dimethoxyphenyl)propane-1,3-diol:hydrogen-peroxide oxidoreductase, EC 1.11.1.14), and recalcitrant dyes by DyP-type peroxidases (reactive-blue-5:hydrogen-peroxide oxidoreductase, EC 1.11.1.19).

Direct supplementation of peroxidases to food was applied for the degradation of carotenoids used for coloring of cheese. The whey resulting from colored cheese production has an orange-yellowish tint, which interferes with further usage of the whey. Recently, a fungal peroxidase of the DyP-type was commercialized for bleaching of this kind of whey fluid under the name MaxiBright (DSM). To generate the hydrogen peroxide required by the peroxidase as a cofactor in situ, a glucose oxidase and a β-galactosidase were employed as auxiliary enzymes [108] (Fig. 3).

4.4.3 Lipoxygenase

Microorganisms, fungi, plants, and their enzymes can be used to synthesize natural flavor compounds [109]. Probably the most important example is the biotechnological production of the highly sought after vanillin [110]. Recently, disrupted

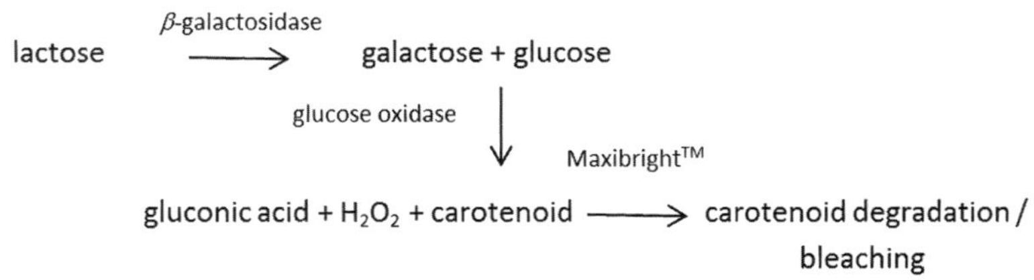

Fig. 3 Bleaching of whey and milk by a multiple enzyme system (modified from [108])

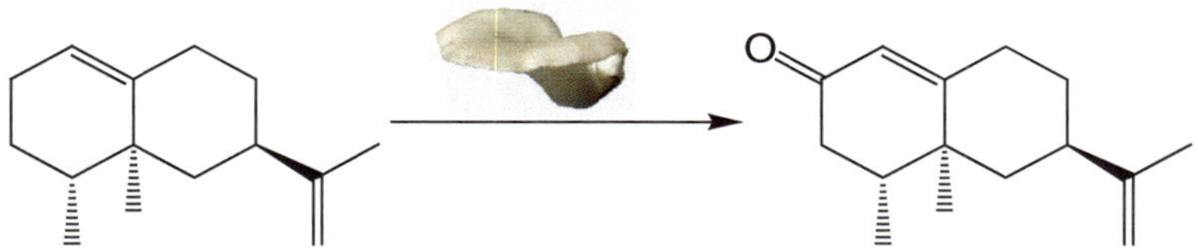

Fig. 4 Oxidation of (+)-valence to the grapefruit flavor (+)-nootkatone by lyophilisates of the basidiomycete *Pleurotus sapidus*

cells of the edible basidiomycete *Pleurotus sapidus* were deployed as a potent biocatalyst for the transformation of (+)-valencene to natural (+)-nootkatone [111] (Fig. 4).

The enzyme responsible for the biotransformation was biochemically characterized and purified, and the enzyme encoding cDNA was amplified from a cDNA library by polymerase chain reaction [112]. The catalytic reaction sequence of the enzyme was further investigated and a lipoxygenase-type oxidation of (+)-valencene via secondary and tertiary hydroperoxides was suggested [113]. In ongoing research, the dioxygenase was heterologously expressed in the cytosol and periplasm of *Escherichia coli* [114]. Only recently, the enzyme was identified as a potent 13*S*-lipoxygenase ($LOX_{Psa}1$; linoleate:oxygen 13-oxidoreductase, EC 1.13.11.12), and the kinetic parameters of the recombinant enzyme were determined by using linoleic acid as the substrate [115].

Acknowledgments This work was supported by the excellence initiative of the Hessian Ministry of Science and Art, which provides a generous grant for the Landes-Offensive zur Entwicklung Wissenschaftlich-Ökonomischer Exzellenz (LOEWE) research focus of insect biotechnology.

References

1. ancienttexts.org (2013) The epic of Gilgamesh: Tablet II. http://www.ancienttexts.org/library/mesopotamian/gilgamesh/tab2.htm. Accessed 18 Mar 2013
2. van Leewenhoeck A (1676) Observations, communicated to the publisher by Mr. Antony van Leewenhoeck, in a Dutch letter of the 9th of Octob. 1676. Here english'd: concerning little animals by him observed in rain-well-sea. and snow water; as also in water wherein pepper had lain infused. Philos Trans R Soc London 12:821–831. doi: 10.1098/rstl.1677.0003
3. Kirchhoff GS (1815) Ueber die Zuckerbildung beim Malzen des Getreides, und beim Bebrühen seines Mehls mit kochendem Wasser. Journal für Chemie und Physik 14:389–398
4. Leuchs EF (1831) Ueber die Verzuckerung des Stärkemehls durch Speichel. Archiv für die gesammte Naturlehre 21:105–107
5. Payen A, Persoz J-F (1833) Mémoire sur la Diastase, les principaux Produits de ses Réactions, et leurs applicatíons aux arts industriels. Annal Chim Phys 53:73–92
6. Duclaux É (1899) Traité de microbiologie. Masson and Co., Paris
7. Berzelius JJ (1836) Quelques Idées sur une nouvelle Force agissant dans les Combinaisons des Corps Organiques. Annal Chim Phys 61:146–151

8. Pasteur ML (1857) Mémoire sur la fermentation appelée lactique. Comptes rendus hebdomadaires des séances de l'Académie des sciences 45:913–916
9. Kühne W (1877) Ueber das Verhalten verschiedener organisirter und sog. ungeformter Fermente. Verhandlungen des naturhistorisch-medicinischen Vereins zu Heidelberg 1:190–193
10. Nobelprize.org (2013) The Nobel Prize in chemistry 1909. http://www.nobelprize.org/nobel_prizes/chemistry/laureates/1909/. Accessed 18 Mar 2013
11. Takamine J (1894) Process of making diastatic enzyme. United States Patent Application US 525823
12. Nobelprize.org (2013) The Nobel Prize in chemistry 1902. http://www.nobelprize.org/nobel_prizes/chemistry/laureates/1902/. Accessed 18 Mar 2013
13. Fischer E (1894) Einfluss der Configuration auf die Wirkung der Enzyme. Ber Dtsch Chem Ges 27:2985–2993
14. Buchner E (1897) Alkoholische Gährung ohne Hefezellen. Ber Dtsch Chem Ges 30:117–124
15. Nobelprize.org (2013) The Nobel Prize in chemistry 1907. http://www.nobelprize.org/nobel_prizes/chemistry/laureates/1907/. Accessed 18 Mar 2013
16. Henri V (1903) Lois générales de l'action des diastases. A. Hermann, Paris
17. Menten L, Michaelis MI (1913) Die Kinetik der Invertinwirkung. Biochem Z 49:333–369
18. Johnson KA, Goody RS (2011) The original Michaelis constant: Translation of the 1913 Michaelis–Menten paper. Biochemistry 50:8264–8269. doi:10.1021/bi201284u
19. Nobelprize.org (2013) The Nobel Prize in physiology or medicine 1945. http://www.nobelprize.org/nobel_prizes/medicine/laureates/1945/. Accessed 18 Mar 2013
20. Sumner JB (1926) The isolation and crystallization of the enzyme urease. J Biol Chem 69:435–441
21. Nobelprize.org (2013) The Nobel Prize in chemistry 1946. http://www.nobelprize.org/nobel_prizes/chemistry/laureates/1946/. Accessed 18 Mar 2013
22. Schellmann JA, Schellman CG (1997) Kaj Ulrik Linderstrøm-Lang (1896–1959). Protein Sci 6:1092–1100. doi:10.1002/pro.5560060516
23. Nobelprize.org (2013) The Nobel Prize in chemistry 1929. http://www.nobelprize.org/nobel_prizes/chemistry/laureates/1929/. Accessed 18 Mar 2013
24. Nobelprize.org (2013) The Nobel Prize in physiology or medicine 1978. http://www.nobelprize.org/nobel_prizes/medicine/laureates/1978/. Accessed 18 Mar 2013
25. Novozymes (2013) Novozymes. http://www.novozymes.com/. Accessed 21 Mar 2013
26. Neet KE, Koshland DE Jr (1966) The conversion of serine at the active site of subtilisin to cysteine: a "chemical mutation". Proc Natl Acad Sci U S A 56:1606–1611
27. DeSantis G, Jones JB (1999) Chemical modification of enzymes for enhanced functionality. Curr Opin Biotechnol 10:324–330. doi:10.1016/S0958-1669(99)80059-7
28. Arnold FH, Volkov AA (1999) Directed evolution of biocatalysts. Curr Opin Chem Biol 3:54–59. doi:10.1016/S1367-5931(99)80010-6
29. Kotzia GA, Platis D, Axarli IA et al (2012) Biocatalysis, enzyme engineering and biotechnology. In: Simpson BK, Nollet LML, Toldrá F et al (eds) Food biochemistry and food processing, 2nd edn. Wiley-Blackwell, Oxford
30. Lorenz P, Eck J (2005) Metagenomics and industrial applications. Nat Rev Microbiol 3:510–516. doi:10.1038/nrmicro1161
31. Novozymes (2012) The Novozymes Report 2011. http://report2011.novozymes.com/Service/Download+report/The-Novozymes-Report-2011.pdf. Accessed 5 Sep 2012
32. AMFEP (2009) List of enzymes October 2009. http://amfep.drupalgardens.com/sites/amfep.drupalgardens.com/files/Amfep-List-of-Commercial-Enzymes.pdf. Accessed 21 Mar 2013
33. DSM (2013) DSM. http://www.dsm.com/. Accessed 21 Mar 2013
34. DuPont (2013) DuPont. http://www.dupont.com/. Accessed 21 Mar 2013
35. Henkel (2013) Home—Henkel. http://www.henkel.com. Accessed 21 Mar 2013

36. Spök A, Proksch M (2012) Lebensmittelenzyme in der EU, 3rd edn. Bundesministerium für Gesundheit, Wien
37. Kirk O, Borchert TV, Fuglsang CC (2002) Industrial enzyme applications. Curr Opin Biotechnol 13:345–351. doi:10.1016/S0958-1669(02)00328-2
38. Sanchez S, Demain AL (2011) Enzymes and bioconversions of industrial, pharmaceutical, and biotechnological significance. Org Process Res Dev 15:224–230. doi:10.1021/op100302x
39. Whitehurst RJ, van Oort M (2009) Enzymes in food technology, 2nd edn. Blackwell Publishing Ltd, Oxford. doi:10.1002/9781444309935
40. Bedford MR, Partridge GG (2010) Enzymes in farm animal nutrition. CABI, Wallingford
41. Son J-H, Ravindran V (2011) Feed enzyme technology: present status and future developments. Recent Pat Food Nutr Agric 3:102–109. doi:10.2174/2212798411103020102
42. Guzmán-Maldonado H, Paredes-López O, Biliaderis CG (1995) Amylolytic enzymes and products derived from starch: a review. Crit Rev Food Sci Nutr 35:373–403. doi:10.1080/10408399509527706
43. Sivaramakrishnan S, Gangadharan D, Nampoothiri KM et al (2006) α-Amylases from microbial sources—an overview on recent developments. Food Technol Biotech 44:173–184
44. Akoh CC, Chang S-W, Lee G-C, Shaw J-F (2008) Biocatalysis for the production of industrial products and functional foods from rice and other agricultural produce. J Agric Food Chem 56:10445–10451. doi:10.1021/jf801928e
45. Kelly RM, Dijkhuizen L, Leemhuis H (2009) Starch and α-glucan acting enzymes, modulating their properties by directed evolution. Bioresour Technol 140:184–193. doi:10.1016/j.jbiotec.2009.01.020
46. Tang S-Y, Le Q-T, Shim J-H et al (2006) Enhancing thermostability of maltogenic amylase from *Bacillus thermoalkalophilus* ET2 by DNA shuffling. FEBS J 273:3335–3345. doi:10.1111/j.1742-4658.2006.05337.x
47. Priyadharshini R, Manoharan S, Hemalatha D, Gunasekaran P (2010) Repeated random mutagenesis of α-amylase from *Bacillus licheniformis* for improved pH performance. J Microbiol Biotechnol 20:1696–1701
48. Collier PD, Cromie DDO, Davies AP (1991) Mechanism of formation of chloropropanols present in protein hydrolysates. J Am Oil Chem Soc 68:785–790. doi:10.1007/BF02662173
49. Linke D, Krings U, Zorn H et al (2010) Peptidases from basidiomycetes. European Patent Application EP2139996 (A1)
50. Haefner S, Knietsch A, Scholten E et al (2005) Biotechnological production and applications of phytases. Appl Microbiol Biotechnol 68:588–597. doi:10.1007/s00253-005-0005-y
51. Pallauf J, Rimbach G (1997) Nutritional significance of phytic acid and phytase. Arch Anim Nutr 50:301–319. doi:10.1080/17450399709386141
52. Lenis NP, Jongbloed AW (1999) New technologies in low pollution swine diets: diet manipulation and use of synthetic amino acids, phytase and phase feeding for reduction of nitrogen and phosphorus excretion and ammonia emission. Asian-australas J Anim Sci 12:305–327
53. Selle PH, Ravindran V (2007) Microbial phytase in poultry nutrition. Anim Feed Sci Technol 135:1–41. doi:10.1016/j.anifeedsci.2006.06.010
54. Cao L, Wang W, Yang C et al (2007) Application of microbial phytase in fish feed. Enzyme Microb Technol 40:497–507. doi:10.1016/j.enzmictec.2007.01.007
55. Kumar V, Sinha AK, Makkar HPS et al (2012) Phytate and phytase in fish nutrition. J Anim Physiol Anim Nutr 96:335–364. doi:10.1111/j.1439-0396.2011.01169.x
56. Kumar V, Sinha AK, Makkar HPS, Becker K (2010) Dietary roles of phytate and phytase in human nutrition: a review. Food Chem 120:945–959. doi:10.1016/j.foodchem.2009.11.052
57. Aquilina G, Bories G, Brantom P et al (2010) Scientific opinion on Ronozyme® P (6-phytase) as feed additive for chickens and turkeys for fattening, laying hens, and piglets (weaned), pigs for fattening and sows (poultry and pigs). EFSA J 8(1862):1–27

58. Roopesh K, Ramachandran S, Nampoothiri KM et al (2005) Comparison of phytase production on wheat bran and oilcakes in solid-state fermentation by *Mucor racemosus*. Bioresour Technol 97:506–511. doi:10.1016/j.biortech.2005.02.046
59. Bogar B, Szakacs G, Linden JC et al (2003) Optimization of phytase production by solid substrate fermentation. J Ind Microbiol Biotechnol 30:183–189.
60. Brinch-Pedersen H, Olesen A, Rasmussen SK, Holm PB (2000) Generation of transgenic wheat (*Triticum aestivum* L.) for constitutive accumulation of an *Aspergillus* phytase. Mol Breed 6:195–206
61. Chen R, Xue G, Chen P et al (2008) Transgenic maize plants expressing a fungal phytase gene. Transgenic Res 17:633–643. doi:10.1007/s11248-007-9138-3
62. Gao XR, Wang GK, Su Q et al (2007) Phytase expression in transgenic soybeans: stable transformation with a vector-less construct. Biotechnol Lett 29:1781–1787. doi: 10.1007/s10529-007-9439-x
63. Hamada A, Yamaguchi K-I, Harada M et al (2006) Recombinant, rice-produced yeast phytase shows the ability to hydrolyze phytate derived from seed-based feed, and extreme stability during ensilage treatment. Biosci Biotechnol Biochem 70:1524–1527. doi: 10.1271/bbb.60039
64. McCann D, Barrett A, Cooper A et al (2007) Food additives and hyperactive behaviour in 3-year-old and 8/9-year-old children in the community: a randomised, double-blinded, placebo-controlled trial. Lancet 370:1560–1567. doi:10.1016/S0140-6736(07)61306-3
65. Ananou S, Maqueda M, Martínez-Bueno M, Valdivia E (2007) Biopreservation, an ecological approach to improve the safety and shelf-life of foods. In: Méndez-Vilas A (ed) Communicating current research and educational topics and trends in applied microbiology, 2007th edn. Formatex Research Center, Badajoz
66. Stoyanova LG, Ustyugova EA, Netrusov AI (2012) Antibacterial metabolites of lactic acid bacteria: their diversity and properties. Appl Biochem Microbiol 48:229–243. doi: 10.1134/S0003683812030143
67. Juneja VK, Dwivedi HP, Yan X (2012) Novel natural food antimicrobials. Ann Rev Food Sci Technol 3:381–403. doi:10.1146/annurev-food-022811-101241
68. Takala TM, Saris PEJ (2007) Nisin: past, present, and future. In: Riley MA, Gillor O (eds) Research and applications in bacteriocins. Horizon Bioscience, Norfolk
69. Callewaert L, Walmagh M, Michiels CW, Lavigne R (2011) Food applications of bacterial cell wall hydrolases. Curr Opin Biotechnol 22:164–171. doi:10.1016/j.copbio.2010.10.012
70. Carini S, Mucchetti G, Neviani E (1985) Lysozyme: activity against *Clostridia* and use in cheese production - a review. Microbiol Aliments Nutr 3:299–320
71. Bartowsky EJ (2009) Bacterial spoilage of wine and approaches to minimize it. Lett Appl Microbiol 48:149–156. doi:10.1111/j.1472-765X.2008.02505.x
72. Makki F, Durance TD (1996) Thermal inactivation of lysozyme as influenced by pH, sucrose and sodium chloride and inactivation and preservative effect in beer. Food Res Int 29:635–645. doi:10.1016/S0963-9969(96)00074-9
73. Takahashi H, Kuramoto S, Miya S et al (2011) Use of commercially available antimicrobial compounds for prevention of *Listeria monocytogenes* growth in ready-to-eat minced tuna and salmon roe during shelf life. J Food Prot 74:994–998. doi:10.4315/0362-028X.JFP-10-406
74. Nattress FM, Yost CK, Baker LP (2001) Evaluation of the ability of lysozyme and nisin to control meat spoilage bacteria. Int J Food Microbiol 70:111–119. doi:10.1016/S0168-1605(01)00531-1
75. Panesar PS, Panesar R, Singh RS et al (2006) Microbial production, immobilization and applications of β-D-galactosidase. J Chem Technol Biotechnol 81:530–543. doi: 10.1002/jctb.1453
76. Wu Y, Yuan S, Chen S et al (2013) Enhancing the production of galacto-oligosaccharides by mutagenesis of *Sulfolobus solfataricus* -galactosidase. Food Chem 138:1588–1595. doi: 10.1016/j.foodchem.2012.11.052

77. Niehaus F, Eck J (2012) Novel beta-galactosidases useful for the production of lactose depleted milk products. European Patent Application EP2530148 (A1)
78. Burin L, Jouppila K, Roos YH et al (2004) Retention of -galactosidase activity as related to Maillard reaction, lactose crystallization, collapse and glass transition in low moisture whey systems. Int Dairy J 14:517–525. doi:10.1016/j.idairyj.2003.11.003
79. Sabioni JG, Pinheiro AJR, Silva DO et al (1984) Control of lactose crystallization in "Dulce de Leche" by beta-D-galactosidase activity from permeabilized *Kluyveromyces lactis* cells. J Dairy Sci 67:2210–2215
80. Panesar PS, Kumari S, Panesar R (2010) Potential applications of immobilized β-galactosidase in food processing industries. Enzyme Res 2010:1–16. doi:10.4061/2010/473137
81. Roy I, Gupta MN (2003) Lactose hydrolysis by Lactozym™ immobilized on cellulose beads in batch and fluidized bed modes. Process Biochem 39:325–332. doi:10.1016/S0032-9592(03)00086-4
82. Holst HH, Lauritzen K (2010) Process for producing lactose-free milk. European Patent Application EP2207428 (A1)
83. Shimakura K, Tonomura Y, Hamada Y et al (2005) Allergenicity of crustacean extractives and its reduction by protease digestion. Food Chem 91:247–253. doi:10.1016/j.foodchem.2003.11.010
84. Hendriksen HV, Kornbrust BA, Østergaard PR, Stringer MA (2009) Evaluating the potential for enzymatic acrylamide mitigation in a range of food products using an asparaginase from *Aspergillus oryzae*. J Agric Food Chem 57:4168–4176. doi:10.1021/jf900174q
85. Friedman M, Levin CE (2008) Review of methods for the reduction of dietary content and toxicity of acrylamide. J Agric Food Chem 56:6113–6140. doi:10.1021/jf0730486
86. Ciesarová Z, Kukurová K, Bednáriková A et al (2009) Improvement of cereal product safety by enzymatic way of acrylamide mitigation. Czech J Food Sci 27:S96–S98
87. Filip S, Fink R, Hribar J, Vidrih R (2010) *Trans* fatty acids in food and their influence on human health. Food Technol Biotech 48:135–142
88. Brown PH, Carvallo FD, Dinwoodie RC et al (1994) Enzymatic method for preparing transesterified oils. United States Patent Application US 5288619 (A)
89. Ptok S, Heseker H (2010) Trans fatty acids. Ernährungs Umschau 57:472–480
90. Ming LO, Ghazali HM, Chiew Let C (1999) Use of enzymatic transesterified palm stearin-sunflower oil blends in the preparation of table margarine formulation. Food Chem 64:83–88. doi:10.1016/S0308-8146(98)00083-1
91. Zhang H, Xu X, Nilsson J et al (2001) Production of margarine fats by enzymatic interesterification with silica-granulated *Thermomyces lanuginosa* lipase in a large-scale study. J Am Oil Chem Soc 78:57–64
92. Farmani J, Hamedi M, Safari M, Madadlou A (2007) Trans-free Iranian vanaspati through enzymatic and chemical transesterification of triple blends of fully hydrogenated soybean, rapeseed and sunflower oils. Food Chem 102:827–833. doi:10.1016/j.foodchem.2006.06.015
93. Johannes C, Majcherczyk A (2000) Natural mediators in the oxidation of polycyclic aromatic hydrocarbons by laccase mediator systems. Appl Environ Microbiol 66:524–528. doi:10.1128/AEM.66.2.524-528.2000
94. Hoegger PJ, Kilaru S, James TY et al (2006) Phylogenetic comparison and classification of laccase and related multicopper oxidase protein sequences. FEBS J 273:2308–2326. doi:10.1111/j.1742-4658.2006.05247.x
95. Arakane Y, Muthukrishnan S, Beeman RW et al (2005) *Laccase* 2 is the phenoloxidase gene required for beetle cuticle tanning. Proc Natl Acad Sci U S A 102:11337–11342. doi:10.1073/pnas.0504982102
96. Claus H (2004) Laccases: structure, reactions, distribution. Micron 35:93–96. doi:10.1016/j.micron.2003.10.029

97. Gavnholt B, Larsen K (2002) Molecular biology of plant laccases in relation to lignin formation. Physiol Plant 116:273–280. doi:10.1034/j.1399-3054.2002.1160301.x
98. Osma JF, Toca-Herrera JL, Rodríguez Couto S (2010) Uses of laccases in the food industry. Enzyme Res 2010: Article ID 918761–8 pages. doi:10.4061/2010/918761
99. Rodríguez Couto S, Toca Herrera JL (2006) Industrial and biotechnological applications of laccases: a review. Biotechnol Adv 24:500–513. doi:10.1016/j.biotechadv.2006.04.003
100. Minussi RC, Pastore GM, Durán N (2002) Potential applications of laccase in the food industry. Trends Food Sci Technol 13:205–216. doi:10.1016/S0924-2244(02)00155-3
101. Sponholz W-R (2000) Suberase: Eine biotechnologische Möglichkeit Korken zu reinigen. Schweiz Z Obst-Weinbau 24:621–625
102. Schroeder M, Pollinger-Zierler B, Aichernig N et al (2008) Enzymatic removal of off-flavors from apple juice. J Agric Food Chem 56:2485–2489. doi:10.1021/jf073303m
103. Giovanelli G, Ravasini G (1993) Apple juice stabilization by combined enzyme-membrane filtration process. Lebensm-Wiss u-Technol 26:1–7
104. Bouwens EC, Trivedi K, van Vliet C, Winkel C (1999) Method of enhancing color in a tea-based foodstuff. United States Patent Application US 5879730
105. Flander L, Rouau X, Morel M-H et al (2008) Effects of laccase and xylanase on the chemical and rheological properties of oat and wheat doughs. J Agric Food Chem 56:5732–5742. doi:10.1021/jf800264a
106. Renzetti S, Courtin CM, Delcour JA, Arendt EK (2010) Oxidative and proteolytic enzyme preparations as promising improvers for oat bread formulations: Rheological, biochemical and microstructural background. Food Chem 119:1465–1473. doi:10.1016/j.foodchem.2009.09.028
107. Selinheimo E, Kruus K, Buchert J et al (2006) Effects of laccase, xylanase and their combination on the rheological properties of wheat doughs. J Cereal Sci 43:152–159. doi:10.1016/j.jcs.2005.08.007
108. Szweda RT, Schmidt K, Zorn H (2013) Bleaching of colored whey and milk by a multiple-enzyme system. Eur Food Res Technol. doi:10.1007/s00217-013-2000-3
109. Berger RG (2009) Biotechnology of flavours—the next generation. Biotechnol Lett 31:1651–1659. doi:10.1007/s10529-009-0083-5
110. Priefert H, Rabenhorst J, Steinbüchel A (2001) Biotechnological production of vanillin. Appl Microbiol Biotechnol 56:296–314. doi:10.1007/s002530100687
111. Zorn H, Fraatz MA et al (2012) Enzymatic synthesis of nootkatone. US Patent Application US 2012/0045806 A1
112. Fraatz MA, Riemer SJL et al (2009) A novel oxygenase from *Pleurotus sapidus* transforms valencene to nootkatone. J Mol Catal B Enzym 61:202–207. doi:10.1016/j.molcatb.2009.07.001
113. Krügener S, Krings U, Zorn H, Berger RG (2010) A dioxygenase of *Pleurotus sapidus* transforms (+)-valencene regio-specifically to (+)-nootkatone via a stereo-specific allylic hydroperoxidation. Bioresour Technol 101:457–462. doi:10.1016/j.biortech.2009.08.087
114. Zelena K, Krings U, Berger RG (2012) Functional expression of a valencene dioxygenase from *Pleurotus sapidus* in *E. coli*. Bioresour Technol 108:231–239. doi:10.1016/j.biortech.2011.12.097
115. Plagemann I, Zelena K, Arendt P et al (2013) $LOX_{Psa}1$, the first recombinant lipoxygenase from a basidiomycete fungus. J Mol Catal B Enzym 87:99–104. doi:10.1016/j.molcatb.2012.11.004

Adv Biochem Eng Biotechnol (2014) 143: 257–295
DOI: 10.1007/10_2013_237

Published Online: 14 August 2013

Recent Developments in Manufacturing Oligosaccharides with Prebiotic Functions

Zoltán Kovács, Eric Benjamins, Konrad Grau, Amad Ur Rehman, Mehrdad Ebrahimi and Peter Czermak

Abstract The market for prebiotics is steadily growing. To satisfy this increasing worldwide demand, the introduction of effective bioprocessing methods and implementation strategies is required. In this chapter, we review recent developments in the manufacture of galactooligosaccharides (GOS) and fructooligosaccharides (FOS). These well-established oligosaccharides (OS) provide several health benefits and have excellent technological properties that make their use as food ingredients especially attractive. The biosyntheses of lactose-based GOS and sucrose-based FOS show similarities in terms of reaction mechanisms and product formation. Both GOS and FOS can be synthesized using whole cells or (partially) purified enzymes in immobilized or free forms. The biocatalysis results in a final product that consists of OS, unreacted disaccharides, and monosaccharides. This incomplete conversion poses a challenge to manufacturers because an enrichment of OS in this mixture adds value to the product. For removing digestible carbohydrates from OS, a variety of bioengineering techniques have been investigated,

Z. Kovács · K. Grau · A. Ur Rehman · M. Ebrahimi · P. Czermak (✉)
Institute of Bioprocess Engineering and Pharmaceutical Technology,
University of Applied Sciences Mittelhessen, Wiesenstrasse 14,
35390 Giessen Germany
e-mail: pczermak@k-state.edupeter.czermak@lcb.chemie.uni-giessen.de

Z. Kovács
e-mail: kovacs.zoltan@kmub.thm.de

Z. Kovács
Department of Food Engineering, Corvinus University of Budapest, Ménesi st 44,
H-1118, Budapest Hungary

E. Benjamins
FrieslandCampina, Amersfoort The Netherlands
e-mail: eric.benjamins@frieslandcampina.com

P. Czermak
Department of Chemical Engineering, Kansas State University, Durland Hall 1005,
Manhattan KS, 66506-5102, USA

P. Czermak
Faculty of Biology and Chemistry, Justus Liebig University, Giessen Germany

including downstream separation technologies, additional bioconversion steps applying enzymes, and selective fermentation strategies. This chapter summarizes the state-of-the-art manufacturing strategies and recent advances in bioprocessing technologies that can lead to new possibilities for manufacturing and purifying sucrose-based FOS and lactose-based GOS.

Contents

1 Introduction

The market for prebiotics is steadily growing. The concept of prebiotics was originally introduced by Roberfroid and Gibson [1]. In 1995, prebiotics were defined as "a nondigestible food ingredient that beneficially affects the host by selectively stimulating the growth and/or activity of one or a limited number of bacteria in the colon, and thus improves host health. This definition has since been revised several times to achieve its current form: "A prebiotic is a selectively fermented ingredient that allows specific changes, both in the composition and/or activity in the gastrointestinal microflora that confers benefits upon host well-being and health"[1]

Prebiotics—food ingredients that are indigestible by humans—have a positive influence on the host organism by specifically stimulating the growth and/or the activities of bacteria or a limited number of bacteria species in the colon; that is, they are substrates for the growth and the metabolism of the probiotic bacteria. This supports the balance in the intestinal ecosystem of the human host and improves health.

[1] Roberfroid, Marcel (2007) Prebiotics: the concept revisited. J Nutr 137(3) http://jn.nutrition.org/content/137/3/830S.abstract and http://jn.nutrition.org/content/137/3/830S.full.pdf+html

The human colon contains an estimated 1,014 different germs. The intestinal microflora consist of about 500 different anaerobic species [2], of which approximately 90 % of the bacterial species have not been characterized yet. Based on that high quantity, it seems obvious that the metabolic activities of these bacteria might have a tremendous influence on physiological and biochemical processes of the human host [3].

Prebiotics must fulfill the following criteria:

- No hydrolysis or absorption in the upper part of the digestive system (mouth, stomach, small intestine)
- Selective substrate for one or more desired bacteria species in the colon and stimulation of that species regarding growth and activation
- Able to positively influence the numeric proportion of different bacteria species in the colon.

The critical aspect of prebiotics is their selective use by certain bacteria in a mixed culture. Of most interest are lactic acid–forming Lactobacteria and Bifidobacteria. It has been confirmed that nondigestible oligosaccharides (NDO) play an important role in the nutrition of humans [4]. There are currently only two groups of substances that completely fulfill the criteria for prebiotics: the fructans FOS, including lactulose and the fructopolysaccharide inulin, and the GOS. FOS are linear carbohydrate chains as a mixture of 3–10 β-(1,2)-glycosidical linked D-fructose units with a terminal glucose rest. Inulin, appearing as the natural storage carbohydrate in more than 3,600 different plants, consists of a mixture of different fructose chains up to a length of 65 fructose units. Also, these fructosyl rests are linked to the fructosyl rests of the saccharose by β-(1,2) connections. Lactulose is a synthetic disaccharide from galactose β-(1,4)-glycosidical linked to fructose gained by isomerization of lactose. The Lobre-de-Bruyn-van-Ekenstein relocation happens in an alkaline milieu or under high temperatures. Alternatively, the lactulose can be formed enzymatically from lactose by β-galactosidase. Commercially, GOS are produced from lactose by enzymatic synthesis using β-galactosidase.

Humans are only able to digest α-(1,4) and α-(1,6) linked oligo- and polysaccharides. The above-mentioned FOS and GOS are not hydrolyzed enzymatically but rather reach the colon more or less unchanged. Here exists an intestinal microflora in which most of the Bifido bacteria, but only a few representatives of other strains, possess a β-fructosidase (EC3.2.1.7) that qualifies them for the hydrolysis of β-(1,2) and β-(1,4) glycosidic connections in the oligosaccharides. Short-chained fatty acids, lactose, hydrogen, methane, and carbon dioxide are metabolic products that lead to an acidic milieu in the colon, which antagonizes the survival and the augmentation of pathogenic bacteria. The hydrogen formed during the fermentation of prebiotic oligosaccharides by the bacterial flora in the colon is absorbed by the blood and exhaled via the lung. The quantity of oligosaccharides arriving indigested in the colon and then undergoing microbial fermentation can be determined by the hydrogen concentration in the exhaled breath [5].

In Japan, which is one of the leading markets for functional food, additional types of carbohydrates have already received the status of prebiotics, been released, and been approved for human consumption. These prebiotics include xylo-oligosaccharides (XOS), formed by a partial enzymatic hydrolysis of xylan, a heteropolysaccharide mainly appearing in wood. They consist of up to eight β-(1,4)-glycosidic linked xylose-molecules. XOS are indigestible by human beings and therefore reach the colon completely. A significant growth of some species of the intestinal microflora has been shown by in vitro fermentations with XOS as substrate [6]. Resistant starch (RS) is not enzymatically degradable by α-amylase in the human duodenum but is fermentable by the intestinal flora in the colon. RS can be formed amylose or amylopectin, appears naturally in plants, and consists of α-(1,4)-linked glucane chains. RS is distinguishable by the fact that, during fermentative degradation into short-chained fatty acids by the intestinal flora, a very high level of butyrate is formed.

Gluco-oligosaccharides (gluco-OS) are formed enzymatically from saccharose and maltose via transglucosylation by glucosyltransferase. Isomalto-oligosaccharides (IMO) are α-(1,6)-linked saccharides that can be gathered by enzymatic hydrolysis of starch by α-amylases, pullulanases, and α-glucosidases. Pectin-oligosaccharides and polydextrose are currently being examined for their suitability as prebiotics. Polydextrose is a synthetic branched polysaccharide formed of α-(1,6)-linked glucose molecules. It has been approved as food ingredient E1200 in the European Union (EU) without any limitation in quantity. Some other substances being discussed as potential prebiotics do not yet fulfill the defined requirements completely. Natural soybean oligosaccharides, gathered by extraction of soybeans, are selectively fermented in the colon. However, approximately 50 % is digested in the human gastrointestinal system and does not reach the colon [6]. Many fermentable carbohydrates (e.g. some types of enzyme-resistant starch) and dietary fibers (e.g. pectins) that are naturally contained in daily nutrition have not yet been evaluated as to whether they are selectively fermented in the colon [7, 8].

The application of prebiotics leads to changes in the makeup and metabolism of intestinal microflora. Some positive impacts on the physiological processes in humans have been proven [9]. The complete bacterial biomass in the colon increases by the strengthened fermentation and stimulates the actions of the bowels. The application of inulin reduces the frequency of constipation for patients with chronic obstipation [10]. Consider also how the composition of the intestinal microflora has changed. The reproduction of potentially pathogenic bacteria may be blocked by Bifidobacteria. That effect results on the one hand from the decrease in pH value due to the forming of short-chained fatty acids and lactate during fermentation; on the other hand, the forming of bactericidal substances by Bifidobacteria has been proven [11].

In the colon, indigestible fermented carbohydrates such as lactulose and FOS stimulate the resorption of calcium, iron, and magnesium from food [12]. Furthermore, the addition of OS and inulin leads to a reduction of the serum cholesterol level. The quantity of triglycerides in the serum and liver activates the liver function and therefore the separation of toxic substances [13, 14]. Bifidobacteria, activated in growth by OS and inulin, might improve the immune reaction of the host—that is, it may have a positive impact on the human immune system, thus possibly decreasing the risk of colon cancer [15, 16].

The prebiotics FOS, GOS, inulin, and lactulose, which are allowed as food additives in Europe, also fulfill the U.S. requirements for being generally recognized as safe (GRAS). Because of their technological characteristics, they are mainly added in food production. Inulin can be used as a fat substitute, for example. Due to processing technologies with high shear rates, the tertiary three-dimensional structure of the molecules can be modified in a manner that they develop fat-similar organoleptic characteristics regarding chew-feel and creaminess. The use of prebiotic oligosaccharides as dietary fibers often results in an improvement in taste and texture [17].

The use of inulin in fat-reduced products provides a stability in emulsions and an easy-to-spread texture. The addition of FOS in light yogurt, for example, improves the mouth-feel and lowers syneresis; when combined with aspartame or acesulfam potassium, a synergistic flavor effect can be reached without increasing the calorie content. In baked goods, FOS can substitute for sugar, increase dietary fibers, and keep products moist. The use of prebiotics as an ingredient with functional properties allows food to be labeled as prebiotic or bifidogenic, calorie reduced, containing dietary fibers, or able to improve the resorption of calcium.

A further important area of use for FOS and GOS is the supplementation of infant food. Oligosaccharides naturally appear in women's milk as special effective prebiotics with a concentration of about 1 g per 100 ml. The structures are very complex and the functional consequences are not completely understood yet [18, 19]. To date, oligosaccharides have not been available in formulated baby food. A clinical study in 2002 [20] has shown that an addition of GOS and long-chained FOS in a ratio of 90–10 % up to a concentration of 0.8 g per 100 ml lead to a similar bifidogenic effect as in breast feeding. Due the influence on the development of the microbial flora, the addition of oligosaccharides also has a positive impact on the immunological development of formula-nourished children [21]. The European Food Safety Authority (EFSA) has not noted any concerns about the addition of an OS mixture in the above-mentioned concentration in commercial baby food [22, 23].

Among the commercially available prebiotic preparations, GOS and FOS are considered to be well-established functional oligosaccharides. The bioprocessing techniques applied for their synthesis and their purification show considerable similarities. In this chapter, we review recent developments in the manufacture and purification of lactose-derived GOS and sucrose-derived FOS. First, we discuss the preparation of GOS- and FOS-containing carbohydrate mixtures via biocatalysis using whole cells and (partially) purified enzymes. In the second part of this chapter, we review the various purification tools employed for separating oligosaccharides from disaccharides and monosaccharides. Finally, we devote special attention to emerging membrane-based processes that can be used for both manufacturing and fractionation purposes.

2 Bioengineering Aspects of Oligosaccharide Synthesis

In this section, we introduce the reader to the fundamentals of oligosaccharide biosynthesis. As previously mentioned, we focus on the synthesis of lactose-based GOS and sucrose-based FOS. As we will show, these systems show similarities in terms of reaction mechanisms and product formation. The biocatalytic reaction results in a carbohydrate mixture consisting of OS, unreacted disaccharides, and monosaccharides as byproducts. Both GOS and FOS can be synthesized either using whole cells or (partially) purified enzymes, immobilized or in free form. A large amount of literature deals with these manufacturing concepts. Here, we briefly summarize the enzymatic reaction mechanisms, highlighting the various technological factors affecting OS yield and purity, and report the state-of-the-art technologies used in commercial production.

2.1 Lactose-Based Galactooligosaccharides

The mechanism of transglycosylation by *b*-galactosidases has been long known [24]. GOS are derived from lactose through this mechanism. GOS are carbohydrates built up from glucose and galactose, according to the formula Gal_n–Glc, where $n = 2$–20; in general, disaccharides with linkages other than Galβ1–4Glc (lactose) are considered GOS as well [25]. The Japanese company Yakult has produced GOS since the early 1990s. Together with their activities on probiotics, they are considered pioneers in the field. Gibson and Roberfroid [26] introduced the concept of prebiotics, which, in combination with a growing interest in prebiotics and the recognition of their functionality, boosted the application of GOS.

The efficiency of the synthesis of GOS by transgalactosidase activity of β-galactosidases depends on the conditions applied as well as on the enzyme of choice. GOS synthesis is a kinetically controlled reaction; therefore, the enzyme characteristics strongly determine the formation of GOS, GOS structures, and the productivity of the enzyme [27]. Lactose, however, only poorly dissolves in water (18.9 g per 100 g at 25 °C, [28]). To achieve high substrate concentrations, elevated temperatures are thus required. Whereas elevated temperatures increase the reaction rate of oligosaccharides formation, these temperatures can be detrimental to the biocatalyst. On the other hand, high sugar concentrations have been shown to have a stabilizing effect on proteins due to preferential hydration of the protein [29]. This enables GOS synthesis at temperatures higher than the optimum in diluted aqueous solutions. A good example is the β-galactosidase derived from *Kluyveromyces lactis*, which has a optimum temperature of ~40 °C. The enzyme activity rapidly decreases with increasing temperature in aqueous solutions. Padilla et al. [30] and Matinez-Villaluenga et al. [31], however, performed the synthesis of GOS using the same enzyme at 50 °C with a lactose concentration of 250 g/L, indicating the stabilization effect of lactose on the enzyme conformation and stability. GOS synthesis at high temperatures using *A. oryzae* β-galactosidase

Table 1 Substrate, temperature, and pH ranges of GOS synthesis by β-galactosidases from different sources

Enzyme source	Lactose concentration (g/L)	Temperature (°C)	pH	GOS yield (%w/w)	References
Bifidobacterium sp.	300–500	37–65	6.0–7.5	20–55	[34–41]
Bacillus circulans	45–400	15–60	4.5–7.0	6–56	[42–55]
Kluyveromyces lactis	200–400	37–50	6.5–7.3	5–50	[31, 45, 55–60]
Aspergillus oryzae	100–475	30–60	4.5–6.5	10–35	[32, 33, 45, 58, 61–67]

was studied by Vera et al. [32] and Huerta et al. [33], using supersaturated or partially dissolved lactose solutions.

Besides temperature, the pH value of the solution is another important factor influencing the reaction. Evidently, the pH dependency of the enzyme activity is greatly determined by the organism from which the enzyme is derived. Table 1 provides examples of well-described β-galactosidases derived from different sources. Additionally, the substrate ranges and GOS yields are also mentioned.

In addition to GOS synthesis with the use of (partially) purified enzyme preparations, GOS synthesis using whole cells was also investigated. The obvious advantage is the fact that the purification step for the enzyme can be omitted. GOS synthesis with whole cells of *Bifidobacterium bifidum* NCIMB 41171 was carried out by Osman et al. [41], yielding 50 % GOS starting with an initial lactose concentration of 390 g/L. Onishi et al. [54] described GOS synthesis with a number of organisms. Their best results in terms of GOS yield were obtained with the yeast *Sterigmatomyces elviae* CBS8119 in a fermentation system, where inhibiting glucose was consumed for cell growth. A total of 232 g/L GOS was produced from 360 g/L lactose after incubation at 30 °C for 60 h. In addition to the previously mentioned whole cell systems, the application of immobilization technology might also be considered. An appealing example is the alginate entrapment of whole cells of *Sporobolomyces singularis*, which enabled the reuse of the cells for 20 consecutive batch productions of GOS [68].

Nowadays, a number of companies produce GOS on an industrial scale. Table 2 provides an overview of commercially available GOS products and the sources of the enzymes that are used to produce these products.

Three manufacturers have obtained the Generally Recognized as Safe (GRAS) status for their products by the U.S. Food and Drug Administration. GOS are applied mainly in infant nutrition, but other food applications are also known. Vivinal GOS (manufactured by FrieslandCampina Domo, The Netherlands) is available either as a syrup (75 % solids) or co-spray dried with whey protein concentrate or maltodextrin. The composition of GOS was extensively studied by Coulier et al. [69], who used a combination of analytical techniques to unravel

Table 2 Commercially available galactooligosaccharide (GOS) products

Product	Manufacturer	GOS fraction (%w/w)	GRAS status notification	Organism/enzyme	References
Vivinal® GOS	FrieslandCampina Domo	59	Yes	*Bacillus circulans*	[69, 70]
Oligomate® 55 N	Yakult Pharmaceuticals	55	Yes	*Sporobolomyces singularis/ Kluyveromyces lactis*	[71]
Purimune™	GTC Nutrition	90[a]	Yes	*Bacillus circulans*	[72]
Cup Oligo	Kowa Company/ Nisshin Sugar	70[a]	No	*Cryptococcus laurentii*	[73]
Bimuno	Clasado	48	No	*Bifidobacterium bifidum* NCIMB 41171	[26, 74]

[a] High GOS content was obtained by additional purification steps

Table 3 Composition of commercially available galactooligosaccharide (GOS) products

Concentration (w/w% on total solids)	Vivinal GOS FrieslandCampina Domo	Oligomate 55 N Yakult Pharmaceuticals	Purimune GTC Nutrition	Bimuno Clasado	Cup Oligo Nissin Sugar
GOS	59–60	55–56	90–92[a]	48	70–75[a]
Lactose	19	12.7	7–10	22	n. r.[b]
Glucose	21	22.4	0–1	18	n. r.[b]
Galactose	1.3	8.6	0–0.5	12	n. r.[b]

[a] High GOS content was obtained by additional purification steps
[b] not reported

structures and linkage types present in Vivinal GOS. Table 3 shows its carbohydrate composition as compared to that of other commercial products.

The Vivinal GOS process is schematically depicted in Fig. 1.

The process starts with dissolving lactose at a high temperature to obtain a substrate solution with a sufficient dry matter content. Subsequently, the solution is cooled to the optimal reaction temperature and transferred to a reactor (stirred batch reactor). After adjustment of the pH using sodium hydroxide, the enzyme is added to the reactor for the conversion of lactose to GOS. After the synthesis reaction, the enzyme is inactivated by heating. After refining and decoloring, citric acid is added and the product is concentrated by an evaporator to obtain a syrup with 75 % total solids. Finally, the product is transferred to the filling station for final packaging.

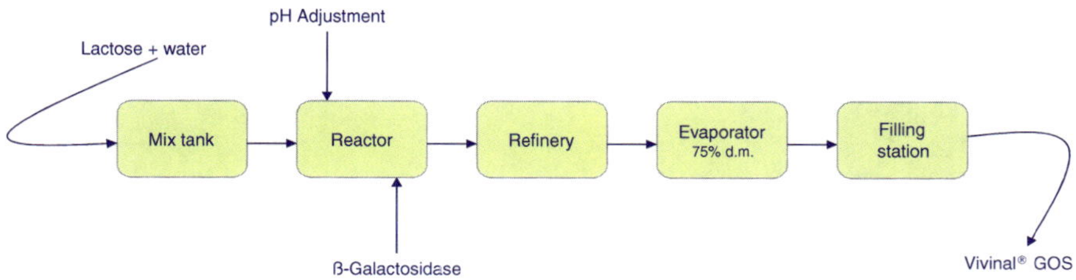

Fig. 1 Flow chart of the galactooligosaccharide (GOS) manufacturing process

2.2 Sucrose-Based Fructooligosaccharides

As discussed in Sect. 1, a wide variety of health benefits have been associated with FOS. Beside their favorable nutritional properties, FOS also have excellent technological properties. They can be thought of as low molecular weight, nonviscous, highly soluble dietary fibers. Being nonreducing sugars, FOS do not undergo a Maillard reaction. When purified, the sweetness of FOS is about 30 % that of sucrose. Moreover, it has a well-balanced sweetness profile with no off-flavor and can mask the aftertaste of artificial sweeteners [75]. It is stable above pH 3 and under 140 °C [76].

FOS can be synthesized either by hydrolysis of inulin (e.g. from chicory) or by enzymatic transfructosylation from sucrose. FOS formed by hydrolysis contain longer fructo-oligomer chains, and not all of the β-(2-2)-linked fructosyl chains end with a terminal glucose [77]. FOS naturally occur in several biological materials and can be extracted from, for instance, caprine milk [78], onion [79], asparagus [80], and banana peel [81]. Here, and in the remainder of this chapter, we focus on sucrose-based FOS biosynthesis.

The FOS-producing enzyme is usually classified as β-D-fructofuranosidase (invertase, EC. 3.2.1.26) or fructosyltransferase (EC. 2.4.1.9). The synthesis is a complex process involving a multitude of sequential reactions leading to the final products. The resulting FOS structures are mainly 1-kestose (GF_2), nystose (GF_3), and fructofuranosyl nystose (GF_4), where *G* and *F* represent the glucosyl and fructosyl moieties of the sucrose molecule, respectively [82]. The byproduct of the conversion is glucose, which has been reported to be the main factor decreasing yield during FOS synthesis.

The network of reaction mechanisms for FOS synthesis has been studied by several investigators [83–86]. Depending on the source of enzyme, the proposed networks differ from each other in their individual reactions and the species produced (e.g. [87–89]).

As an example of a typical FOS reaction mechanism, we report the network proposed by Nishizawa et al. [83] and the simulated reaction kinetics for an FTase obtained from *Aspergillus niger* ATCC 20611. The rate equations for the individual reactions were obtained by a two-substrate random bi–bi model with noncompetitive inhibition by glucose. This model assumes that hydrolysis

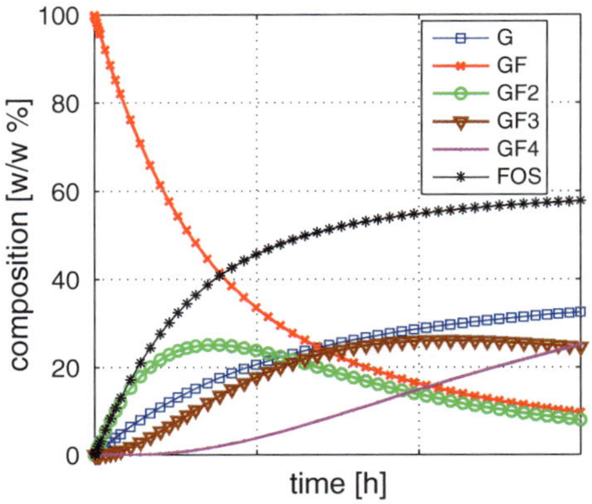

Fig. 2 Simulated FOS kinetics based on the reaction network proposed by Nishizawa et al. [83] illustrating the change in saccharide composition over time for a typical batch reaction

reactions of GF, GF_2, GF_3, and GF_4 do not occur, and formation of GF_5 is negligible. The proposed mechanism for the reaction is as follows:

$$GF + GF \xrightarrow{r_1} GF_2 + G$$

$$GF_2 + GF_2 \xrightarrow{r_2} GF_3 + GF$$

$$GF_3 + GF_3 \xrightarrow{r_3} GF_4 + GF_2$$

$$GF + GF_2 \xrightarrow{r_4} GF_3 + G$$

$$GF + GF_3 \xrightarrow{r_5} GF_4 + G$$

$$GF_2 + GF_3 \xrightarrow{r_6} GF_4 + GF$$

Fig. 2 shows an example of the reaction kinetics obtained by numerical simulation using the reaction network above [83].

We note that the type of glucose inhibition depends also on the source of enzyme. As opposed to the noncompetitive inhibition found for FTase originating from *Aspergillus niger* ATCC 20611 [83], competitive inhibition by glucose has been observed using a FTase obtained from *Rhodotorula* sp. [84], *Aspergillus japonicus* [85], and *Aureobasidium pullulans* [86]. The major challenge in FOS production is to achieve a high yield and a high purity of FOS. The amount and nature of the FOS formed in the enzymatically catalyzed process depends upon several factors, including the source of the enzyme, the concentration, and nature of the substrate, and the reaction conditions.

Sucrose-based FOS can also be produced from sucrose present in different sources, such as agrowastes and cheap byproducts. These include molasses from beet processing [90], aqueous extracts from date byproducts [91], logan syrup [92], cassava wastes [93], cereal bran, corn products, sugarcane bagasse, and byproducts of coffee and tea processing [94]. Isolated enzymes from cells can be applied in both soluble and immobilized forms to produce OS. Table 4 provides an overview

Table 4 Fructooligosaccharide (FOS) yield and sucrose conversion using free enzymes in stirred-tank reactor

Source of enzyme	Initial sucrose concentration [g/L]	Reaction conditions	Sucrose conversion [%w/w]	FOS yield [%w/w]	Fructose liberated [%w/w]	References
Aspergillus sp. 27H	615	40 °C, pH 5.5	52	61	Traces	[95]
Aspergillus sp. N74	550	60 °C, pH 5.5	90	50	n. r.[a]	[96]
Aspergillus oryzae	600	55 °C, pH 6 (0.05 M citrate/phosphate)	86	57	Traces	[97]
Aspergillus oryzae MTCC 5154	600	55 °C	80	54	n. r.[a]	[98]
Aureobasidium pullulans	600	55 °C, pH 5 (sodium acetate)	n. r.[a]	59	n. r.[a]	[99]
Bacillus macerans EG-6	200	50 °C, pH 5 (sodium acetate)	93	39	10	[100]
Bacillus macerans EG-6	500	37 °C, pH 6 (phosphate)	98	42	9	[89]
Fusarium oxysporum 172464	500	26 °C, pH 5 (sodium acetate)	50	30	30	[101]
Penicillium rugulosum	775	55 °C, pH 5.5 (sodium acetate)	98	84	1.4	[102]
Rhodotorula sp. LEB-V10	775	50 °C, pH 4.5 (50 mM sodium acetate)	91	54	2	[103]
Rhizopus stolonifer LAU 07	600	55 °C, pH 5.2 (0.1 M citric acid/K_2HPO_4)	n. r.[a]	34	n. r.[a]	[93]
Pectinex Ultra SP-L, *Aspergillus aculeatus*	400–600	50–65 °C, pH 5–6	70–89	53–69	0–6	[90, 104–109]

[a] not reported

on literature data on FOS synthesis on sucrose in stirred tank reactor using free enzymes.

In Table 4, the FOS yield and the corresponding sucrose conversion is given as follows. The FOS yield, as a percentage of all the sugars present in the media, is calculated as the value (%) of the FOS present in the final product divided by the sum of all the other carbohydrate components present in the media (FOS, GF, G, and F). The degree of sucrose conversion is defined as a conversion of sucrose to FOS and monosaccharides in percentage—that is, the initial quantity of sucrose minus the sucrose present in the final product divided by the initial sucrose. In the rest of this chapter, we use these definitions to refer to OS yield and substrate conversion.

For the production of lactose-based GOS, a high concentration of substrate is also required for efficient FOS synthesis. At low sucrose concentrations, FTase shows mainly hydrolyzing activity (invertase), whereas at high concentration the transfructosylating activity is more pronounced. In Table 4, small values of generated fructose indicate that the reported enzymes have a restricted hydrolyzing capability at the given reaction conditions.

As shown in Table 4, Pectinex Ultra SP-L (Novozyme A/S, Denmark) seems to be a popular choice among investigators. This liquid enzyme preparation is classified as polygalacturonase by its manufacturer and mainly used in fruit juice processing industries. The FTase, which is responsible for FOS synthesis, represents a relatively small fraction of the total protein content of the crude enzyme preparation, as shown by isolation and purification studies [110, 111].

Isolated enzymes from cells can be applied in both soluble and immobilized forms to produce FOS. Table 4 summarizes the results of batch FOS production with soluble enzymes. The main drawback of using free enzymes in a batch reactor is that the biocatalysts have to be inactivated or removed from the resulting FOS-containing mixture prior to its application in food formulas. In contrast to that, continuous techniques allow the reuse of enzymes. Continuous production with soluble enzymes is typically done with ultrafiltration-assisted enzyme reactors, as discussed later in Sect. 4.3.1. An alternative to this technology involves enzyme immobilization.

Studies have reported on FOS production using immobilized enzymes in calcium alginate beads [112], methacrylamide-based polymeric beads [113, 114], epoxy-activated acrylic beads (Eupergit C) [115], epoxy-activated polymethacrylate carriers (Sepabeads EC-EP5) [90], porous glass [116], anionic ion exchange resin (Amberlite IRA 900 Cl) [117], and different polymeric and ceramic membrane filters (see Sect. 4.2). In general, immobilized FTase is found to be more stable to changes in pH and temperature than free FTase (e.g. [115]) and an increased operational stability is observed (e.g. reaching a half-life of 275 days [114]).

Immobilization has been recognized as an effective bioengineering tool for retaining enzymes in reactors, enhancing enzyme stability, and enabling a continuous operation [118]. The main technical problems associated with packed-bed reactors are microbial contamination, adsorption of feed components, and channeling [119]. To overcome these problems, periodic washing, regular pasteurization, and changing the feed direction is required. Although a large body of literature deals with immobilized enzymes, these issues are far less investigated for FOS production.

To obtain purified or partially purified enzymes from fermentation broth, a sequence of various downstream processing steps is required. The employment of expensive isolation techniques is not necessary when using resting or living whole cells as biocatalysts instead of isolated enzymes.

Bacterial and fungal species are used in microbial FOS production. A large and growing body of literature has been devoted to find microbes with high FTase activity. These include *Lactobacillus reutri* [120], *Penicillium citrinum* [88], *Candida* sp., *Rhodotorula* sp., *Cryptococcus* sp., *Rhodotorula* sp. [121], *Aspergillus niger* [122, 123], *Aspergillus oryzae* [123], *Gluconobacter oxydans* [124], *Aspergillus japonicus* [124–126], *Aspergillus phoenicis* [127], *Aureobasidium pullulans* [123, 124, 128–131], and *Penicillium expansum* [132, 133].

Microbial fermentations have been performed by using suspended cell systems and immobilized cell systems. Generally, submerged fermentation or solid-state fermentation is employed to produce FOS with suspended cell systems [134]. Very recently, Aziani et al. [127] have reported the production of FOS by *Aspergillus phoenicis* forming biofilm on polyethylene as an inert support. Also, a considerable amount of literature has been published on immobilization of whole cells on different lignocellulosic materials (including brewer's spent grain, wheat straw, corn cobs, coffee husks, cork oak, and loofa sponge) [125], synthetic fiber, polyurethane foam, stainless steel sponge [133], entrapped in gluten [135] and calcium alginate [126, 130]. Microbial production of FOS has been extensively reviewed by several authors. We refer the reader here to the studies of Prapulla et al. [136] and Sangeetha et al. [134].

The market for FOS is already substantial and is growing rapidly. Currently, a number of companies produce FOS on industrial scale. Nishizawa et al. [83] summarized the manufacturing process of Neosugar G and Neosugar P, the two FOS formulations produced commercially by Meiji Seika (Japan). Neosugar G is produced in a conventional batch process using free enzymes obtained from *Aspergillus niger* ATCC 20611. The reaction is performed in a stirred tank reactor with a sucrose solution of high concentration (50–60 % (w/w)) at pH 5.5–6.0 and at 50–60 °C. The resulting carbohydrate mixture containing FOS, glucose, and residual sucrose is then heated up to 90 °C for 30 min in order to deactivate the enzyme. The reaction mixture is cooled to less than 50 °C, clarified by filtration, and deionized by ion-exchange resin column. The purified reaction mixture is concentrated to 75 % (w/w) by evaporation. The resulting product, Neosugar G, consists of about 55–60 % FOS on a total carbohydrate basis. Neosugar P, in which the FOS content is more than 95 % on a total carbohydrate basis, is obtained from Neosugar G by removing residual sugar and glucose by simulated moving-bed chromatography.

Continuous production of FOS on an industrial scale has also been reported [137]. Meiji Seika Kaisha (Tokyo, Japan) developed a process in 1983 for FOS production with packed bed reactors using immobilized *A. niger* cells entrapped in calcium alginate gel. Later, Cheil Foods and Chemicals (Seoul, Korea) also developed a continuous process, and two pieces of 1-m^3 packed bed reactors using immobilized cells of *A. pullulans* were put into operation in 1990. The stability of the immobilized cells in this system is reported to be about 3 months at 50 °C [137].

3 Strategies for High OS Content

The current production technologies of both FOS and GOS have a common and significant weakness, namely the incomplete conversion. The bioreaction actually results in a mixture of carbohydrates consisting of OS, remaining disaccharides (i.e. nonreacting substrates), and monosaccharides as byproducts. OS yields rarely exceed 50–60 % due to byproduct inhibition. Most typical yields are between 30 and 50 % [138]. Enrichment of OS in this mixture adds value to the product. Obviously, the removal of disaccharide and monosaccharide fractions could expand the use of purified OS in the food and pharmaceutical industries. For this reason, a number of bioprocess engineering techniques have been investigated, including various downstream separation technologies, additional bioconversion steps applying enzymes, and selective fermentation process steps, as described in the following sections. The membrane-based techniques are reviewed in Sect. 4.

3.1 Purification Techniques

A variety of downstream unit operation tools have been tested for removing digestible carbohydrates from OS. Liquid chromatography has been long used on large scale in the sugar industry for the separation of glucose from sucrose and sucrose from molasses. Its employment for OS separation seems then to be a straightforward choice. We should note, however, that a number of competitive techniques have been recently proposed. In fact, membrane cascades (discussed in Sect. 4.1) might compare favorably with simulated moving bed chromatography. For example, activated charcoal treatment has been shown to be comparable with liquid chromatography in terms of yield and purity. Also, supercritical extraction and precipitation with ethanol are potential candidates in purifying OS. Recent developments on these technologies are summarized in the following sections.

3.1.1 Liquid Chromatography

Analytical chromatography is a well-established method for quantitative analysis of GOS- and FOS-containing carbohydrate mixtures (e.g. [139, 140]). The high cost of chromatographic purification, however, causes considerable resistance to the widespread use of this separation technique in manufacturing processes. The design of an industrial-scale chromatographic separation system starts with the selection of a suitable adsorbent material [141]. The most widely used type of adsorbent for saccharides are cation-exchange resins.

Most of the literature available on the chromatographic separation of FOS or GOS from monosaccharides and disaccharides deals with purification for analytical purposes. Preparative elution chromatography for the separation of FOS has been investigated by Vaňková and Polakovič [142]. For this purpose, true process-size particles of a commercial cation-exchange resin (Amberlite CR1320Ca) were used in two columns (23 cm^3 and 324 cm^3 packing volume). A maximal FOS yield of ~86 % and a selectivity of 82 % was reported for optimal superficial velocity and column load at 60 °C. It was also shown that the column load has a great influence on the separation efficiency, with values larger than 20 % leading to minimal selectivity.

Chromatographic processing can be done in a batch (elution) mode, with integrated or side-stream recycling, in simulated moving bed chromatography (SMB) mode, or in various hybrid modes [143]. The continuous separation of FOS using a rotating annular chromatograph with a rotating feed nozzle and product collectors was investigated by Takahashi and Goto [144]. It was demonstrated that FOS can be continuously separated from mono- and disaccharides to obtain lower calorie sweetener.

SMB is the state-of-the-art technology for industrial sugar purification [143, 145, 146]. The SMB is basically a binary separator that presents three main advantages over batch chromatography: (i) saving significant amounts of eluent, (ii) enhancing productivity, and because it is a continuous process, (iii) simplifying the connection to associated unit operations [147]. SMB chromatography is generally known as a high-cost separation technique. The total costs associated with SMB break down mainly to resin cost, column cost, pump cost, pumping costs, costs associated with heating or cooling, and the cost of eluent [148]. Of these, the resin and the column costs are the greatest contributors to the total cost, together reaching 80–95 % of the total costs [148].

Vaňková and Polakovič [145] carried out design simulations of an SMB chromatography unit for the separation of FOS from a mixture containing about 40 % (w/w) digestible sugars. The technological goal was to produce a raffinate with a FOS purity of 90 % and a FOS yield higher than 95 %. The design of the SMB unit was optimized, taking into account the switch time and the feed, raffinate, and extract flow rates as optimization parameters. Other performance criteria were the product dilution, the eluent consumption both in the raffinate and extract, and the yield and purity of mono- and disaccharides in the extract. The design was carried out for a 12-column SMB system consisting of identical columns of defined dimensions. A sensitivity study showed that the designed process was very sensitive to a change in the switch time and recycle flow rate. These operating parameters should be kept in a very narrow range; otherwise, the values of key performance parameters—yield and purity—decrease considerably. It was found that the performance criteria of the SMB system exceeds by far those calculated for a battery of identical batch columns operated in parallel.

The same research group [149] provided an analysis of a complete process for a plant with an annual production of 10,000 tons of FOS. The flow chart includes three sections: (i) the production of FTase, (ii) its immobilization, and (iii) the (partial) enzymatic conversion of sucrose to FOS followed by the purification of the resulting carbohydrate mixture. In this latter section, SMB is proposed for the purification of FOS. The technological objective was set as lowering the weight percentage of monosaccharides and disaccharides in the final product to less than 5 %. The SMB produces 23,600 tons/year of purified FOS syrup, which is then spray-dried. The system consists of four chromatographic columns connected in series, each of them being filled with a cation exchange resin and having a volume of 3.4 m^3. The switch time is 22 min. The feed flow rate was 3.38 m^3/h, the flow rate of water (which is used as eluent) was 2.7 m^3/h, and the flow rates of extract and raffinate were 3.04 m^3/h and 2.96 m^3/h, respectively. The equipment investment cost for the SMB column is estimated as 453,000 EUR, and the cost of the resin was assumed to be 235,000 EUR.

Liquid chromatography is a well-established technology. It is proven to be a fast and efficient technique for separation FOS- and GOS-containing carbohydrate mixtures. High selectivity, purities, and yields may compensate for the relatively high costs that are associated with the implementation of this technology.

3.1.2 Activated Charcoal Treatment

Recent investigations [150–152] have shown that a treatment with an activated charcoal fixed-bed column is a good alternative for sugar separation. Activated charcoal is rather inexpensive as compared to chromatography resins, has a good sorption capacity due to its large surface area and volume, and has been reported to be easily regenerated with ethanol [153].

Generally, sugars are selectively adsorbed onto activated charcoal. OS are adsorbed more strongly than monosaccharides and disaccharides. Using adequate desorption strength by applying water/ethanol mixtures, the selective recovery of different sugar fractions is possible. As described in Ref. [152], the purification process is typically performed in three steps: (i) adsorption of sugars onto the activated charcoal; (ii) column washing with water; and (iii) stepwise desorption of sugar fractions using water/ethanol mixtures with increasing concentration of ethanol. Kuhn and Filho [151] applied an activated charcoal fixed-bed column using ethanol 15 % (v/v) as eluent at 50 °C to purify FOS from a mixture of sugars obtained by enzymatic synthesis. The final purification degree and recovery of FOS were approximately 80 and 97.8 %, respectively.

Another step toward handling more complex systems has been made by Nobre et al. [152]. They tested the performance of an activated charcoal column on FOS-containing fermentation broth. The broth was first microfiltered to remove cells, then residual proteins were removed with a centrifugal filter. The resulting mixtures containing 50.6 % (w/w) of FOS were then purified with an activated charcoal column. The equilibrium loading on the column was reported to be 222 mg of FOS per gram of activated charcoal. Fractions rich in FOS were obtained by desorption using ethanol percentages between 10 % and 40 % (v/v). The purity of FOS in these fractions was found to be 92.9 % (w/w), representing 74.5 % (w/w) of FOS from the broth. This excellent study also demonstrates that the FOS fractions gained by desorption are free of the ions originally present in the fermentation broth.

The influence of the concentration of ethanolic water solutions used for desorption on selectivity in GOS recovery was studied by Hernández et al. [150]. Ethanolic solutions of 8 % led to a high recovery of GOS (90 %), but 20 % of disaccharides were also recovered. Solutions with 10 % ethanol allowed almost complete removal of disaccharides, but only 53 % of GOS trisaccharides were recovered.

In summary, activated charcoal treatment is an inexpensive and efficient process to purify OS, even from complex process liquids such as fermentation broths. Therefore, it is a potential candidate for larger scale applications.

3.1.3 Ethanol Precipitation

Exploiting differences in the solubility of components to be separated is a commonly used purification technique in downstream processing. Recently, Sen et al. [154] studied the purification of GOS from a mixture consisting of GOS, lactose, glucose, and galactose by ethanol precipitation. GOS was enriched 2.3-fold in the precipitate formed in a solution of 90 % (v/v) ethanol with 28 g/L of total saccharides at 40 °C. The corresponding recovery of GOS was 47 %. Performing two such precipitations sequentially reduced the monosaccharides from 48 % (w/w) of the total saccharides to 4 % (w/w) and increased the percentage of GOS from 15 % (w/w) in the feedstock to 75 % (w/w) in the product.

Simplicity and cost-effectiveness during scale-up are the main advantages associated with this technique. It can be used as an alternative to the competing technologies or in combination with them. As the authors suggest, it could be used, for example, as a step to enrich and concentrate GOS prior to chromatography [154].

3.1.4 Supercritical Extraction

Montañés et al. published a series of papers [155–157] on the feasibility of supercritical fluid extraction for purifying carbohydrate mixtures containing GOS, lactose, galactose, and glucose. Although pure supercritical carbon dioxide (SC-CO_2) as a solvent is not suitable to extract sugars, some polar cosolvents may considerably enhance the solubility of carbohydrates [155]. Using SC-CO_2 solvent with ethanol/water cosolvent under suitable extraction conditions (including temperature, pressure, and cosolvent flow rate) allowed the almost complete removal of monosaccharides and disaccharides from the mixture and led to a residue in the extraction cell, where mainly tri saccharides and tetrasaccharides were present with approximately 75 % w/w of purity and 94 % recovery [156].

Another concept was proposed by the same group in 2010 [157]. This novel method involves isomerization of the reducing oligosaccharides prior to supercritical extraction. The isomerization of lactose, glucose, and galactose is carried out using complexating agents such as sodium aluminate or boric acid. Subsequent to the reaction, the pH of the solution is neutralized with sulfuric acid, inducing the precipitation of aluminates (or borates). Then, supercritical extraction is carried out in three subsequent steps to remove monosaccharides, disaccharides, and trisaccharides in each extraction step, respectively. Under optimized conditions, an almost complete removal of monosaccharides and disaccharides was achieved, leading to pure oligosaccharides as result of the extraction process. In addition, a quasi-complete elimination of aluminates (and borates) from the prebiotic carbohydrate mixture was obtained. The concentration of aluminum in the extracts is reported to be below the safe limit for consumers [157].

Supercritical fluid purification of carbohydrate mixtures is a high-yield/high-purity technology. A considerable drawback of this technique might be the relatively high investment and operating costs that are generally associated with this technology. To the best of our knowledge, no reports are available in the published literature on the economic evaluation of large-scale GOS purification with supercritical technology.

3.2 Mixed-Enzyme Systems

Several investigators have attempted to enhance the purity of OS using mixed-enzyme systems. The synthesis of saccharide mixtures with high-OS content is based on the concept of eliminating the byproduct that inhibits the reaction; thus, substrate utilization can be maximized. The strategies and main findings of these investigations are summarized in this section.

As discussed in Sect. 2.2, the activity of this FOS-producing enzyme is severely inhibited by glucose, which is a byproduct of the reaction. Yun et al. [158] have examined an enzymatic method using two kinds of enzymes to enhance the conversion of FOS by eliminating glucose. The batch production of high-content FOS from sucrose employing a mixed-enzyme system of β-fructofuranosidase and glucose oxidase was investigated. The reaction was performed in a single-step under the optimal (compromised) reaction conditions for the mixed-enzyme system. Under optimum conditions (pH 5.5, 40 °C, 0.7 L/min oxygen flow rate, 700 g/L substrate concentration), high-content fructo-oligosaccharides up to 98 % were obtained with complete consumption of sucrose and glucose by the mixed-enzyme system.

The performance of the same mixed-enzyme system, consisting of β-fructofuranosidase (obtained from *Aspergillus japonicus*) and commercial glucose oxidase (Gluzyme, Novo Nordisk), was investigated by Sheu et al. [159]. However, they performed the reaction in an aerated stirred tank reactor controlled at pH 5.5 by the addition of $CaCO_3$ slurry. This procedure allows the precipitation of gluconic acid (which is obtained by the conversion of glucose by glucose oxidase) and the formation of calcium gluconate in the solution. The system produced more than 90 % (w/w) FOS on a dry weight basis; the remainder was glucose, sucrose, and a small amount of calcium gluconate. The drawbacks of this procedure are that glucose oxidase is expensive and it has a limited stability at high temperature. Moreover, the production is performed in batch fashion. To overcome these problems, Sheu et al. [160] proposed a continuous process using a mixed-cell system (see Sect. 4.3.2).

Another interesting concept has been proposed by Tanriseven and Gokmen [109]. They obtained FOS from sucrose using the commercial enzyme preparation Pectinex Ultra SP-L (Novozyme A/S, Denmark). Then, the sugar mixture was further processed by *Leuconostoc mesenteroides* B-512 FM dextransucrase to convert all the unreacted sucrose to isomaltooligosaccharides, which also enhance the activity of *Bifidobacteria*. Thus, the final product is composed of FOS, isomaltooligosaccharides (isomaltose through isomaltodecaose), glucose, and fructose.

The synthesis of GOS was also investigated with a mixed-enzyme system consisting of *Bacillus* β-galactosidase and glucose oxidase. Cheng et al. [161] performed a single-step batch reaction following the method proposed for FOS production by Sheu et al. [159]. However, less promising results were obtained. The pH was controlled at 5.0 by adding 40 % (w/w) $CaCO_3$ in the aerated tank reactor. The total GOS peaked at 5 h and then decreased. Authors explain this effect with a possible shift of the reaction equilibrium of β-galactosidase to hydrolytic activity in the absence of glucose. A large amount of lactose remained in the product, resulting in a low-content GOS (less than 53 % on a dry weight basis). The difference in the yields between GOS and FOS, as interpreted by the authors, might result from the nature of the enzymes. In addition, galactose, a byproduct of the reaction, might be a competitive inhibitor of β-galactosidase.

Splechtna et al. [162] also presented a combined method of enzymatic treatment and chromatographic separation steps to synthesize a product containing 97 % GOS, 1.2 % lactose, and 2.1 % monosaccharides. The lactose present in a GOS-containing carbohydrate mixture was first oxidized into lactobionic acid and then removed by anion exchange chromatography. The selective enzymatic oxidation of lactose was achieved by using fungal cellobiose dehydrogenase and a redox mediator (2,6-dichloro-indophenol). During the reaction, the redox mediator was reduced and reoxidized continuously by laccase in the presence of molecular oxygen. NaOH was added during the reaction as required to maintain a constant pH value of 5.0. Combined with the addition of reactants, this caused a maximum 1.3-fold dilution of the original sugar mixture. Subsequent to lactose hydrolysis, a near-complete removal of monosaccharides was achieved in a single step of cation exchange chromatography. The overall GOS yield of the multistep process related to original lactose is only 25 %. However, lactobionic acid, the byproduct of the conversion, is known as a valuable component for the food and pharmaceutical industry.

3.3 Selective Fermentation

The basic idea behind selective fermentation is to remove monosaccharides and disaccharides from OS with the assistance of microorganisms. Oda and Ouchi [163], early investigators of this method, constructed a sucrose-fermenting strain, incapable of hydrolyzing FOS, by cross-breeding a baking yeast strain and a laboratory strain. Their investigation was motivated by the fact that commercial baker's yeast hydrolyzes FOS. Due to the very low fermenting ability, FOS-containing white bread was produced with the new strain.

Investigations conducted by Crittenden and Playne [164] showed that a complete removal of glucose, fructose, and sucrose present in OS mixtures was possible with *Zymomonas mobilis* fermentation. Using encapsulated cells in alginate beads, various OS mixtures (fructo-, malto-, isomalto-, gentio-, and inulin-oligosaccharides) containing total carbohydrate concentrations of 300 g/L were fermented in batch reactors for 12 h without nutrient additions and pH control. The fermentation end products were ethanol and carbon dioxide, and no degradation of the oligosaccharides in the mixtures was observed. Although batch fermentations were performed in their investigations, the authors note that *Z. mobilis* purification has the potential to be adapted to continuous processes.

Li et al. [165] challenged *Saccharomyces cerevisiae* and *Kluyveromyces lactis* cultures entrapped in calcium alginate with a carbohydrate mixture consisting of GOS (29 % w/w), galactose (5 % w/w), glucose (18 % w/w), and lactose (48 % w/w). Results indicated that a complete and rapid removal of monosaccharides can be achieved by *S. cerevisiae*, whereas no lactose and only a small fraction (1.4 %) of GOS is digested. In contrast, *K. lactis* uses both monosaccharides and disaccharides, yet with a considerably slower consumption rate. Eighteen hours of fermentation resulted in a 19.9 % reduction in the original GOS content and a nearly complete removal of the monosaccharides and lactose. In contrast with the excellent stability of immobilized *S. erevisiae*, a rapid stability loss for entrapped *K. lactis* cells was observed.

Similar results of fermentation with *S. cerevisiae* have been reported by Goulas et al. [34]. A carbohydrate mixture was successfully purified without any adverse effect on the GOS content. The glucose was almost completely metabolized, mainly to ethanol and CO_2, whereas galactose concentration was slightly reduced. The reason for low consumption of galactose is explained because the galactose uptake by *S. cereviciae* is only inducible by galactose as sole carbon source; that is, the activation of digestion mechanism requires the cultivation of yeast in a medium where galactose is the only carbon source. It is also reported that the obtained high ethanol concentration (7.4 % v/v) caused ethanol toxicity to the yeast.

Cheng et al. [161] investigated the action of *Kluyveromyces marxianus* on a GOS syrup. During fermentation, monosaccharides and disaccharides were depleted, resulting in up to ~97 % on a dry weight basis of high-content GOS, with a yields of 31 %. Most of the trisaccharides were consumed by the cells, and a ~30 g/L concentration of ethanol was reached at the end of the fermentation.

Selective fermentation is a relatively new concept. Although a number of promising results are available, it is still considered to be an unexplored area and the possible advantages of this technology are not yet fully exploited. Investigations show that removing digestible sugars from low-purity OS mixtures is technically feasible with selectively fermenting cultures. This type of purification can be performed at a low cost and on an industrial scale. This technology, however, suffers from some weaknesses that limit its spread and the applicability of the purified OS product. First, the substrate-based OS yield of the overall production process is low due to the amount of digested sugars. The fraction of digested sugars represents ~30–70 % (w/w) of the initial carbohydrate content; thus, the utilization of the products of carbon conversion has to be addressed in order to be able to develop an economically viable process. Second, the final OS product consists of the metabolic products of microbial activity and the remaining ingredients of growth media. These components (e.g. ethanol, organic acids) alter product composition, nutrition, and taste. Thus, without further purification, the resulting product can only be used in a limited number of food formulas.

4 Membrane-Based Techniques

Membranes, according to their roles in OS production, can be generally categorized as follows: (i) membranes as separation tools to fractionate mixtures of OS, disaccharides, and monosaccharides; (ii) membranes as porous matrices for immobilizing enzymes; and (iii) membranes as attachments of reactors that use free cells or enzymes. These membrane-assisted technologies are critically reviewed below.

4.1 Carbohydrate Fractionation

Pressure-driven membrane filtration is known as one of the most feasible downstream bioseparation tools. Its advantages include low energy requirements, easy control of operation, and easy scale-up [166]. Several investigations [150, 167–172] have aimed at fractionating carbohydrate mixtures consisting of OS, disaccharides, and monosaccharides using nanofiltration (NF) membranes. NF separation of carbohydrates is primarilyy based on coupled diffusion and convection because the electrostatic effects between the membrane and such noncharged solutes are negligible. The requirement of an ideal separation is to have a

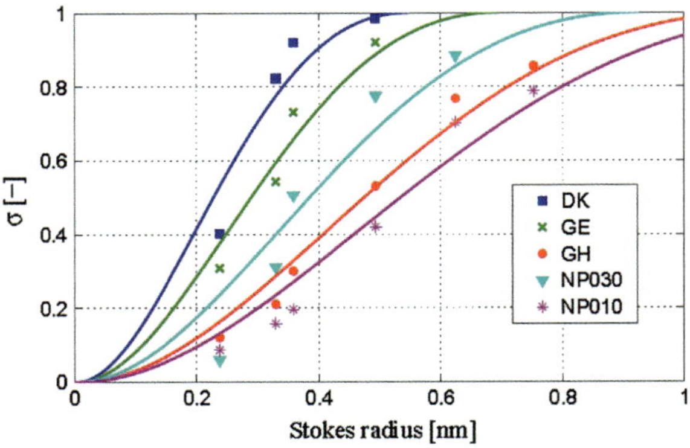

Fig. 3 Relationship between reflection coefficient and Stokes radii of noncharged solutes for some commercial NF membranes (DK, GE, GH from GE W&P Technologies, US; NP030 and NP010 from Microdyn-Nadir, Germany). Symbols denote measured values; *solid lines* illustrate model predictions using Eq. 1. (Figure adopted from [174])

membrane that completely retains OS but allows the free passage of disaccharides and monosaccharides. Such an idealized membrane, however, does not exist. Applying the theory of membrane transport, it can be shown that NF membranes possess poor permselectivity for such carbohydrates. The steric-hindrance pore model [173] allows the prediction of solute rejection, *s*, of a membrane of which pore-size distribution is uniform and consists of cylindrical and nonconnected pores. This model reads as

$$\sigma = 1 - \left(1 + \frac{16}{9}\lambda^2\right)\left(1 - \lambda^2\right)\left[2 - (1 - \lambda)^2\right] \tag{1}$$

where *l* is the ratio of solute to pore radius.

Operational parameters of NF and chemical features of saccharides that may also affect membrane separation are summarized in Ref. [166]. A first approximation of NF separation behavior, however, can be provided by using the formula above. Figure 3 shows the reflection coefficient *s* (e.g. maximal achievable rejection) as a function of Stokes radii of uncharged solutes calculated with Eq. 1 for different commercial NF membranes.

The Stokes radius of solutes in question (e.g. G, GF, GF2, GF3, GF4) ranges between approximately 0.3 and 0.7 nm. Thus, it is obvious from the figure that no sharp cutoff can be achieved with NF membranes. OS losses can only be avoided when a dense membrane with a relatively high rejection for monosaccharides and disaccharides is selected. Alternatively, an NF membrane can be chosen that allows the free passage of monosaccharides and disaccharides but has a non-complete OS rejection. Thus, there is a trade-off between yield and purity. Another disadvantage of NF when working with high concentration of polarizing species is that it requires high pressure to achieve a moderate permeate flux, making the process energy intensive.

NF fractionation might be improved with diafiltration, in which a diluant is supplied to the feed solution in order to "wash out" the low molecular weight solutes. In this case, however, the consumed wash-water causes additional costs, and the NF permeate becomes diluted. Typically, an OS-rich carbohydrate solution can be obtained as final retentate that consists of a considerable amount of remaining monosaccharides and disaccharides. Accordingly, the diluted permeate is rich in monosaccharides and disaccharides and contains considerable amounts of OS. For example, Feng et al. [171] achieved a final product with an OS purity of 55 % (~1.5 times higher of that in the starting material) and an OS yield of 70 %. Another batchwise membrane-based technique has been investigated by Kuhn et al. [175]. They have proposed a two-stage nanofiltration/diafiltration process to purify FOS from the byproduct glucose and sucrose remaining in the reaction mixture. The 5-h batch purification process resulted in an increase in purity from 55 to 90 % in FOS with a yield of ~80 %.

The trade-off between yield and purity, the high energy consumption, and the considerable amount of diluant required for the separation put the significance of NF technology operating in batch configuration into question. A novel and promising tool for large-scale continuous purification of OS is the so-called membrane cascades technology. Lightfoot [176] demonstrated by numerical calculations that membrane cascades might replace chromatography in separating molecules having similar molecular weights.

Membrane cascades address the previously mentioned problems associated with NF technology. They reduce the consumption of diluant and improve the product purity without compromising product yield [177]. This is, of course, achieved at the expense of higher investment and operating costs. In a membrane cascade, the fractionation is achieved through a series of membrane stages. The operating principle of a countercurrent membrane cascade is similar to that of distillation. The feed solution is continuously supplied to the feed stage of a multistage membrane system. The permeate of this stage is then fed to the next stage, and the retentate is sent to the previous stage. The retentate of the first stage is the stripped product stream enriched in the less permeable solute. The permeate of the last stage is enriched in the most permeable solute.

Vanneste et al. [148] studied membrane cascades for the purification of monosaccharides and oligosacharides and showed that cascades can be designed to reach the same final product specifications as with SMB chromatography. They have compared the separation of raffinose from sucrose with industrial-scale SMB and an equivalent membrane cascade system. Their technoeconomic calculations indicate that the competitiveness of membrane cascades over SMB chromatography increases with increasing plant size. It has been shown that the cost of a six-stage membrane cascade (3.6 million EUR), designed for a productivity of 464 tons per year, is approximately the half of the calculated cost of a SMB installation (6.8 million EUR).

To the best of our knowledge, no reports are available yet in the published literature on the performance of membrane cascades on purifying GOS- and FOS-containing mixtures. We note, however, that the molecular weights of raffinose and sucrose (504 and 342 g/mol, respectively) and their assumed rejections (0.8 and 0.56, respectively) fall into the range which is of interest for GOS and FOS fractionation. Thus, based on the work of Vanneste et al. [148], we can conclude that this technology might be a promising alternative to chromatography for large-scale continuous processes.

4.2 Enzyme Immobilization onto Membranes

Membranes are thin layers of porous materials providing a large surface area for enzyme immobilization. The operating principle of biocatalytic membrane technology is straightforward: the substrate is forced through the membrane by applying transmembrane pressure and reacts during permeation through the membrane. This forced-flow reduces diffusive limitations that may occur in conventional column reactors using beads. Immobilization onto membrane supports allows production in continuous operation mode and may result in some technological benefits, including enhanced enzyme stability and resistance against pH, temperature, and ionic strength of the environment. The major drawbacks associated with biocatalytic membranes may include low immobilization efficiency, loss of enzyme activity as compared to free enzymes, reduced flux performance due to immobilization, fouling of membranes caused by components of the process solution, limited capability for enzyme reimmobilization, and troublesome membrane flux recovery (cleaning and sterilization without causing losses in its biocatalytic activity). Moreover, immobilization is advised to be carried out with highly purified enzymes, which poses an additional cost factor. As compared to a batch process using free enzymes, the obvious benefit of biocatalytic membrane technology is the reuse of enzymes.

It needs to be carefully investigated whether these advantages compensate for the possible drawbacks. In recent years, a number of studies have been published on biocatalytic membranes for GOS and FOS production. A brief survey summarizing the most important findings is given here.

A forced-flow membrane reactor system for transfructosylation has been investigated using ceramic membranes by Nishizawa et al. [178]. β-Fructofuranosidase from *Aspergillus niger* ATCC 20611 was immobilized chemically to the inner surface of a 0.5-mm ceramic membrane activated by a silane-coupling reagent. The FOS conversion was determined as a function of residence time. Results indicate that the ratio of FOS to total saccharides was more than 55 % w/w at a residence time of 10 s. This value is only slightly lower than that in a conventional batch process (~60 % w/w). The volumetric productivity obtained by employing a short residence time (11 s) was found to be 3.87 kg/m^3 per second, which is 560 times higher than that in the reported batch system. Based on the long-term operational test, the half-life of the immobilized enzyme was estimated to be 35 days.

A recent study of Sen et al. [179] provides a comparison on the applicability of enzyme-immobilized membranes using polyamide thin-film composite nanofiltration and both polyether sulfone and cellulose triacetate ultrafiltration membranes as matrices for immobilization. Both adsorption and cross-linking based immobilization techniques were employed in this investigation. The best results on GOS yield (~30 % w/w) were achieved with the NF membrane.

One of the main concerns attributed to membrane immobilization is the long-term operational stability of the resulting catalytic membranes. Recently, Palai et al. [180] tested the repeated usability of biocatalytic membranes. Their study deals with the immobilization of β-galactosidase on polyvinylidene difluoride (PVDF) membrane by cross-linking with glutaraldehyde. The biocatalytic activity of the same membrane was tested in consecutive runs. Between runs, the membrane was stored in a phosphate buffer at pH 6. The study indicates that enzyme activity fell below 57 % after the second run. We note that a similar trend can be observed when storing soluble enzymes in a buffer solution instead of a stabilizing solution. Thus, an interesting question is that of how far such a dramatic loss in activity can be avoided when using a stabilizing solution. In case of free enzymes, glycerol and potassium chloride, for example, are common stabilizers, and the process solution itself works well due to its high sugar content.

Similar results in respect to membrane reusability have been reported by Jochems et al. [181] and Gülec [182]. Jochems et al. [181] used a 13.8-kDa cutoff polymer-ceramic hybrid membrane for immobilizing β-galactosidase by adsorption. After 14 days of storage in 50 mM Tris–HCl buffer at pH 7, the immobilized enzyme showed 41 % of its initial activity. In his 2013 study, Gülec [182] reported on the immobilization of β-galactosidase from *Kluyveromyces lactis* onto plasma-modified cellulose acetate membrane surfaces by adsorption and covalent binding. The membranes were kept dry and conditioned with phosphate buffer before each run. After 10 consecutive runs, residual activity of enzymes immobilized by adsorption dropped to zero, while that of covalently immobilized enzymes decreased to ~30–70 %, depending on the type of technique used for covalent binding. This is in a good agreement with the findings of Ulbricht and Papra [183], who noted that irreversible immobilization of β-galactosidase onto ultrafiltration membranes might be better suited for industrial applications than adsorption techniques, because cross-linking yields a better enzyme stability.

Although adsorption-based immobilization may compare unfavorably with cross-linking in terms of enzyme stability, it allows a loose binding of the enzyme and, thus, an enhanced reusability of the membrane matrix. However, this loose binding may cause enzyme leakage, especially when dealing with process solutions containing salts. Adsorption onto membranes can also be done by using commercially available membrane adsorbers, as proposed by Engel et al. [184–186]. A Mustang Q (Pall GmbH, Germany) anion exchange polyether sulfone membrane with a pore size of 0.8 mm was used as matrix for immobilization. One of the rapidly growing field of applications of membrane chromatography is the separation of proteins, and in past decades many efforts have been put into the development of superior membrane adsorbers. By selecting the proper chromatography membrane for the protein of question and adjusting the appropriate conditions such as pH and temperature, efficient adsorption can be achieved. The results obtained with a 20 % (w/w) lactose solution show a lactose conversion of 82 % and a GOS yield of 24 % at 1-h reaction time.

To conclude, immobilization of enzymes onto membranes for GOS and FOS production is technically feasible. A wide range of techniques are available to perform immobilization onto different polymeric and ceramic (or hybrid) membranes. In addition, there is an increasing number of publications on this issue,

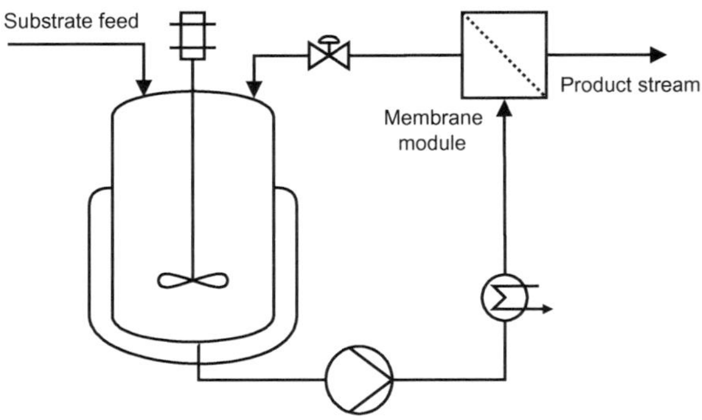

Fig. 4 Schematic diagram of the continuous stirred tank reactor equipped with an external ultrafiltration module for enzyme retention

most of them published within the past few years. To the best of our knowledge, all investigations have been carried out at laboratory scale and reports on the economics of this technology for GOS/FOS production are not available in the published literature.

4.3 Membrane-Assisted Biocatalytic Reactors

In Sect. 4, we discussed how membrane filtration can be used as a downstream fractionation tool for a carbohydrate mixture subsequent to the biosynthesis. Here, we review the membrane filtration technologies that can be integrated into the process for enhancing the biocatalytic performance of the reaction step. Generally, microfilters can be used for retaining cells cultivated in a fermenter and ultrafilters allow the recovery of enzymes, whereas nanofilters have the potential to eliminate the lower molecular weight fractions from the solution during biocatalysis. These coupled processes are briefly summarized below.

4.3.1 Ultrafiltration-Assisted Enzymatic Reactor

Ultrafiltration-assisted (UF) enzymatic reactor (EMR) represents a specific mode for running a continuous processes in which enzymes are separated from the resulting OS with the help of an ultrafilter. It couples the separation process with the enzymatic reaction. One of the main advantages of this technology is its simple configuration. A schematic of a EMR is shown in Fig. 4. In UF-EMR, fresh substrate is supplied to the reactor, and the enzyme-free product stream is continuously removed.

In the UF model, an inhomogeneity in the local concentration of biocatalysts arises due to the accumulation of retained species at the membrane wall. Rejected proteins form a boundary layer adjacent to the membrane. The build-up of this so-called concentration polarization layer depends on several factors, most importantly, protein load, applied pressure, and hydrodynamic conditions [187, 188],

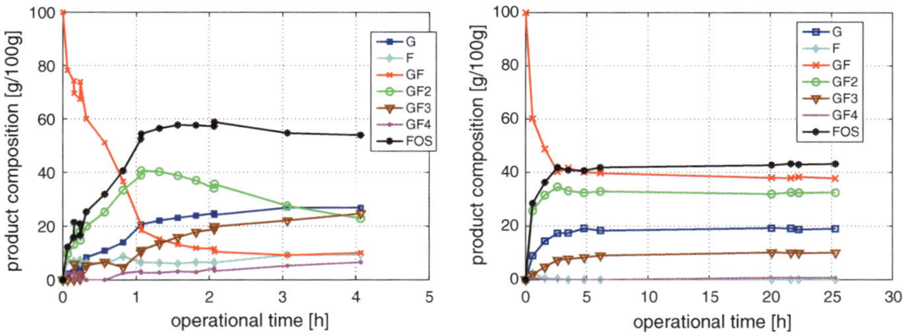

Fig. 5 Saccharide composition vs. processing time in a conventional batch (*left side*) and ultrafiltration-assisted enzymatic membrane reactor operating at 1-h residence time (*right side*). Reaction conditions: Pectinex Ultra SP-L (Novozyme A/S, Denmark), 7.5 g/100 g crude enzyme dosage, 40 g/100 g substrate concentration, 50 °C, pH 5.6 adjusted by citrate/phosphate buffer. Process conditions: 20-kDa ceramic membrane, 0.5 bar, 0.42 m/s cross-flow velocity, 1.0 h residence time

which are crucial parameters in the design of UF-EMR. Inappropriate selection of these operating parameters might result in a low volumetric productivity of UF-EMR and the aggregation and denaturation of proteins [189, 190].

Also, fouling of the membrane may occur during operation. Fouling is generally a complex mechanism that may involve adsorption, pore blocking, layer-by-layer deposition, and, in some cases, gel formation [191]. Fouling leads to a (often partially irreversible) loss of membrane flux [192] and, as a consequence, to troublesome cleaning [193]. It may also result in pronounced enzyme activity losses during long-term operation.

Although it is often stated that product inhibition of enzyme catalyzed reactions can be diminished if membranes are integrated with the reactors for continuous or semicontinuous product removal (e.g. [166, 194, 195]), this is not the case in UF-EMR producing sucrose-based FOS and lactose-based GOS. In fact, both experimental data and simulation results show that the OS fraction in UF-EMR permeate cannot reach values as high as in a batch process. In Fig. 5, we show a set of typical experimental data from our laboratory on FOS synthesis in a batch reactor and in UF-EMR. Although the concentration of OS obtained in UF-EMR is lower than in batch, UF-EMR offers a better productivity (i.e. the total quantity of OS synthesized in a given time by the same amount of enzyme) [196–198].

Czermak et al. [199] described the continuous production of GOS from pure lactose (350 g/L) in a UF-EMR using commercial enzyme preparations. The GOS concentrations in the product stream are reported to be between 25 and 40 % (w/w) on total carbohydrate basis, depending on the applied residence time. Other investigators studied GOS synthesis on whey and deproteinated whey [200, 201], reaching GOS concentrations of 9–22 % (w/w) on a total carbohydrate basis. The main factors influencing GOS yield are the source of enzyme, enzyme dosage, residence time, pH (and buffer composition), temperature, and substrate concentration.

Enzyme activity decay during operation constitutes a problem that has been far less investigated [195]. Sustainable operating performance requires a narrow control of fouling and, in some cases, the addition of fresh enzymes to compensate unavoidable activity losses [202–204]. In fact, Petzelbauer et al. [205] identified enzyme adsorption onto the membrane as one of the major factors of enzyme inactivation. They studied GOS synthesis in UF-EMR using two thermostable β-glycosidases at 70 °C. The half-life times of the β-glycosidases from *Sulofulfolobus solfataricus* and *Pyrococcus furiosus* are reported to be 5 and 7 days, respectively. It was also shown in a 2011 study [206] that the activity of β-galactosidase from *Bacillus circulans* drops to 23 % after 5 h of reaction time in an UF-EMR used for GOS production.

4.3.2 Microfiltration-Assisted Bioreactor

The principle of microfiltration-assisted bioreactor (MBR) is the same as of UF-EMR, with the exception that MBR uses free cells instead of enzymes. Obviously, a MF membrane with a pore size greater than an UF membrane can be employed to retain whole cells.

An elegant technique has been proposed by Sheu et al. [160] for continuously producing high-content FOS mixtures with a MBR. In the MBR, simultaneous reactions take place by using fungal mycelia with FTase activity and living bacterial cells with glucose dehydrogenase activity. The MBR is aerated and stirred at a constant temperature, with calcium carbonate slurry added to control the pH. The gluconic acid generated during the reactions is rendered insoluble, and high-purity FOS is recovered by means of a MF module. Meanwhile, sucrose solution is fed continuously into the bioreactor, forming a system capable of continuously producing 80 % (w/w) FOS on dry mass basis with only a small amount of calcium gluconate in the filtrate. It has been shown that the system was stable for 7 days with a dilution rate of 0.04/h and a volumetric productivity for total FOS more than 160 g/L per hour.

In 2011, Avalakki et al. [207] issued a US patent on a process for production of high yield of pure GOS by MBR. Whole cells of *B. singularis* and *Saccharomyces* sp. are mixed in a reactor equipped with a cross-flow hollow-fiber MF module. The 24-h fermentation is carried out with a 30 % (w/w) lactose solution at 30 °C. The proposed MBR operates in a discontinuous mode. When the purity of GOS reaches more than 90 %, the volume in the reactor is reduced to half by filtrating the broth through the MF membrane. Then, a fresh lactose solution is added into the tank to adjust the starting volume. This cycle is repeated until a 10 % drop in the purity of GOS is observed. The patent claims that a more than 90 % pure GOS was obtained in the reactor in 39 repeated cycles wherein the cell biomass was reused in the GOS production. GOS yield on lactose basis was not indicated.

4.3.3 Nanofiltration-Assisted Enzymatic Reactor

The concept of a nanofiltration-coupled enzyme reactor (NF-EMR) was proposed by Nishizawa et al. [83]. The configuration of the NF-EMR is practically the same as for UF-EMR (see Fig. 4). The applied membrane, however, is different. Here, an NF membrane is selected through which glucose permeates but sucrose and FOS do not permeate.

It is known that the transfructosylation reaction is inhibited by the glucose that is formed during the reaction. Thus, a higher reaction conversion can be achieved by the simultaneous removal of produced glucose. Nishizawa et al. [83] made a screening test with five commercial membranes, and NF-45 (Sanko Shokai, Japan) was selected as best-performing membrane for the given separation task. At a 4-MPa transmembrane pressure, the observed rejection values of this membrane were 74 %, 98 %, >99 %, >99 %, and >99 % for glucose, sucrose, GF_2, GF_3, and GF_4, respectively.

The authors claim that after 12 h of operation, the FOS fraction in the reactor was 93 %, which is much higher than in the product obtained in conventional batch operating mode (55–60 %), and residual sucrose percentage was 5 %. The amount of diafiltration buffer, which is used to compensate permeate losses and to keep the feed volume constant, can be estimated as ~1.5-fold of the feed volume.

It should be noted that an NF-EMR has to operate under a considerably higher pressure than an UF-EMR. The requirement of a positive permeate flux is to overcome the osmotic pressure generated by the species rejected by the NF membrane. At the surface of an NF membrane, the polarizing species (i.e. GF, GF_2, GF_3, GF_4, and the rejected fraction of glucose) form a boundary layer of high concentration that may affect enzyme stability. Long-term performance and possible enzyme activity losses were not investigated in Ref. [83].

5 Conclusion

In this chapter, we reviewed the recent developments in the manufacture and purification of GOS and FOS. These well-established OS provide several health benefits because they selectively support the growth of beneficial microflora of the colon and thus improve health. In addition, they also have excellent technological properties such as low viscosity, high pH and temperature stability, high solubility, and fine taste. These unique properties make their use as food ingredients especially attractive.

The biosynthesis processes of lactose-based GOS and sucrose-based FOS show similarities in terms of reaction mechanisms and product formation. Both GOS and FOS can be synthesized either using whole cells or (partially) purified enzymes, in immobilized or free form. A large body of literature has been devoted to study these various biocatalytic strategies; today, a number of companies produce OS on industrial scale.

A considerable drawback of biocatalysis is that the reaction actually results in a carbohydrate mixture consisting of OS, unreacted disaccharides, and monosaccharides. The incomplete conversion poses a challenge to manufacturers because an enrichment of OS in this mixture adds value to the product. For removing digestible carbohydrates from OS, a variety of bioengineering techniques have been investigated. These include downstream separation technologies, additional bioconversion steps applying enzymes, and selective fermentation strategies.

Among the downstream separation technologies, liquid chromatography has been long used on large scale in the sugar industry, and its employment for OS separation seems to be a straightforward choice. However, a number of competitive techniques have been recently proposed in this relatively expensive technology. In fact, activated charcoal treatment and membrane cascades might compare favorably with simulated moving-bed chromatography in terms of purity and yield. Also, supercritical extraction and precipitation with ethanol are potential candidates in purifying OS.

Another approach to enhance the purity of OS is based on using mixed-enzyme systems to eliminate the inhibiting byproducts from the reaction mixture and, thus, to maximize substrate conversion. The batch production of high-content FOS is successfully realized by employing a mixed-enzyme system of β-fructofuranosidase and glucose oxidase. High-content FOS up to 98 % can be obtained in this way with complete consumption of sucrose and glucose. In contrast to that, the same mixed-enzyme system performs less well for GOS synthesis, leading to a relatively low-content GOS product (less than 53 % on dry weight basis). In the case of GOS-containing mixtures, the removal of digestible sugars is possible through a combined method of enzymatic treatment with laccase and chromatographic separation steps.

Selective fermentation is a relatively new concept. Although a number of promising results are available, it is still considered to be an unexplored area and the possible advantages of this technology are not yet fully exploited. This type of purification is technically feasible and can be performed at a low cost and on an industrial scale. However, the substrate-based OS yield of the overall production process is low due to the high amount of digested sugars. Because the fraction of digested sugars typically represents $\sim$30–70 % w/w of the initial carbohydrate content, the utilization of the products of carbon conversion has to be addressed in order to be able to develop an economically viable process. Moreover, the final OS product consists of the metabolic products of microbial activity and remaining ingredients of growth media. These components alter product quality; thus, without further purification, the resulting product can only be used in a limited number of food formulas.

In this chapter, we devoted special attention to membrane-based processes as emerging techniques used for both manufacturing and fractionation purposes. Membranes, according to their roles in OS production, can be generally categorized as (i) membranes as separation tools to fractionate mixtures of OS, disaccharides, and monosaccharides; (ii) membranes as porous matrices for immobilizing enzymes; and (iii) membranes as attachments of reactors that use free cells or enzymes.

Membrane filtration—or more precisely, nanofiltration—can be used to fractionate carbohydrate mixtures obtained from the biosynthesis step. Membrane filtration is typically associated with low energy requirements, easy control of operation, and easy scale-up. NF membranes, however, show poor permselectivity for the carbohydrates in question due to the small differences in their relative molecular sizes. This problem can be addressed with cascade arrangement of multiple nanofiltration units. Recent studies based on theoretical calculations suggest that membrane cascades technology may be an alternative to chromatography for large-scale continuous fractionation of carbohydrates in the future.

There is a wide range of available techniques to immobilize enzymes onto different polymeric and ceramic (or hybrid) membranes for GOS and FOS production. Although an increasing number of publications prove this technology to be technically feasible, investigations so far are restricted to the laboratory scale and no reports on the economics of this technology are available.

Membrane filtration can also be coupled with biosynthesis for enhancing the biocatalytic performance of the reaction step. Generally, microfilters can be used for retaining cells cultivated in a fermenter and ultrafilters allow the recovery of enzymes, whereas nanofilters have the potential to eliminate the lower molecular weight fractions from the solution during biocatalysis. Such membrane-assisted enzyme reactors allow the integration of the separation process with the biocatalytic reaction into a single step and enable a continuous production of OS mixture that is free of biocatalysts. The stability decay of biocatalysts during operation, however, constitutes a problem that has so far been less investigated.

The market for prebiotics is steadily increasing. To satisfy this growing worldwide demand, the introduction of effective bioprocessing methods and implementation strategies is required. In this chapter, we have critically reviewed the state-of-the-art manufacturing strategies and the recent advances in bioprocessing technologies that can open new possibilities for manufacturing sucrose-based FOS and lactose-based GOS.

Acknowledgments We thank the Hessen State Ministry of Higher Education, Research and the Arts for the financial support within the Hessen Initiative for Scientific and Economic Excellence (LOEWE Program). The first author is grateful for the Marie Curie FP7 Integration Grant provided by the 7th European Union Framework Programme (PCIG11-GA-2012-322219).

References

1. Gibson GR, Roberfroid MB (1995) Dietary modulation of the human colonic microbiota: introducing the concept of prebiotics. J Nutr 125(6):1401–1412. http://jn.nutrition.org/content/125/6/140.short and http://jn.nutrition.org/content/125/6/140.full.pdf+html
2. Hammes W, Weiss N, Holzapfel W (1992) The genera *Lactobacillus* and *Carnobacterium*, vol II., The ProcaryotesSpringer, Berlin

3. Van Loo J, Cummings J (1999) Br J Nutr 81:121–132
4. Cummings J, Roberfroid M (1997) Eur J Clin Nutr 51(7):417
5. Rumessen J, Gudmand-Hoyer E (1998) Am J Clin Nutr 68(2):357
6. Hopkins M, Cummings J, Macfarlane G (1998) J Appl Microbiol 85(2):381
7. Wang X, Gibson G (1993) J Appl Bact 75(4):373
8. Kleessen B, Sykura B, Zunft H, Blaut M (1997) Am J Clin Nutr 65:1397
9. Musatto S, Mancilha I (2007) Carbohydr Polym 68:587
10. Den Hond E, Geypens B, Ghoos Y (2000) Nutr Res 20(5):731
11. Gibson G, Wang X (1994) J Appl Microbiol 77:412
12. Igarashi M (1994) Bifidus 7:139
13. Schoterman H (2001) Galactooligosaccharides: properties and health aspects., Advanced dietary fibre technologyBlackwell Science, Oxford
14. Swennen K, Courtin C, Delcour J (2006) Crit Rev Food Sci Nutr 46:459
15. Sekine K, Ohta J (1995) Biol Pharmaceut Bull 18:148
16. Reddy B (1998) Br J Nutr 80:219
17. Franck A, Coussement P (1997) Food ingredients and analysis international 51
18. Kunz C, Rudloff S (1993) Acta Pädiatr 82:903
19. Boehm G, Stahl B (2003) *Oligosaccharides*, functional dairy, products edn. Woodhead Publishing Ltd. Cambridge UK, Cambridge UK
20. Moro G, Minoli I (2002) J Pediatr Gastroenterol Nutr 34:291
21. Eiwegger T, Stahl B, Schmitt J, Boehm G, Gerstmayr M, Pichler J, Dehlin KE, Loibichler C, Urbanek R, Szépfalusiz (2004) Human milk–derived oligosaccharides stimulate cytokine production of cord blood T-cells in vitro. Pediatr Res 56(4):536–540
22. EFSA (2010a) Consolidated list of Article 13 health claims of references received by EFSA-Part 3 (internet). Parma: Scientific Panel on Dietetic Products, Nutrition and Allergies Unit. URL http://www.efsa.europa.eu/de/ndaclaims13/docs/ndaart13ref01.pdf
23. EFSA (2010b) Consolidated list of Article 13 health claims of references received by EFSA-Part 3 (internet). Parma: Scientific Panel on Dietetic Products, Nutrition and Allergies Unit. URL http://www.efsa.europa.eu/de/ndaclaims13/docs/ndaart13ref03.pdf
24. Wallenfels K, Malhotra OP (1961) Adv Carbohydr Chem 16:239
25. Voragen AGJ (1998) Trends Food Sci Technol 8–9:328
26. Gibson GR, Roberfroid MB (1995) J Nutr 125:1401
27. Torres DPM, do Pilar M, Goncalves F, Teixeira JA, Rodrigues LR (2010) Compr Rev Food Sci Food Safety 9:438
28. Machadoa JJB, Coutinho JA, Macedo EA (2000) Fluid Phase Equilibria 173:121
29. Arakawa T, Timasheff SN (1982) Biochemistry 21(25):6536
30. Padilla B, Ruiz-Matute AI, Belloch C, Cardelle-Cobas A, Corzo N, Manzanares P (2012) J Agric Food Chem 60:5134
31. Martinez-Villaluenga C, Cardelle-Cobas A, Olano A, Corzo N, Villamiel M, Jimeno ML (2008) J Agric Food Chem 56:557
32. Vera C, Guerrero C, Conejeros R, Illanes A (2012) Enzym Microb Technol 50:188
33. Huerta LM, Vera C, Guerrero C, Wilson L, Illanes A (2011) Process Biochem 46:245
34. Goulas A, Tzortzis G, Gibson GR (2007) Int Dairy J 17:648
35. Hsu CA, Lee SL, Chou CC (2007) J Agric Food Chem 55:2225
36. Joergensen F, Hansen OC, Stougaard P (2001) Appl Microbiol Biotechnol 57:647
37. Rabiu BA, Jay AJ, Gibson GR, Rastall RA (2001) Appl Environ Microbiol 67(6):2526
38. Hung MN, Lee BH (2002) Appl Microbiol Biotechnol 58:439
39. Roy D, Daoudi L, Azaola A (2002) J Indus Microbiol Biotechnol 29:281
40. Goulas T, Goulas A, Tzortzis G, Gibson GR (2009) Appl Microbiol Biotechnol 84:899
41. Osman A, Tzortzis G, Rastall RA, Charalampopoulosa D (2010) J Biotechnol 150:140
42. Gosling A, Stevens GW, Barber AR, Kentish SE, Gras SL (2011) J Agric Food Chem 59:3366
43. Kamerke C, Pattky M, Huhn C, Elling L (2012) J Mol Catal B Enzym 79:27
44. Mozaffar Z, Nakanishi K, Matsuno R, Kamikubo T (1984) Agric Biol Chem 48(12):3053

45. Boon MA, Janssen AEM, Van't Riet K (2000) Enzym Microb Technol 26:271
46. Vetere A, Paoletti S (1996) FEBS Lett 399:203
47. Yanahira S, Kobayashi T, Suguri T, Nakakoshi M, Miura S, Ishikawa H, Nakajima I (1995) Biosci Biotechnol Biochem 59(6):1021
48. Das R, Sen D, Sarkar A, Bhattacharyya S, Bhattacharjee C (2011) Indus Eng Chem Res 50:806
49. Li W, Sun Y, Ye H, Zeng X (2010) Eur Food Res Technol 231:55
50. Li W, Xiang X, Tang S, Hu B, Tian L, Sun Y, Ye H, Zeng X (2009) J Agric Food Chem 57:3927
51. Mozaffar Z, Nakanishi K, Matsuno R (1985) J Food Sci 50(6):1602
52. Pruksasri S (2007) Production and separation of galacto-oligosaccharides from lactose by *b*-galactosidase immobilized on nanofiltration membranes
53. Usui T, Kubota S, Ohi H (1993) Carbohydr Res 244(2):315
54. Onishi N, Yamashiro A, Yokozeki K (1995) Appl Environ Microbiol 11:4022
55. Rodriguez-Colinas B, de Abreu MA, Fernandez-Arrojo L, de Beer R, Poveda A, Jimenez-Barbero J, Haltrich D, Olmo AOB, Fernandez-Lobato M, Plou FJ (2011) J Agric Food Chem 59:10477
56. Montilla A, Corzo N, Lano A (2012) Milchwissenschaft 67(1):14
57. Burvall A, Asp NG, Dahlqvist A (1979) Food Chem 4(4):243
58. Adamczak M, Charubin D, Bednarski W (2009) Chem Pap 63(2):111
59. Chockchaisawasdee S, Athanasopoulos VI, Niranjan K, Rastall RA (2005) Biotechnol Bioeng 89(4):434
60. cová KP, Curda L, sún DM, Dryáková A, Diblíková L (2010) J Food Eng 99(4):479
61. Iwasaki KI, Nakajimab M, Nakao SC (1996) Process Biochem 31(1):69
62. Reuter S, Nygaard AR, Zimmermann W (1999) Enzym Microb Technol 25:509
63. Bankova E, Bakalova N, Petrova S, Kolev D (2006) Biotechnol Biotechnol Equip 20(3):114
64. Toba T, Yokota A, Adachi S (1985) Food Chem 16(2):147
65. Vera C, Guerrero C, Illanes A (2011) Carbohydr Res 346:745
66. Chen C, Hsu C, Chiang B (2002) Process Biochem 38:801
67. Cheng CC, Yu MC, Cheng TC, Sheu DC, Duan KJ, Tai WL (2006) Biotechnol Lett 28:793
68. Sakai T, Tsuji H, Shibata S, Hayakawa K, Matsumoto K (2008) J Gen Appl Microbiol 54:285
69. Coulier L, Timmermans J, Bas R, Van Den Dool R, Haaksman I, Klarenbeek B, Slaghek T, Van Dongen W (2009) J Agric Food Chem 57(18):8488. doi: 10.1021/jf902549e. URL http://pubs.acs.org/doi/abs/10.1021/jf902549e
70. US Food and Drug Administration. Agency response letter gras notice no. grn 000236
71. US Food and Drug Administration. Agency response letter gras notice no. grn 000334
72. US Food and Drug Administration. Agency response letter gras notice no. grn 000286
73. Tzortzis G, Vulevic J (2009) Galacto-oligosaccharide prebiotics. Springer, New York, pp 207–244 (Prebiotics and Probiotics - Science and Technology)
74. Silk DBA, Davis A, Vulevic J, Tzortzis G, Gibson GR (2009) Aliment Pharmacol Ther 29:508
75. Mabel M, Sangeetha P, Platel K, Srinivasan K, Prapulla S (2008) Carbohydr Res 343(1):56. doi: 10.1016/j.carres.2007.10.012. URL http://www.sciencedirect.com/science/article/pii/S0008621507004387
76. Khan R (1993) Low-calorie foods and food ingredients. Springer, London
77. Crittenden R, Playne M (1996) Trends Food Sci Technol 7(11):353. doi: 10.1016/S0924-2244(96)10038-8. URL http://www.sciencedirect.com/science/article/pii/S0924224496100388
78. Martinez-Ferez A, Guadix A, Guadix EM (2006) J Membr Sci 276(1–2):23. doi: 10.1016/j.memsci.2005.09.027. URL http://www.sciencedirect.com/science/article/pii/S0376738805006678
79. Shiomi N (1978) J Facul Agric 58:4
80. Shiomi N, Yamada J, Izawa M (1976) Agric Biol Chem 40(3):567

81. Kurtoglu G, Yildiz S (2011) Gazi Univ J Sci 24(4):877
82. Monsan P, Paul F (1995) FEMS Microbiol Rev 16(2–3):187. doi:10.1111/j.1574-6976.1995.tb00165.x
83. Nishizawa K, Nakajima M, Nabetani H (2001-02-01) Food Sci Technol Res 7(1):39. doi:10.3136/fstr.7.39
84. Alvarado M, Maugeri F (2007) J Biotechnol 131:S91–S92
85. Duan K, Chen J, Sheu D (1994) Enzym Microb Technol 16(5):334
86. Jung K, Yun J, Kang K, Lim J, Lee J (1989) Enzyme Microb Technol 11:491
87. Kilian S, Sutherland F, Meyer P, Preez J (1996) Biotechnol Lett 18:975. doi:10.1007/BF00154633. URL http://dx.doi.org/10.1007/BF00154633
88. Park M, Lim J, Kim J, Park S, Kim S (2005) Biotechnol Lett 27:127. doi:10.1007/s10529-004-7339-x. URL http://dx.doi.org/10.1007/s10529-004-7339-x
89. Kim BW, Choi JW, Yun JW (1998) Biotechnol Lett 20(11):1031. URL http://www.ingentaconnect.com/content/klu/bile/1998/00000020/00000011/00177101
90. Ghazi I, Fernandez-Arrojo L, Gomez De Segura A, Alcalde M, Plou FJ, Ballesteros A (2006) J Agric Food Chem 54(8):2964. doi:10.1021/jf053023b
91. Smaali I, Jazzar S, Soussi A, Muzard M, Aubry N, Marzouki MN (2012) Biotechnol Bioprocess Eng 17:385. doi:10.1007/s12257-011-0388-9
92. Surin S, Seesuriyachan P, Thakeow P, Phimolsiripol Y (2012) J Appl Sci 12(11):1118. doi:10.3923/jas.2012.1118.1123
93. Lateef A, KANA EBG (2012) Roman Biotechnol Lett 17(3):7309. URL http://ebooks.unibuc.ro/biologie/RBL/rbl3vol17/11.pdf
94. Sangeetha P, Ramesh M, Prapulla S (2004) Appl Microbiol Biotechnol 65:530. doi:10.1007/s00253-004-1618-2. URL http://dx.doi.org/10.1007/s00253-004-1618-2
95. Fernandez RC, Maresma BG, Juarez A, Martinez J (2004) J Chem Technol Biotechnol 79:268. doi: 10.1002/jctb.967. URL http://onlinelibrary.wiley.com/doi/10.1002/jctb.967/abstract
96. Sanchez OF, Rodriguez AM, Silva E, Caicedo L (2010) Food Bioprocess Technol 3(4):662. doi:10.1007/s11947-008-0121-7
97. Park YK, Almeida MM (1991) World J Microbiol Biotechnol 7(3):331. doi:10.1007/BF00329399
98. Mabel MJ, Sangeetha PT, Platel K, Srinivasanb K, Prapulla SG (2008) Carbohydr Res 343(1):55. doi: 10.1016/j.carres.2007.10.012. URL http://www.sciencedirect.com/science/article/pii/S0008621507004387
99. Lateef A, Oloke JK, Prapulla SG (2007) Turkish J Biol 31(3):147
100. Park JP, Oh TK, Yun JW (2001) Process Biochem 37(5):471. doi:10.1016/S0032-9592(01)00237-0. URL http://www.sciencedirect.com/science/article/pii/S0032959201002370
101. Patel V, Saunders G, Bucke C (1994) Biotechnol Lett 16(11):1139
102. Barthomeuf C, Pourrat H (1995) Biotechnol Lett 17(9):914
103. Kuhn RC, Filho FM, New Biotechnology (2010) 27(6):862. doi: 10.1016/j.nbt.2010.05.008. URL http://www.sciencedirect.com/science/article/pii/S1871678410004437
104. Hang YD, Woodams EE (1995) Biotechnol Lett 17(7):741
105. Hang YD, Woodams EE (1996) LWT Food Sci Technol **29**(5–6):578. doi:10.1006/fstl.1996.0089. URL http://www.sciencedirect.com/science/article/pii/S0023643896900894
106. Nemukula A, Mutanda T, Wilhelmi BS, Whiteley CG (2009) Bioresour Technol 100(6):2040. doi: 10.1016/j.biortech.2008.10.022. URL http://www.sciencedirect.com/science/article/pii/S096085240800878X
107. Tanriseven A, Aslan Y (2005) Enzyme Microb Technol 36(4):550. doi:10.1016/j.enzmictec.2004.12.001. URL http://www.sciencedirect.com/science/article/pii/S014102290400362X

108. Ghazi I, Arrojo LF, Arellano HG, Ferrer M, Ballesteros A, Plou FJ (2007) J Biotechnol 128(1):204. doi: 10.1016/j.jbiotec.2006.09.017. URL http://www.sciencedirect.com/science/article/pii/S0168165606007905
109. Tanriseven A, Gokmen F (1999) Biotechnol Tech 13(3):207
110. Nemukula A, Mutanda T, Wilhelmi B, Whiteley C (2009) Bioresour Technol 100(6):2040. doi:10.1016/j.biortech.2008.10.022
111. Ghazi I, Fernandez-Arrojo L, Garcia-Arellano H, Ferrer M, Ballesteros A, Plou FJ (2007) J Biotechnol 128(1):204. doi:10.1016/j.jbiotec.2006.09.017
112. Yun J, Jung K, Oh J, Lee J (1990) Appl Biochem Biotechnol 24–25:299. doi: 10.1007/BF02920254. URL http://dx.doi.org/10.1007/BF02920254
113. Chiang CJ, Lee WC, Sheu DC, Duan KJ (1997) Biotechnol Prog 13(5):577. doi: 10.1021/bp970067z. URL http://dx.doi.org/10.1021/bp970067z
114. Hayashi S, Tubouchi M, Takasaki Y, Imada K (1994) Biotechnol Lett 16:227. doi: 10.1007/BF00134616. URL http://dx.doi.org/10.1007/BF00134616
115. Tanriseven A, Aslan Y (2005) Enzym Microb Technol 36(4):550. doi: 10.1016/j.enzmictec.2004.12.001
116. Hayashi S, Kinoshita J, Nonoguchi M, Takasaki Y, Imada K (1991) Biotechnol Lett 13:395. doi: 10.1007/BF01030989. URL http://dx.doi.org/10.1007/BF01030989
117. Csanádi Z, Sisak C (2008) Hung J Indus Chem 36(1–2):23
118. Clark DS (1994) Trends Biotechnol 12(11):439. doi: 10.1016/0167-7799(94)90018-3. URL http://www.sciencedirect.com/science/article/pii/0167779994900183
119. Panesar PS, Panesar R, Singh RS, Kennedy JF, Kumar H (2006) J Chem Technol Biotechnol 81(4):530. doi: 10.1002/jctb.1453. URL http://dx.doi.org/10.1002/jctb.1453
120. van Hijum S, van Geel-Schutten G, Rahaoui H, van der Maarel M, Dijkhuizen L (2002) Appl Environ Microbiol 68(9):4390. doi: 10.1128/AEM.68.9.4390-4398.2002
121. Maugeri F, Hernalsteens S (2007) J Mol Catal B Enzym 49(1–4):43. doi: 10.1016/j.molcatb.2007.08.001. URL http://www.sciencedirect.com/science/article/pii/S1381117707001580
122. Hidaka H, Hirayama M, Sumi N (1988) Agric Biol Chem 52:1181. URL http://ci.nii.ac.jp/naid/110006323785
123. Fungsin B, Saman P, Meeploy M, Chatanon L, Srichuai A, Sukcharurn J, Artjariyasripong S (2010) In: The 8th international symposium on biocontrol and biotechnology. Pattaya
124. Sheu DC, Duan KJ, Cheng CY, Bi JL, Chen JY (2002) Biotechnol Prog 18:1282. doi: 10.1021/bp020081y. URL http://onlinelibrary.wiley.com/doi/10.1021/bp020081y/abstract
125. Mussatto SI, Aguilar CN, Rodrigues LR, Teixeira JA (2009) J Mol Catal B Enzym 59:76
126. Cruz R, Cruz VD, Belini MZ, Belote JG, Vieira CR (1998) Bioresour Technol 65(1–2):139. doi: 10.1016/S0960-8524(98)00005-4. URL http://www.sciencedirect.com/science/article/pii/S0960852498000054
127. Aziani G, Terenzi H, Jorge J, Guimaraes L (2012) Food Technol Biotechnol 50(1):40
128. Antosova M, Polakovie M, Slovinska M, Madlova A, Illeova V, Bales V (2002) Chem Pap 56(6):394
129. Dominguez A, Nobre C, Rodrigues LR, Peres A, Torres D, Rocha I, Lima N, Teixeira J (2012) Carbohydr Polym 89(4):1174. doi: http://dx.doi.org/10.1016/j.carbpol.2012.03.091
130. Yun JW, Jung KH, Oh JW, Lee JH (1990) Appl Biochem Biotechnol 24/25:299
131. Shin H, Baig S, Lee S, Suh D, Kwon S, Lim Y, Lee J (2004) Bioresour Technol 93(1):59. doi:10.1016/j.biortech.2003.10.008
132. Prata M, Mussatto S, Rodrigues L, Teixeira J (2010) Biotechnol Lett 32(6):837. doi: 10.1007/s10529-010-0231-y
133. Mussatto S, Prata M, Rodrigues L, Teixeira J (2012) Eur Food Res Technol 235:13. doi: 10.1007/s00217-012-1728-5
134. Sangeetha P, Ramesh M, Prapulla S (2005) Trends Food Sci Technol 16(10):442. doi: 10.1016/j.tifs.2005.05.003. URL http://www.sciencedirect.com/science/article/pii/S0924224405001445

135. Chien CS, Lee WC, Lin TJ (2001) Enzym Microb Technol 29(4–5):252. doi: 10.1016/S0141-0229(01)00384-2. URL http://www.sciencedirect.com/science/article/pii/S0141022901003842
136. Prapulla SG, Subhaprada V, Karanth NG, (2000) Microbial production of oligosaccharides : a review. Advance in Applied Microbiology vol. 47. Academic Press, pp 299–343. doi: 10.1016/S0065-2164(00)47008-5. URL http://www.sciencedirect.com/science/article/pii/S0065216400470085
137. Yun JW (1996) Enzym Microb Technol 19(2):107. doi:10.1016/0141-0229(95)00188-3
138. Gosling A, Stevens GW, Barber AR, Kentish SE, Gras SL (2010) Food Chem 121(2):307. doi:10.1016/j.foodchem.2009.12.063
139. Shiomi N, Onodera S, Chatterton NJ, Harrison PA (1991) Agric Biol Chem 55(5):1427
140. Silva MTMV, Gomes P, Rodrigues A (2012) In: Inamuddin D, Luqman M (eds) Ion exchange technology II. Springer, The Netherlands, pp 109–135
141. Chilamkurthi S, Willemsen JH, van der Wielen LA, Poiesz E, Ottens M (2012) J Chromatogr A 1239(0):22. doi: 10.1016/j.chroma.2012.03.042. URL http://www.sciencedirect.com/science/article/pii/S0021967312004554
142. Vaňková K, Polakovič M (2010) Process Biochem 45(8):1325. doi:10.1016/j.procbio.2010.04.025
143. Pynnonen B (1998) J Chromatogr A 827(2):143. doi: 10.1016/S0021-9673(98)00732-8. URL http://www.sciencedirect.com/science/article/pii/S0021967398007328
144. Takahashi Y, Goto S (1994) Sep Sci Technol 29(10):1311. doi: 10.1080/01496399408006942. URL http://www.tandfonline.com/doi/abs/10.1080/01496399408006942
145. Vaňková K, Polakovič M (2012) Chem Eng Technol 35(1):161. doi:10.1002/ceat.201100254
146. da Silva EAB, de Souza AAU, de Souza SGU, Rodrigues AE (2006) Chem Eng J 118(3):167. doi: 10.1016/j.cej.2006.02.007. URL http://www.sciencedirect.com/science/article/pii/S1385894706000763
147. Nicoud RM 2000) In: Ahuja S (ed) Handbook of bioseparations, separation science and technology, vol. 2. Academic Press, pp 475–509. doi: 10.1016/S0149-6395(00)80060-4. URL http://www.sciencedirect.com/science/article/pii/S0149639500800604
148. Vanneste J, Ron SD, Vandecruys S, Soare SA, Darvishmanesh S, der Bruggen BV (2011) Sep Purif Technol 80(3):600. doi: 10.1016/j.seppur.2011.06.016. URL http://www.sciencedirect.com/science/article/pii/S1383586611003546
149. Vaňková K, Onderková Z, Antosová M, Polakovič M (2008) Chem Pap 62:375. doi: 10.2478/s11696-008-0034-y. URL http://dx.doi.org/10.2478/s11696-008-0034-y
150. Hernández O, Ruiz-Matute AI, Olano A, Moreno FJ, Sanz ML (2009) Int Dairy J 19(9):531. doi: 10.1016/j.idairyj.2009.03.002. URL http://www.sciencedirect.com/science/article/pii/S0958694609000521
151. Kuhn RC, Filho FM (2010) New Biotechnol 27(6):862. Papers from Symbiosis - The 14th European congress on biotechnology (Part 1), Barcelona, Sept 2009. doi: 10.1016/j.nbt.2010.05.008. URL http://www.sciencedirect.com/science/article/pii/S1871678410004437
152. Nobre C, Teixeira J, Rodrigues L (2012) New Biotechnol 29(3):395. doi:10.1016/j.nbt.2011.11.006
153. Chinn D, King CJ (1999) Indus Eng Chem Res 38(10):3738. doi: 10.1021/ie990286k. URL http://pubs.acs.org/doi/abs/10.1021/ie990286k
154. Sen D, Gosling A, Stevens GW, Bhattacharya PK, Barber AR, Kentish SE, Bhattacharjee C, Gras SL (2011) Food Chem 128(3):773. doi: 10.1016/j.foodchem.2011.03.076. URL http://www.sciencedirect.com/science/article/pii/S0308814611004584
155. Nés FM, Fornari T, Stateva RP, Olano A, nez EI (2009) J Supercrit Fluids 49(1):16. doi: 10.1016/j.supflu.2008.11.014. URL http://www.sciencedirect.com/science/article/pii/S0896844608003859
156. Nés FM, Olano A, Reglero G, nez EI, Fornari T (2009) Sep Purif Technol 66(2):383. doi: 10.1016/j.seppur.2008.12.006. URL http://www.sciencedirect.com/science/article/pii/S1383586608005091

157. n és FM, Fornari T, Olano A, n ez EI (2010) J Supercrit Fluids 53(1–3):25. Selected papers from the 9th international symposium on supercritical fluids (ISSF 2009) - new trends in supercritical fluids: energy, materials, processing, Arcachon, France, May 18-20 2009. doi: 10.1016/j.supflu.2010.02.011. URL http://www.sciencedirect.com/science/article/pii/S0896844610000811
158. Yun JW, Lee MG, Song SK (1994) J Ferment Bioeng 77(2):159. doi: 10.1016/0922-338X(94)90316-6
159. Sheu D, Lio P, Chen S, Lin C, Duan K (2001) Biotechnol Lett 23:1499. doi: 10.1023/A:1011689531625. URL http://dx.doi.org/10.1023/A:1011689531625
160. Sheu DC, Duan KJ, Cheng CY, Bi JL, Chen JY (2002) Biotechnol Prog 18(6):1282. doi: 10.1021/bp020081y. URL http://dx.doi.org/10.1021/bp020081y
161. Cheng CC, Yu MC, Cheng TC, Sheu DC, Duan KJ, Tai WL (2006) Biotechnol Lett 28:793. doi: 10.1007/s10529-006-9002-1. URL http://dx.doi.org/10.1007/s10529-006-9002-1
162. Splechtna B, Petzelbauer I, Baminger U, Haltrich D, Kulbe KD, Nidetzky B (2001) Enzym Microb Technol 29(6-7):434. doi: 10.1016/S0141-0229(01)00412-4. URL http://www.sciencedirect.com/science/article/pii/S0141022901004124
163. Oda Y, Ouchi K (1991) Enzym Microb Technol 13(6):495. doi:10.1016/0141-0229(91)90008-X
164. Crittenden R, Playne M (2002) Appl Microbiol Biotechnol 58:297. doi: 10.1007/s00253-001-0886-3. URL http://dx.doi.org/10.1007/s00253-001-0886-3
165. Li Z, Xiao M, Lu L, Li Y (2008) Process Biochem 43(8):896. doi: 10.1016/j.procbio.2008.04.016. URL http://www.sciencedirect.com/science/article/pii/S1359511308001281
166. Pinelo M, Jonsson G, Meyer AS (2009) Sep Purif Technol 70(1):1. doi: 10.1016/j.seppur.2009.08.010. URL http://www.sciencedirect.com/science/article/pii/S1383586609003608
167. Grandison A, Goulas A, Rastall R, Songklanakarin (2002) J Sci Technol 24(Supplement):915. URL http://centaur.reading.ac.uk/13347/
168. Kuhn RC, Palacio L, Prádanos P, Hernández A, Filho FM (2011) Desalin Water Treat 27(1–3):18. doi:10.5004/dwt.2011.2038
169. Li W, Li J, Chen T, Chen C (2004) J Membr Sci 245(1–2):123. doi:10.1016/j.memsci.2004.07.021
170. Goulas AK, Kapasakalidis PG, Sinclair HR, Rastall RA, Grandison AS (2002) J Membr Sci 209(1):321. doi: 10.1016/S0376-7388(02)00362-9. URL http://www.sciencedirect.com/science/article/pii/S0376738802003629
171. Feng Y, Chang X, Wang W, Ma R (2009) J Taiwan Inst Chem Eng 40(3):326. Festschrift Issue In honor of Professor Yi Hua Ma. doi: 10.1016/j.jtice.2008.12.003. URL http://www.sciencedirect.com/science/article/pii/S1876107008001958
172. Botelho-Cunha VA, Mateus M, Petrus JC, de Pinho MN (2010) Biochem Eng J 50(1–2):29. doi: 10.1016/j.bej.2010.03.001. URL http://www.sciencedirect.com/science/article/pii/S1369703X1000080X
173. Nakao S, Kimura S (1982) J Chem Eng Japan 15(2):200
174. Kovács Z, Samhaber W (2008) Membrántechnika 12(2):22
175. Kuhn R, Filho FM, Silva V, Palacio L, Hernández A, Prádanos P (2010) J Membr Sci 365:356
176. Lightfoot EN (2005) Sep Sci Technol 40(4):739. doi: 10.1081/SS-200047994. URL http://www.tandfonline.com/doi/abs/10.1081/SS-200047994
177. Siew WE, Livingston AG, Ates C, Merschaert A (2013) Sep Purif Technol 102(0):1. doi: 10.1016/j.seppur.2012.09.017. URL http://www.sciencedirect.com/science/article/pii/S1383586612004923
178. Nishizawa K, Nakajima M, Nabetani H (2000) Biotechnol Bioeng 68(1):92
179. Sen D, Sarkar A, Gosling A, Gras SL, Stevens GW, Kentish SE, Bhattacharya P, Barber AR, Bhattacharjee C (2011) J Membr Sci 378(1–2):471. Membranes for a Sustainable

Future Section. doi: 10.1016/j.memsci.2011.05.032. URL http://www.sciencedirect.com/science/article/pii/S0376738811003711

180. Palai T, Bhattacharya PK (2013) J Biosci Bioeng 115(6):668. doi: http://dx.doi.org/10.1016/j.jbiosc.2012.12.014. URL http://www.sciencedirect.com/science/article/pii/S1389172312005282
181. Jochems P, Satyawali Y, Roy SV, Doyen W, Diels L, Dejonghe W, Enzym Microb Technol 49(6–7):580 (2011). Special Issue on Papers presented at the 14th international biotechnology symposium and exhibition (IBS2010). doi: 10.1016/j.enzmictec.2011.06.010. URL http://www.sciencedirect.com/science/article/pii/S014102291100127X
182. Güleç HA (2013) Colloids Surf B Biointerfaces 104(0):83. doi: 10.1016/j.colsurfb.2012.11.039. URL http://www.sciencedirect.com/science/article/pii/S092777651200687X
183. Ulbricht M, Papra A (1997) Enzym Microb Technol 20(1):61
184. Engel L, Ebrahimi M, Czermak P (2008) Desalination 224(1–3):46. Issues 1 and 2: 11th Aachener Membran Kolloquium, 28-29 March 2007, Aachen, Issue 3: Aqua 2006, 2nd international conference on water science and technology - integrated management of water resources, November 2006, Athens. doi: 10.1016/j.desal.2007.04.078. URL http://www.sciencedirect.com/science/article/pii/S0011916408000295
185. Engel L, Schneider P, Ebrahimi M, Czermak P (2007) Open Food Sci J 1:17
186. Ebrahimi M, Engel L, Peter S, Grau K, Czermak P (2006) Desalination 200(1–3):509. Euromembrane 2006. doi: 10.1016/j.desal.2006.03.415. URL http://www.sciencedirect.com/science/article/pii/S0011916406007193
187. Mignard D, Glass D (2001) J Membr Sci 186(1):133. doi:10.1016/S0376-7388(00)00661-X
188. Prádanos P, Hernández A, Calvo J, Tejerina F (1996) J Membr Sci 114(1):115. doi: 10.1016/0376-7388(95)00324-X
189. Meireles M, Aimar P, Sanchez V (1991) Biotechnol Bioeng 38:528
190. van Reis R, Goodrich EM, Yson CL, Frautschy LN, Whiteley R, Zydney AL (1997) J Membr Sci 130(1–2):123 10.1016/S0376-7388(97)00012-4
191. Bacchin P, Aimar P, Field R (2006) J Membr Sci 281(1–2):42 10.1016/j.memsci.2006.04.014
192. Kim KJ, Sun P, Chen V, Wiley DE, Fane AG (1993) J Membr Sci 80(1):241. doi: 10.1016/0376-7388(93)85148-P
193. Field R, Hughes D, Cui Z, Tirlapur U (2008) Desalination 227(1–3):132. doi: 10.1016/j.desal.2007.08.004
194. Giorno L, Drioli E (2000) Trends Biotechnol 18(8):339. doi: 10.1016/S0167-7799(00)01472-4. URL http://www.sciencedirect.com/science/article/pii/S0167779900014724
195. Rios G, Belleville M, Paolucci D, Sanchez J (2004) J Membr Sci 242(1–2):189. Membrane Engineering Special Issue. doi: 10.1016/j.memsci.2003.06.004. URL http://www.sciencedirect.com/science/article/pii/S0376738804003230
196. Gonzalez R, Ebrahimi M, Czermak P (2009) Open Food Sci J 3:1. doi:10.2174/1874256400903010001
197. Czermak P, Ebrahimi M, Kandzia S, Klein K, Sawatzki G (2002) Chemie Ingenieur Technik 74(5):645. doi: 10.1002/1522-2640(200205)74:5<645::AID-CITE645>3.0.CO;2-W. URL http://dx.doi.org/10.1002/1522-2640(200205)74:5<645::AID-CITE645>3.0.CO;2-W
198. Ebrahimi M, Gonzalez R (2006) Czermak P Desalination 200(1–3):686. Euromembrane 2006. doi: 10.1016/j.desal.2006.03.468. URL http://www.sciencedirect.com/science/article/pii/S0011916406006874
199. Czermak P, Ebrahimi M, Grau K, Netz S, Sawatzki G, Pfromm P (2004) J Membr Sci 232(8):85
200. Foda MI, Lopez-Leiva M (2000) Process Biochem 35(6):581. doi: 10.1016/S0032-9592(99)00108-9
201. Pocedičová K, Čurda L, Mišún D, Dryáková A, Diblíková L (2010) J Food Eng 99(4):479. doi:10.1016/j.jfoodeng.2010.02.001

202. Maria G (2012) Comput Chem Eng 36:325. doi:10.1016/j.compchemeng.2011.06.006
203. Bélafi-Bakó K, Eszterle M, Kiss K, Nemestóthy N, Gubicza L (2007) J Food Eng 78(2):438. doi: 10.1016/j.jfoodeng.2005.10.012. URL http://www.sciencedirect.com/science/article/pii/S026087740500703X
204. Olano-Martin E, Mountzouris K, Gibson G, Rastall R (2001) J Food Sci 66(7):966. doi: 10.1111/j.1365-2621.2001.tb08220.x. URL http://dx.doi.org/10.1111/j.1365-2621.2001.tb08220.x
205. Petzelbauer I, Splechtna B, Nidetzky B (2002) Biotechnol Bioeng 77(4):394. doi: 10.1002/bit.10106. URL http://dx.doi.org/10.1002/bit.10106
206. Das R, Sen D, Sarkar A, Bhattacharyya S, Bhattacharjee C (2011) Indus Eng Chem Res 50(2):806. doi: 10.1021/ie1016333. URL http://pubs.acs.org/doi/abs/10.1021/ie1016333
207. Avalakki UK, Maheswaran P Saravanan R (2011) Process for production of galactooligosaccharides (gos)

Adv Biochem Eng Biotechnol (2014) 143: 297–301
DOI: 10.1007/978-3-662-43761-2

Index

Printed by Printforce, the Netherlands